Springer Texts in Education

More information about this series at http://www.springer.com/series/13812

Andy Liu

S.M.A.R.T. Circle Overview

Andy Liu
Professor Emeritus
Department of Mathematical
  and Statistical Sciences
University of Alberta
Edmonton, AB
Canada

ISSN 2366-7672　　　　　　　ISSN 2366-7680　(electronic)
Springer Texts in Education
ISBN 978-3-319-56822-5　　　ISBN 978-3-319-56823-2　(eBook)
https://doi.org/10.1007/978-3-319-56823-2

Library of Congress Control Number: 2017946035

© Springer International Publishing AG 2018
This work is subject to copyright. All rights are reserved by the Publisher, whether the whole or part of the material is concerned, specifically the rights of translation, reprinting, reuse of illustrations, recitation, broadcasting, reproduction on microfilms or in any other physical way, and transmission or information storage and retrieval, electronic adaptation, computer software, or by similar or dissimilar methodology now known or hereafter developed.
The use of general descriptive names, registered names, trademarks, service marks, etc. in this publication does not imply, even in the absence of a specific statement, that such names are exempt from the relevant protective laws and regulations and therefore free for general use.
The publisher, the authors and the editors are safe to assume that the advice and information in this book are believed to be true and accurate at the date of publication. Neither the publisher nor the authors or the editors give a warranty, express or implied, with respect to the material contained herein or for any errors or omissions that may have been made. The publisher remains neutral with regard to jurisdictional claims in published maps and institutional affiliations.

Printed on acid-free paper

This Springer imprint is published by Springer Nature
The registered company is Springer International Publishing AG
The registered company address is: Gewerbestrasse 11, 6330 Cham, Switzerland

This book is dedicated to my teachers and mentors

Harvey L. Abbott

William G. Brown

Martin Gardner

Solomon W. Golomb

Murray S. Klamkin

George Sicherman

Raymond M. Smullyan

Preface
A Brief History of the S.M.A.R.T.Circle

A most beneficial side effect of the collapse of the former Soviet Union in 1992 was the migration of the Mathematical Circles across the Atlantic to the United States. Mathematical Circles, originated in Hungary during the nineteenth century, are a glorious tradition in Eastern Europe. They are organizations which discover and nurture young mathematical talents through meaningful extra-curricular activities.

The process took a few years, leading to the formation in 1998 of the Berkeley Mathematical Circle. With the support of the Mathematical Sciences Research Institute, the movement has caught fire in the United States, culminating in the formation of a Special Interest Group in the Mathematical Association of America under the leadership of Tatiana Shubin of San Jose State University.

Unbeknown to this community, a Mathematical Circle had existed in North America almost two decades earlier. The ultimate inspiration was still of Soviet origin, but the migration took place across the Pacific, via the People's Republic of China in the form of their Youth Palaces. This was the S.M.A.R.T. Circle in Edmonton, Canada, founded in 1981. The acronym stood for Saturday Mathematical Activities, Recreations & Tutorials.

I was born in China during the over-time sudden-death period of the Second World War, but moved to Hong Kong at age six. Thus I had never attended any session of any Youth Palace. However, I followed reports of their activities, and this fueled my interest in mathematics. The first mathematics book I had was a Chinese translation of Boris Kordemski's *Moscow Puzzles*, which was on their recommended reading list. An English version is now an inexpensive Dover paperback. Later, I acquired Chinese translations of several wonderful books by Yakov Perelman. Dover has published his *Figures for Fun* in English.

I came to Canada at age twenty, and eventually got a tenure-track position at the University of Alberta in 1980. That fall, I was invited to a general meeting of the Edmonton Chapter of the Association for Bright Children. My comment was that their activities seemed heavily biased towards the Fine Arts. Having put my foot in my mouth, I was obliged to take some concrete action. The next spring, the S.M.A.R.T. Circle was born.

During the first year, the members ranged from Grade 3 to Grade 6, because of the clientele of the A.B.C. However, to do meaningful mathematical activities, I preferred the children to be a bit more mature. So the Grade level rose by one each year, until in 1985, the members ranged from Grade 7 to Grade 10. Many of them stayed throughout this period.

As we moved away from the normal age of the clientele of the A.B.C., the Circle practically became an independent operation. This also became necessary because in 1983, we received a grant of $1,500 from the University of Alberta, arranged by Vice-President Academic **Amy Zelmer**. With the money, I built up a Circle Library containing mathematical books, games and puzzles. This was the only funding the Circle had received in its thirty-two year history.

We met on the University of Alberta campus from 2:00 pm to 3:00 pm every Saturday in October, November, February and March. A second classroom adjacent to the meeting room was open from 1:30 pm to 3:30 pm as the Circle Library. **Adrian Ashley**, a former Circle member, was hired at $5 an hour to look after it. There was a comedy of error in that for a while, his salary came out of the Student Union cafeteria account! They soon put a stop to that, but never bothered to claim readjustment.

Because of the members' tender ages, most came with their parents, and some parents stayed in Circle Library during the session. Members also had half an hour before and half an hour after the session to browse through. Sometimes, some younger members' attention span wandered during the session, and they would drift to the Circle Library for a few minutes.

In 1986, the three-year period of the grant ran out. As I closed the account, I turned the Circle Library over to the Faculty of Education. Then I started building a replacement out of my own pocket. Meanwhile, the A.B.C. had acquired new headquarters in the form of a house, where the basement was set up as a classroom. The Circle was invited to move its operation there. As a result, I restarted the session for A.B.C. members from Grade 3 to Grade 6 again. This went from 1:00 pm to 2:00 pm while the existing session for the older children ran from 2:30 pm to 3:30 pm. We had quite a few sibling pairs. Sometimes, one was in class in the basement while the other waited upstairs and played with mathematical games and puzzles from the new Circle Library. Sometimes, they sat in the same session despite any disparity in age.

In 1991, this arrangement came to an end, and the Circle moved back to the university campus. Only the Grade 7 to Grade 10 session survived the move. The meeting time was once again from 2:00 pm to 3:00 pm. A section at the back of the classroom was reserved for the Circle Library.

In 1996, there was a reverse migration of the Circle movement back across the Pacific, to Taiwan. My friend Wen-Hsien Sun of Taipei started the Chiu Chang Mathematical Circle, initially based on my model and using much of the material I had accumulated over a decade and a half. Both Circles closed in 2012, though mine was reincarnated as the J.A.M.E.S. Circle, standing for Junior Alberta Mathematics for Eager Students. It is run by my former student **Ryan Morrill**.

The activities of the S.M.A.R.T. Circle may be loosely classified into the following overlapping categories:

1. **Mathematical Conversations**;
2. **Mathematical Competitions**;
3. **Mathematical Congregations**;
4. **Mathematical Celebrations**.

The first two activities are the mainstay of our Circle right from the beginning. The last two did not emerge until the second half of our Circle's thirty-two year history. All four activities are described in this book, each in a separate part, along with a lot of resource material.

Andy Liu,
Edmonton, Alberta, 2017.

Acknowledgement

I am very excited that **Springer-Verlag**, an institution in mathematics publishing, agrees to publish this volume. I am most grateful to their staff for encouragement and support, in particular, to Jan Holland, Bernadette Ohmer and Anne Comment. The technical team of Suganya Manoharan at Scientific Publishing Services, Trichy, India, has made significant contributions to the layout of the book.

Table of Contents

	Preface	vii
	Acknowledgement	xi
	Table of Contents	xiii
Part I	**Mathematical Conversations**	**1**
Chapter 1	**Three Sample Projects : Counting Problems**	**3**
Section 1	River-Crossing with Alibaba	3
Section 2	Martian Citizenship Quiz	8
Section 3	Rook Paths	18
Chapter 2	**A Sample Minicourse : Tessellations**	**27**
Section 1	Platonic and Archimedean Tilings	27
Section 2	From Tessellations to Rectifications	39
Section 3	Frieze and Wallpaper Patterns	48
Part II	**Mathematical Competitions**	**61**
Chapter 3	**Past Papers of the Edmonton Junior Hig Mathematics Invitational**	**65**
Section 1	Problems	65
Section 2	Solutions	82
Chapter 4	**International Mathematics Tournament of the Towns : Selected Problems**	**115**
Section 1	A Problem on Area	115
Section 2	A Problem on Communication	120
Section 3	A Problem on Divisors	124
Section 4	A Problem on Complex Numbers in Geometry	128
Section 5	A Problem on Polyhedra	133
Section 6	A Problem on Sequences	142
Section 7	A Problem on Lottery	146
Section 8	A Problem on Angles	149
Section 9	A Problem on Balance	154
Section 10	A Problem on Magic Tricks	159
Section 11	A Problem on Polyomino Dissections	163
Section 12	A Problem on Chess Tournaments	167
Section 13	A Problem on Electrical Networks	171
Section 14	A Problem on Card Signaling	174
Section 15	A Problem on Tangent Circles	178
Section 16	A Problem on Graph Algorithms	180
Section 17	A Problem on Fixed Points	185
Section 18	A problem on Information Extraction	191

Part III Mathematical Congregations		**201**
Appendix A Canadian Geography		208
Appendix B Mathematical Jeopardy		209
Appendix C Answers		210
Chapter 5 "From Earth to Moon" Sample Contests		**211**
Section 1 Problems		211
Section 2 Answers		223
Section 3 Solutions		225
Chapter 6 Past Papers of the Peking University Mathematics Invitational for Youths		**259**
Section 1 Problems		259
Section 2 Solutions		275
Part IV Mathematical Celebrations		**297**
Chapter 7 Sample SNAP Math Fair Projects		**301**
Section 1 Problems		301
Section 2 Solutions		321
Chapter 8 Sample GAME Math Unfair Games		**333**
Section 1 Problems		333
Section 2 Solutions		341

PART ONE
MATHEMATICAL CONVERSATIONS

The Mathematical Conversations are the heart and soul of the Circle.

Partly because of the word "Tutorials" in the acronym S.M.A.R.T., many people mistook the Circle for a tutoring class for weak students. Quite the opposite is intended. It is designed to serve bright students who are not sufficiently challenged in their regular classrooms. For many such students, the usual outlet was acceleration. However, this simply compounds the problem. This ensures that a bored student now will be bored for more years to come. Acceleration is also not without dangers. While the material glossed over is largely routine, not having gone through the usual drill and practice does leave gaps in the background of the bright students.

The alternative is not to push the students further ahead, but to broaden their horizon by exposing them to topics and ideas not usually encountered in the classroom. This is then the primary mission of the S.M.A.R.T. Circle. Material related to computing science is particularly appropriate. It is necessary to counter the erroneous perception that with the advent of computers, we need to know less mathematics. Quite the contrary is the actual case.

The Circle consists of a Fall Session and a Winter Session each academic year. September is the beginning of a new academic year, and is a particularly busy month for both myself and the students. Thus the Fall Session starts at the beginning of October and runs till the end of November. December is also unsuitable because of university examinations, and the proximity to Christmas holidays. The Winter Session skips over January and stops before April for similar reasons and runs only in February and March. It is possible to do something in May and June, but I reserve that period for my own travels.

We meet every Saturday during those four months, from two to three in the afternoon. When the Circle was first formed, a new topic was introduced each session. There is no shortage of material. North America is blessed with the writings of the late *Martin Gardner*, whose *Mathematical Games* columns in the magazine *Scientific American* enticed and nurtured young mathematicians for over a quarter of a century. When he retired, the mantle passed through several hands until it landed on the lap of **Dennis Shasha** of the Courant Institute, New York. Dennis is an accomplished author in his own right, and his expert handling of the column had revitalized it until *Scientific American* inexplicably decided to take it off the printed pages and exiled it to cyberspace. The logician **Raymond Smullyan** also offers many delightful and profound problems and paradoxes.

The most beneficial effect of the Circle on the students is that they begin to explore further into the topics covered, either independently or in small groups. On many occasions, they astonished me by making elegant generalizations, providing insightful comments and finding original results. Over the thirty-two-year history of the Circle, over fifty of our students' projects have been published in scientific and education journals. This is by far the most successful aspect of the Circle. We provide three samples in Chapter 1. For more of these projects, see the companion volume *The S.M.A.R.T. Circle — Projects*.

As time went on, the students expressed a desire to learn something in greater depth, and in a more coherent manner. This led to my giving, in addition to discussion topics, one minicourse per session, at a very leisurely pace. There was some overlap of material presented from week to week since not all Circle members can be present at every meeting. The topics had to be chosen carefully. They should generate sufficient interest without getting bogged down in details. They were prepared well in advance. The sessions are largely structured group explorations, with the work of the students summarized in the edited lecture notes distributed at the end. We provide a sample in Chapter 2. For more of these minicourses, see the companion volume *The S.M.A.R.T. Circle — Minicourses*.

Chapter One
Three Sample Projects
COUNTING PROBLEMS

Section 1. River-Crossing with Alibaba

Among the Forty Thieves, the Chief Thief is ranked 0, the first mate is ranked 1, and so on down to the bottom mate who is ranked 39.

Fleeing from justice, they get to the near shore of a river, which they must cross. Fortunately, they find a boat which requires two people to row and is large enough to carry all of them. So it seems that the problem is solved.

Unfortunately, the thieves are snobbish. Two thieves whose ranks differ by more than 1 refuse to go into the boat at the same time. So they are faced with a mission impossible, because at most two of them can be in the boat at a time, and the same two who cross over must come back together.

Along comes Alibaba, and the thieves appeal to him for help.

He says, "As it happens, I live on the far shore of the river. Make me an honorary thief, and give me equal rank with the Chief Thief, namely 0. I will lead you across the river. I myself will also follow your snobbish rule."

Thief 0 asks, "Are you sure? How many crossing will your plan require, counting crossings in both directions?"

"Of course I am sure, and I can do it in the minimum number of crossings."

So the thieves agree to Alibaba's condition.

"It's easiest for me to explain things," Alibaba beings, "if we consider the general problem, where there are $n+1$ of you, with ranks from 0 to n. We will let $f(n)$ be the minimum number of crossings required for the task. In your case, we are interested in $f(39)$. What would $f(1)$ be?"

Thief 39 says, "That is easy. For n=1, there will only be three thieves, with ranks 0, 0 and 1. They can cross over together, so that f(1)=1 since it cannot be less."

Thief 1 then says, "For n=2, there will be four thieves, with ranks 0, 0, 1 and 2. They can do it in five crossings. Let me draw a chart and show you how."

Crossing	Ranks of thieves		
Number	on Near Shore	in Boat	on Far Shore
First	2	0,0,1	0,0,1
Second	0,1,2	0,1	0
Third	0	1,2	0,1,2
Fourth	0,0,1	0,1	2
Fifth		0,0,1	0,0,1,2

"Good," Alibaba nods. "You have proved that $f(2) \leq 5$. Can't it be less?"

The thieves are stumped. After a little while, Thief 0 says, "Whatever $f(2)$ is, it must be an odd number."

"Why is that?" several thieves asked at the same time.

"You see," explains Thief 0. "The crossings must be alternately from the near shore and back to the near shore. Since the first and the last crossings are both from the near shore, $f(n)$ will always be an odd number."

"I understand," Thief 1 says. "So if $f(2) < 5$, it can only be 1 or 3. Each crossing from the near shore will get at most three people over, while each crossing back to the near shore will bring back at least two people. Thus we cannot have $f(2) = 1$. Since $2 \times 3 - 2 = 4$, we may barely manage to have $f(2) = 3$."

"You can't have $f(2) = 3$," says Thief 39. "In a crossing with three thieves, their ranks must 0, 0 and 1. This means that the second mate can never gets across. So we indeed have $f(2) = 5$.

"Wow!" says Thief 0. "That is good reasoning. However, it will take us forever to get up to $f(39)$."

"Be patient," says Alibaba. "It may not be necessary to work out all the values of $f(n)$ up to $n = 39$. However, it may help to study a few more cases, and see if we can spot a general pattern."

The thieves take out their pencils and paper and work for some time.

"For $n = 3$," Thief 1 says, "there will be five thieves, with ranks 0, 0, 1, 2 and 3. They can do this in 9 crossings."

He draws a chart similar to the one he has drawn earlier.

Crossing	Ranks of thieves		
Number	on Near Shore	in Boat	on Far Shore
First	2,3	0,0,1	0,0,1
Second	0,1,2,3	0,1	0
Third	0,3	1,2	0,1,2
Fourth	0,0,1,3	0,1	2
Fifth	3	0,0,1	0,0,1,2
Sixth	1,2,3	1,2	0,0
Seventh	1	2,3	0,0,2,3
Eighth	0,0,1	0,0	2,3
Ninth		0,0,1	0,0,1,2,3

"Look," said Thief 39, "after the eighth crossing, if we ignore the people on the far shore, we are back to the same situation as when $n = 1$. The task can be accomplished with 1 more crossing."

"For $n = 4$," Thief 0 says, "there will be six thieves with ranks 0, 0, 1, 2, 3 and 4. We can do this in 13 crossings."

He draws a chart which, apart from being longer, is not all that different.

Crossing	Ranks of thieves		
Number	on Near Shore	in Boat	on Far Shore
First	2,3,4	0,0,1	0,0,1
Second	0,1,2,3,4	0,1	0
Third	0,3,4	1,2	0,1,2
Fourth	0,0,1,3,4	0,1	2
Fifth	3,4	0,0,1	0,0,1,2
Sixth	1,2,3,4	1,2	0,0
Seventh	1,2	3,4	0,0,3,4
Eighth	0,0,1,2	0,0	3,4
Ninth	2	0,0,1	0,0,1,3,4
Tenth	0,1,2	0,1	0,3,4
Eleventh	0	1,2	0,1,2,3,4
Twelfth	0,0,1	0,1	3,4
Thirteenth		0,0,1	0,0,1,2,3,4

"Once again," remarked Thief 39, "after the eighth crossing, if we ignore the people on the far shore, we are back to the same situation as when $n = 2$. The task can be accomplished with 5 more crossings."

"The general pattern becomes clear, doesn't it?" Alibaba remarks. "In the first 8 crossings, we get the last two people over to the far shore. Moreover, they are the only people there. I will draw a chart to show that this can always be done, and I do not think the sub-task cannot be done with few crossings."

Crossing	Ranks of thieves		
Number	on Near Shore	in Boat	on Far Shore
First	$2,3,\ldots,n-2,n-1,n$	$0,0,1$	$0,0,1$
Second	$0,1,2,3,\ldots,n-2,n-1,n$	$0,1$	0
Third	$0,3,\ldots,n-2,n-1,n$	$1,2$	$0,1,2$
Fourth	$0,0,1,3,\ldots,n-2,n-1,n$	$0,1$	2
Fifth	$3,\ldots,n-2,n-1,n$	$0,0,1$	$0,0,1,2$
Sixth	$1,2,3,\ldots,n-2,n-1,n$	$1,2$	$0,0$
Seventh	$1,2,3,\ldots,n-2$	$n-1,n$	$0,0,n-1,n$
Eighth	$0,0,1,2,3,\ldots,n-2$	$0,0$	$n-1,n$

Thief 39 observes, "During these eight crossings, the thieves with rank 3, ..., $n-2$ remain on the near shore."

"Taking your word that the sub-task cannot be done in less than 8 crossings, I think we have $f(n) = f(n-2) + 8$ for $n \geq 3$. What does that tell us about $f(n)$?"

"Recall that $f(1) = 1$ and $f(2) = 5$," Alibaba says. "If n is odd, say, $n = 2k - 1$ for some positive integer k, we can work out $f(2k-1)$ this way."

$$\begin{aligned} f(2k-1) &= f(2k-3) + 8 \\ &= f(2k-5) + 8 \times 2 \\ &= \cdots \\ &= f(1) + 8(k-1) \\ &= 4(2k-1) - 3. \end{aligned}$$

"What happens if n is even," Thief 1 asks, "say $n = 2k$ for some positive integer k?"

Thief 39 says, "I know. It is not really different from what Alibaba has done."

$$\begin{aligned} f(2k) &= f(2k-2) + 8 \\ &= f(2k-4) + 8 \times 2 \\ &= \cdots \\ &= f(2) + 8(k-1) \\ &= 4(2k) - 3. \end{aligned}$$

"In summary," Alibaba concludes, "$f(n) = 4n - 3$ for all n."

"This means that $f(39) = 153$. Good heavens, that will take forever, especially since we have already spent so much time working this out. We will be caught for sure."

Fortunately, having worked together on the problem makes the thieves feel closer to one another. They shelf their snobbishness for now and cross over together along with Alibaba.

After bidding the Forty Thieves farewell on the far shore, Alibaba goes home to sleep. Along the way, he keeps thinking about the river crossing problem again. He is convinced that $f(n) = 4n-3$ is the correct answer, but two things bothers him. First, is 8 indeed the minimum number of crossings required to get the bottom two thieves across? Second, and this is more serious, is performing this sub-task the best way to go? Is there a different scheme which uses a smaller number of crossings than $4n-3$?

He does not sleep well that night. Then suddenly, he sits up in his bed. An idea has come to him, which makes his two worries irrelevant. Without bothering to get dressed, he goes to his desk and tries to jot down this idea before he loses track of it.

"The key is Thief 1's proof that f(2)=5," he writes. "His main argument is that the gain is 1 when three thieves cross over together, except the last time when the gain is 3. To get $n+2$ thieves across, we must of course gain $n+2$. So three thieves must cross over together n times. Obviously, this cannot happen in every crossing from the near shore, because thieves 0, 0 and 1 must all come back. Two of them can come back right away, and one of them can cross over next time to fetch the third. So this can happen every other crossing from the near shore."

Alibaba gets himself a cup of tea and continues his writing.

"So there must be at least 3 other crossings between two crossings from the near shore with three thieves. This means that I have to add $n-1$ sets of 3 crossing to the total. Ah ha! I am right. The minimum total is indeed $n + 3(n-1) = 4n - 3$."

Section 2. The Martian Citizenship Quiz

A serious malfunctioning of its central drive forced the spaceship *Enterprise* to crash-land on Mars. Although there were no casualty, not even an injury, further damages to the spaceship were sustained, and there was no hope of getting off under its own power. Outside technical assistance was required.

Two hours later, a Martian delegation was observed approaching the spaceship. So the crew went out to greet them. None of them spoke a word of the Martian language, and the Martians were not known to be conversant in Earthling languages.

A member of the Martian delegation spoke to their leader and then came forward.

"Greetings, Commander Spock," he said in the Vulcan language.

Like Spock, he had pointed ears.

"Are you a fellow Vulcan?"

"My name is Spark. I am a farmer in Vulcan, Alberta. I spotted your pointed ears and asked my leader if I could approach you. He was glad. You see, he has no idea how he is going to communicate with whomever the unexpected visitors may be. That is why he brings me along, together with an assortment of migrants and aliens."

"How do you come to be so far from home?"

"My spaceship was blown completely off course during an asteroid storm a year ago, and was destroyed when it crash-landed here. The Martians fished me out of the debris. With their advanced medical knowledge, they were able to make a sole survivor out of me."

"Have you managed to learn the Martian language?"

"I have picked up a few words here and there, mostly swear words. They are no help to me when I try to pass the Martian Citizenship Quiz."

"We are much luckier because the *Enterprise* is very well-built. However, we desperately need technical assistance. Would your leader agree to that?"

"I am afraid he will not. The Martians render technical assistance only to their own citizens. Are there any Martian citizens on your spaceship?"

"No, but just now, you mentioned the Martian Citizenship Quiz. Can anyone apply?"

"Yes. I have been doing that everyday this past year. There is a quiz consisting of 30 true-or-false questions. To pass the quiz, you have to answer all 30 questions, and all answers must be correct."

"What happens if you fail? Obviously they do not shoot you, or you wouldn't be here."

"They tell you how many questions you've got right, but not which ones. I have never been able to get more than 15 right."

"Do I understand that if you fail, you can apply again?"

"Yes, any number of times, but at most once a day. You will always get exactly the same quiz."

"So if you change all your answers from one day to the next, you must get at least 15 right at least once."

"I haven't thought of that, but this doesn't really help, does it?"

"No. Could you inform your leader that one of us wish to apply for Martian citizenship?"

"Sure," said Spark.

He returned to the Martian delegation to speak to the leader. He came back right away.

"Is it all right?" asked Spock.

"Yes, and he has offered to take you back to town. It is a two-hour trip from here. We saw the smoke when you crash-landed, and it took us this long to get out here."

"That is very kind of him. Not knowing a word of the Martian language, it will take me quite a few days to ace the quiz. Is there a cheap hotel?"

"There is only one hotel. It used to be called the *Fiery Planet*, but was renamed the *Martian Garden* in honor of **Martin Gardner**. It is most definitely not cheap."

"How bad is it? Tell me the worst."

"Let me put it this way. I work there as a bell hop. What I have made during the past year would not be enough for me to stay there for one night as a guest."

"*Good galaxies!* Well, can you wait while I pack a small carrying bag for the personal stuff I would need?"

"Sure," said Spark.

So Spock and the crew went back inside the *Enterprise*. An emergency meeting was convened. It was essential that the hotel bill be kept to a minimum.

Captain Kirk said to Spock, "In the first test, you should answer True for all 30 questions. Then you will know how many answers should be True and how many should be False."

"Yes, in the m-th test, $2 \leq m \leq 30$, I will change from test 1 only my answer to the m-th question. I will then know the correct answers to these 29 questions. From test 1, I know the number of correct answers that are supposed to be True, and I can deduce the correct answer to the first question as well. I will get Martian citizenship on the 31st day."

Engineer Scott said, "Let the number of correct answers in a particular attempt be denoted by k. If in the first question, $k = 0$ or 30, there is no problem. If $k = 1$ or 29, we can sort the odd one out by a binary search. If $k = 2$ or 28, we can still use a refined binary search and keep it well under 30 days. Beyond that, it is not clear if we can improve on your scheme."

Everyone promised to work something out, and send it to Spock by ether-net.

When he had finished packing, Spock came back out.

"I am ready," he said to Spark.

"I wish you good luck. If you are successful and get help to have your spaceship repaired, would you take me back to Earth?"

After he had settled into the luxurious room in the *Martian Garden*, Spock turned on his ether-net receptor and began pondering some more about the problem himself. It did not take long before the first transmission arrived. It was the joint work by Kirk and Scott, which improved on Spock's own scheme by one day.

* * * * * * * * * * *

We claim that Spock can get Martian citizenship on his 30th attempt. In the first test, he answers True for all 30 questions. Suppose he gets $k = 15$. In the second test, Spock changes the answers in test 1 to questions 2, 3 and 4. Then $k = 12, 18, 14$ or 16. In the first two cases, Spock knows the correct answers to questions 2, 3 and 4, and has enough tests left to sort out the remaining questions. Hence we may assume by symmetry that $k = 14$. This means that the correct answers to two of questions 2, 3 and 4 are True, and the other one False. Spock then changes the answers from test 1 to questions $2m - 1$ and $2m$ in the m-th test, $3 \leq m \leq 15$. If in the mth test, $k = 13$ or 17, then he knows the correct answers to questions $2m - 1$ and $2m$. Thus we may assume that he gets $k = 15$ in each test. Thus one correct answer is True and other False in each of these 13 pairs of questions. Moreover, Spock now knows that the correct answer to question 1 is False. So far, he has used 15 tests.

In test 16, Spock changes the answers from test 1 to questions 2, 3 and 5, and in test 17 to questions 2, 4 and 5. The following chart shows he can deduce the correct answers to questions 2 to 6. He has just enough tests left to sort out the remaining pairs.

Correct Answer to Question					Value of k in	
2	3	4	5	6	Test 16	Test 17
T	T	F	T	F	12	14
T	F	T	T	F	14	12
F	T	T	T	F	14	14
T	T	F	F	T	14	16
T	F	T	F	T	16	14
F	T	T	F	T	16	16

Suppose that in the first test, Spock gets $k = a$ where $3 \leq a \leq 27$ and $a \neq 15$. He changes the answers from test 1 to questions $2m - 1$ and $2m$ in the m-th test, $2 \leq m \leq 15$. In each of these 14 tests, he will get either $k = a$ or $k = a \pm 2$. In the latter case, he will know the correct answers to questions $2m - 1$ and $2m$. In the former case, he will know that one of these two answers is True and the other is False. Spock will also have similar knowledge about questions 1 and 2 since he knows the value of a. Because $a \neq 15$, Spock must know the correct answers to one pair of questions. Hence he only needs at most 14 more questions to sort out the remaining pairs.

$$* * * * * * * * * * *$$

The next transmission, which arrived several hours later, was the joint work by Doctor McCoy and Nurse Chapel.

$$* * * * * * * * * * *$$

In the first test, we answer True for all 30 questions. Suppose we get $k = a$, where $3 \leq a \leq 27$. We divide on the questions into seven groups of 4, with 2 left over. We claim that we only need three more attempts to find the correct answers to all 4 questions in each group. We use another question to find the correct answer to the 29th question. From the value of a, we will also know the correct answer to the last question. We have used $1 + 7 \times 3 + 1 = 23$ tests so far, and we will pass the quiz on the 24th attempt.

We now justify our claim. Let the four questions be 1, 2, 3 and 4. In the next test, we change the answers to questions 1 and 2. We will have $k = a$ or $k = a \pm 2$. In the latter case, we will know the correct answers to the first two questions. In the former case, we change the answers to questions 2, 3 and 4 in the third test. We may have $k = a \pm 1$ or $k = a \pm 3$. In the fourth test, we change the answersto to questions 1 and 3. Here we may have $k = a$ or $k = a \pm 2$. From these data, we can deduce the correct answers to the first four questions, as shown in the chart below.

Value of k in the		Correct Answer to Question			
Third Test	Fourth Test	1	2	3	4
$a - 3$	a	False	True	True	True
	$a - 2$	False	True	False	True
$a - 1$	a	False	True	True	False
	$a + 2$	True	False	True	True
	$a - 2$	True	False	True	False
$a + 1$	a	True	False	False	True
	$a + 2$	False	True	False	False
$a + 3$	a	True	False	False	False

* * * * * * * * * * *

Just as Spock was calling it a day and preparing to go to bed, the final transmission arrived. It was the joint work of Helmsman Sulu and Officer Uhura.

* * * * * * * * * * *

In the first test, we answer True for all 30 questions. Suppose we get $k = a$, where $3 \leq a \leq 27$. We divide on the questions into four groups of 7, with 2 left over. We claim that we only need five more attempts to find the correct answers to all 7 questions in each group. We use another question to find the correct answer to the 29th question. From the value of a, we will also know the correct answer to the last question. We have used $1 + 4 \times 5 + 1 = 22$ tests so far, and we will pass the quiz on the 23rd attempt.

We now justify our claim. Let the 7 questions be 1, 2, 3, 4, 5, 6 and 7. In the next three tests, we change the answers of the following three triples of questions: (2,3,4), (3,1,5) and (1,2,6). On each attempt, we have $k = a \pm 3$ or $k = a \pm 1$. We consider two cases.

The Martian Citizenship Quiz

Case 1. We have $k = a \pm 3$ at least once.

By symmetry, we may assume that we have $k = a - 3$ for (2,3,4). This means that the answers to questions 2, 3 and 4 are all True. We cannot have $k = a + 3$ for either (3,1,5) or (1,2,6). Suppose we also have $k = a - 3$ for (3,1,5). Then the answers to questions 1 and 5 are also True, and the answer to question 6 is True if $k = a - 3$ for (1,2,6), and False if $k = a - 1$ for (1,2,6). Similarly, we know everything about questions 1 to 6 if we have $k = a - 3$ for (1,2,6). If $k = a - 3$ only for (2,3,4), we have three subcases.

Subcase 1(a). We have $k = a - 1$ for (3,1,5) and (1,2,6).
Then either the answer is True for question 1 and False for each of questions 5 and 6, or the other way round. A fourth test will settle the issue.

Subcase 1(b). We have $k = a + 1$ for (3,1,5) and (1,2,6).
Then the answer to each of questions 1, 5 and 6 is False.

Subcase 1(c). We have $k = a - 1$ for one of (3,1,5) and (1,2,6) and $k = a + 1$ for the other.
We may assume by symmetry that $k = a - 1$ for (3,1,5) and $k = a + 1$ for (1,2,6). Then the answer is False for each of questions 1 and 6 and True for question 5.

An additional test, the fifth at the worst, will tell us the answer to question 7.

Case 2. We have $k = a \pm 1$ all three times.

We have two subcases.

Subcase 2(a). The value of k is not the same for all of (2,3,4), (3,1,5) and (1,2,6).
We may assume by symmetry that $k = a - 1$ for (2,3,4) and (3,1,5), and $k = a + 1$ for (1,2,6). Then there are three possibilities, as summarized in the following table.

True	Falese
2,3,5	1,4,6
1,3,4	2,5,6
3,4,5,6	1,2

In the fourth test, we change the answers to the triple (2,3,5) of questions. We cannot have $k = a + 3$. If $k = a - 3$, we have the first possibility. If $k = a + 1$, we have the second possibility. If $k = a - 1$, we have the third possibility. In all cases, a fifth test will tell us the answer to question 7.

Subcase 2(b). The value of k is the same for all of (2,3,4), (3,1,5) and (1,2,6).

We may assume by symmetry that $k = a - 1$ in all three cases. Then there are four possibilities, as summarized in the following table.

True	False
2,3,5,6	1,4
3,1,6,4	2,5
1,2,4,5	3,6
1,2,3	4,5,6

In the fourth test, we change the answers to the triple (2,3,5) of questions. We cannot have $k = a + 3$. If $k = a - 3$, we have the first possibility. If $k = a + 1$, we have the second possibility. In either case, a fifth test will tell us the answer to question 7. If $k = a - 1$, it can either be the third or the fourth possibilities. In the fifth test, we change the answers to the triple (4,5,7) of questions. If now $k = a - 3$, we have the third possibility and the answer to question 7 is True. If $k = a - 1$, we still have the third possibility but the answer to question 7 is False. If $k = a + 1$, we have the fourth possibility and the answer to question 7 is True. Finally, if $k = a + 3$, we still have the fourth possibility but the answer to question 7 is False.

* * * * * * * * * * * *

His mind at peace, Spcok went to bed, dreaming about the problem. Suddenly, in the middle of a deep sleep, he sat up, turned on the light, went to the desk and began writing.

* * * * * * * * * * * *

In the first test, we answer True for all 30 questions. Suppose we are told that $k = a$, then we know that a answers should be True and $30 - a$ answers should be False. If $a = 0$ or 30, there is no problem. If $k = 1$ or 29, we can sort the odd one out by a binary search. If $k = 2$ or 28, we can still use a refined binary search and keep it well under 20 attempts.

Suppose $3 \leq a \leq 27$. We divide the questions into three groups of 9, with 3 left over. We claim that we only need six more attempts to find the correct answers to all 9 questions in each group. We use another two tests to find the correct answer to the 28th and the 29th questions. From the value of a, we will also know the correct answer to the last question. We have used $1 + 3 \times 6 + 2 = 21$ tests so far, and we will pass the quiz on the 22nd attempt.

We now justify our claim. Let the 9 questions be 1, 2, 3, 4, 5, 6, 7, 8 and 9. In the next four tests, we change the answers for (1,2,3,8), (1,2,4,7), (1,3,4,6) and (2,3,4,5). On each attempt, we have $k = a \pm 4$, $k = a \pm 2$ or $k = a$. We consider six cases.

Case 1. We have $k = a \pm 4$ at least once.

By symmetry, we may assume that we have $k = a - 4$ for (1,2,3,8). In the fifth test, we determine the correct answer for 4. This will also yield the correct answers for 6, 7 and 8. In the sixth test, we determine the correct answer for 9.

Case 2. We have $k = a$ all four times.

The correct answers to the pair (1,5) are the same. This is also true of each of the pairs (2,6), (3,7) and (4,8). In the fifth test, we change the answers for (1,2,5). In the sixth test, we change the answers for (3,7,9). We consider two subcases.

Subcase 2(a). $k = a \pm 3$ in the fifth test.
By symmetry, we may assume that $k = a - 3$. Then 1, 2, 5 and 6 are True while 3, 4, 7 and 8 are False. We cannot have $k = a - 3$ or $k = a - 1$ in the sixth test. If $k = a + 1$, then 9 is true. If $k = a + 3$ instead, 9 is False.

Subcase 2(b). $k = a \pm 1$ in the fifth test.
By symmetry, we may assume that $k = a - 1$. Then 1 and 5 are true while 2 and 6 are False. In the sixth test, if $k = a - 3$, then 3, 7 and 9 are True while 4 and 8 are False. If $k = a - 1$, then 3 and 7 are True while 4, 8 and 9 are False. If $k = a + 1$, 4, 8 and 9 are True while 3 and 7 are False. If $k = a + 3$, then 4 and 8 are True while 3, 7 and 9 are False.

In all subsequent cases, we do not have $k = a \pm 4$ and we have $k = a \pm 2$ at least once. By symmetry, we may assume that we have $k = a - 2$ at least once.

Case 3. We have $k = a - 2$ exactly once.

By symmetry, we assume that this occurs for (1,2,3,8). In the fifth test, we change the answers for (1,2,5). We cannot have $k = a + 3$. There are three subcases.

Subcase 3(a). $k = a - 3$.
Then 1, 2, 5 and 8 are True while 3, 4 and 7 are False. From the value of k for (1,3,4,6), we can deduce the correct answer for 6. In the sixth test, we determine the correct answer for 9.

Subcase 3(b). $k = a - 1$.
It is easy to check that 2 and 5 cannot both be True. Hence 1 is True. In the sixth test, we change the answers for (5,8,9). There are four sub-subcases.

Sub-subcase 3(b$_1$). $k = a - 3$.
Then 1, 3, 5, 8 and 9 are True while 2, 4 and 6 are False. From the value of k for (1,2,4,7), we can deduce the correct answer for 7.

Sub-subcase 3(b_2). $k = a - 1$.
Then 8 is True while one of 5 and 9 is True. If $k = a + 2$ for (1,2,4,7), then 1, 3, 5 and 8 are True while 2, 4, 6, 7 and 9 are False. If $k = a$ for (1,2,4,7), then 1, 2, 8 and 9 are True while 3, 4, 5 and 7 are False. From the value of k for (1,3,4,6), we can deduce the correct answer for 6.

Sub-subcase 3(b_3). $k = a + 1$.
Then 5 is False and one of 8 and 9 is False. If $k = a$ for (2,3,4,5), then 1, 2, 3 and 9 are True while 4, 5, 6, 7 and 8 are False. If $k = a + 2$ for (2,3,4,5), then 1, 2 and 8 are True while 3, 4, 5, 7 and 9 are False. From the value of k for (1,3,4,6), we can deduce the correct answer for 6.

Sub-subcase 3(b_4). $k = a + 3$.
Then 1, 2 and 3 are True while 4, 5, 6, 7, 8 and 9 are False.

Subcase 3(c). $k = a + 1$.
One of 1 and 2 is True and the other is False. Hence 3 and 8 are True while 5 is False. Since $k > a - 1$ for both (1,3,4,6) and (2,3,4,5), 4 must be False. From the value of k for (2,3,4,5), we can determine which of 1 and 2 is True. From the values of k for (1,2,4,7) and (1,3,4,6), we can deduce the correct answers for 6 and 7. In the sixth test, we determine the correct answer for 9.

Case 4. We have $k = a - 2$ exactly twice.

By symmetry, we assume that this occurs for (1,2,3,8) and (2,3,4,5). In the fifth test, we change the answers for (1,2,5). We cannot have $k = a + 3$. There are three subcases.

Subcase 4(a). $k = a - 3$.
Then 1, 2, 3 and 5 are True while 4, 6, 7 and 8 are False.

Subcase 4(b). $k = a - 1$.
It is easy to check that 1 and 2 cannot both be False, and neither can 2 and 5. So 2 is False and 1 and 5 are True. Hence 4 and 8 are True while 3 is False. From the values of k for (1,2,4,7) and (1,3,4,6), we can deduce the correct answers for 6 and 7.

Subcase 4(c). $k = a + 1$.
It is easy to check that 1 and 2 cannot both be True, and neither can 2 and 5. So 2 is True and 1 and 5 are False. Hence 3, 4 and 8 are True while 6 and 7 are False.

In each subcase, we determine the correct answer for 9 in the sixth test.

Case 5. We have $k = a - 2$ exactly thrice.

By symmetry, we assume that this occurs for (1,2,3,8), (1,2,4,7) and (1,3,4,6). In the fifth test, we change the answers for (1,2,5). We cannot have $k = a + 3$. There are three subcases.

Subcase 5(a). $k = a - 3$.
Then 1, 2, 3, 5 and 6 are True while 4, 7 and 8 are False.

Subcase 5(b). $k = a - 1$.
It is easy to check that 1 and 2 cannot both be True, and neither can 2 and 5. So 2 is False and 1 and 5 are True. Hence 3, 4 and 8 are True while 6 and 7 are False.
Subcase 5(c). $k = a + 1$.
It is easy to check that 1 and 2 cannot both be False, and neither can 1 and 5. So 1 is True and 2 and 5 are False. Hence 3, 4, 7 and 8 are True while 6 is False.
In each subcase, we determine the correct answer for 9 in the sixth test.

Case 6. We have $k = a - 2$ all four times.

In the fifth test, we change the answers for (1,2,5). We cannot have $k = a + 3$. There are three subcases.
Subcase 6(a). $k = a - 3$.
Then 1, 2, 5 and 6 are True. Either 3 and 7 are True while 4 and 8 are False, or the other way round. In the sixth test, we change the answers for (3,7,9). That will tell us everything.
Subcase 6(b). $k = a - 1$.
It is easy to check that 2 and 5 cannot both be True. If 1 and 2 are True, then 3 and 4 are also True while 5, 6, 7 and 8 are False. If 1 and 5 are true, then 3, 4, 7 and 8 are also True while 2 and 6 are False. In the sixth test, we change the answers for (3,7,9). That will tell us everything.
Subcase 6(c). $k = a + 1$.
It is easy to check that 1 and 2 cannot both be False, and neither can 2 and 5. So 2 is True and 1 and 5 are False. Hence 3, 4, 6, 7 and 8 are all True. In the sixth test, we determine the correct answer for 9.

* * * * * * * * * * * *

"All's well that ends well," said an elated Spark as he stood next to Spock while the *Enterprise* took off on its homeward journey.

Section 3. Rook Paths

We consider some problems on Rook paths.

Problem 1.
There is a Rook on the square at the bottom left corner of a $2 \times n$ chessboard. It may pass through each square at most once, and may not pass through all four squares which form a 2×2 subboard. What is the number of different Rook paths which ends on any square of the rightmost column?

By symmetry, we may assume that the destination is the square at the top right corner, while the starting square is
   (a) at the bottom left corner;
   (b) at the top left corner.

Call the numbers of such paths a_n and b_n respectively. Below are some initial data.

n	0	1	2	3	4	5	6	7	8	9	10
a_n	0	1	2	3	4	6	10	17	28	45	72
b_n	1	1	1	2	4	7	11	17	27	44	72

We take $a_0 = 0$ and $b_0 = 1$ since on an empty board, there is no room to move from the bottom row to the top row, but no movement is needed if we are already on the top row.

Some patterns are easy to spot.
Observation A. If $n \equiv 1$ or $4 \pmod 6$, then $a_n = b_n$.
Observation B. If $n \equiv 2$ or $3 \pmod 6$, then $a_n = b_n + 1$.
Observation C. If $n \equiv 5$ or $0 \pmod 6$, then $a_n = b_n - 1$.
We shall justify these results using a recursive approach.

Suppose the Rook is on the square at the bottom left corner. Its first move may be to the right, and the number of ways of continuing from there is a_{n-1}. If the first move is upward, then the next two moves must be to the right. The number of ways of continuing from there is b_{n-2}.

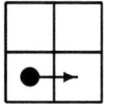 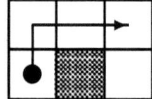

Figure 1.1

We have
$$a_n = a_{n-1} + b_{n-2}, \tag{1}$$
$$b_n = b_{n-1} + a_{n-2}. \tag{2}$$
The second recurrence relation can be proved in an analogous manner.

From (1), we have $b_{n-2} = a_n - a_{n-1}$. Substituting into (2), we have $a_{n+2} - a_{n+1} = a_{n+1} - a_n + a_{n-2}$. This is equivalent to

$$a_n = 2a_{n-1} - a_{n-2} + a_{n-4}.$$

By symmetry, we also have

$$b_n = 2b_{n-1} - b_{n-2} + b_{n-4}.$$

We can now justify our earlier observations by simultaneous induction. The basis can easily be verified from the initial data. Suppose the results hold up to some positive integer n. Let $n \equiv 1 \pmod{6}$. Then $n \equiv 0$, $n - 1 \equiv 5$ and $n - 3 \equiv 3 \pmod{6}$. Hence we have

$$\begin{aligned} a_{n+1} &= 2a_n - a_{n-1} + a_{n-3} \\ &= 2(b_n - 1) - (b_{n-1} - 1) + (b_{n-3} + 1) \\ &= 2b_n - b_{n-1} + b_{n-3} \\ &= b_{n+1}. \end{aligned}$$

The other five cases can be handled in an analogous manner.

A more striking pattern emerges when we consider $t_n = a_n + b_n$.

n	0	1	2	3	4	5	6	7	8	9	10
a_n	0	1	2	3	4	6	10	17	28	45	72
b_n	1	1	1	2	4	7	11	17	27	44	72
t_n	1	2	3	5	8	13	21	34	55	89	144

Note that $t_n = a_n + b_n = (a_{n-1} + b_{n-2}) + (b_{n-1} + a_{n-2}) = t_{n-1} + t_{n-2}$. Hence $\{t_n\}$ satisfies the same recurrence relation as that of the famous Fibonacci sequence.

To explain this rather surprising phenomenon, we use an alternative approach to the problem. Note that the Rook can never move to the left, and vertical moves cannot take place in adjacent columns. Moreover, the path is entirely determined by the vertical moves. The number of vertical moves in any path counted in $\{a_n\}$ must be odd, and the number of vertical moves in any path counted in $\{b_n\}$ must be even.

Suppose we wish to choose k of n vertical columns without choosing two adjacent columns. Denote the chosen columns by 1s and the remaining columns by 0s. This yields a binary sequences of length n with k 1s such that no two are adjacent. If we remove one 0 between every pair of 1s, we obtain a binary sequence of length $n - k + 1$ with k 1s.

Since the process is reversible, we have a one-to-one correspondence between the two classes of binary sequences. The number of binary sequences of either kind is $\binom{n-k+1}{k}$. It follows that

$$a_n = \binom{n}{1} + \binom{n-2}{3} + \binom{n-4}{5} + \cdots,$$

$$b_n = \binom{n+1}{0} + \binom{n-1}{2} + \binom{n-3}{4} + \cdots$$

$$t_n = \binom{n+1}{0} + \binom{n}{1} + \binom{n-1}{2} + \cdots.$$

The diagram below shows that each t_n is the sum of all entries of some diagonal in Pascal's Triangle.

```
                1
              1   1
            1   2   1        t_0 = 1
          1   3   3   1      t_1 = 1 + 1 = 2
        1   4   6   4   1    t_3 = 2 + 1 = 3
      1   5  10  10   5   1  t_4 = 1 + 3 + 1 = 5
                              t_5 = 3 + 4 + 1 = 8
```

Now t_n is the number of binary sequences of length n, with no two 1s adjacent. First we count those in which the last term is 0. Then the first $n-1$ terms can be any binary sequence of length $n-1$, with no two 1s adjacent. Their number is t_{n-1}. Now we count those in which the last term is 1. Then the second last term must be 0, and the first $n-2$ terms can be any binary sequence of length $n-2$, with no two 1s adjacent. Their number is t_{n-2}. It follows that $t_n = t_{n-1} + t_{n-2}$.

Problem 2a.
There is a Rook on the square at the bottom left corner of a $3 \times n$ chessboard. It may pass through each square at most once, and may not pass through all four squares which form a 2×2 subboard. What is the number of different Rook paths which ends on any square of the rightmost column?

By symmetry, we may assume that the destination is the square at the top right corner, while the starting square is
  (c) at the bottom left corner;
  (d) at the top left corner;
  (e) in the middle of the leftmost column.
Call the numbers of such paths c_n, d_n and e_n respectively. Below are some initial data.

Rook Paths

n	1	2	3	4	5	6	7	8	9	10
c_n	1	3	6	10	18	38	86	190	403	837
d_n	1	1	3	9	22	46	91	183	383	819
e_n	1	2	4	8	17	36	76	160	337	710

There are no clear patterns in these sequences. All we can do is to establish some recurrence relations for them.

Suppose the Rook is on the square at the bottom left corner. Its first move may be to the right, and the number of ways of continuing from there is c_{n-1}. Suppose the first move is upward. If the second move is also upward, then the next two moves must be to the right. The number of ways of continuing from there is d_{n-2}. Suppose the second move is to the right. If the third move is also to the right, the number of ways of continuing from there is e_{n-2}. If it is upward, then the next two moves must be to the right. The number of ways of continuing from there is d_{n-3}.

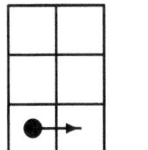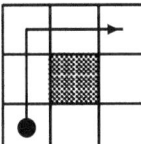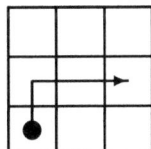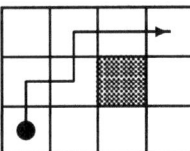

Figure 1.2

We have
$$c_n = c_{n-1} + d_{n-2} + e_{n-2} + d_{n-3}, \tag{3}$$
$$d_n = d_{n-1} + c_{n-2} + e_{n-2} + c_{n-3}, \tag{4}$$
$$e_n = e_{n-1} + c_{n-2} + d_{n-2}. \tag{5}$$

The other two recurrence relations can be proved in an analogous manner.

Combining (3), (4) and (5), we have

$$\begin{aligned} e_{n+2} - e_{n+1} &= c_n + d_n \\ &= (c_{n-1}+d_{n-1}) + (c_{n-2}+d_{n-2}) + 2e_{n-2} + (c_{n-3}+d_{n-3}) \\ &= (e_{n+1} - e_n) + (e_n - e_{n-1}) + 2e_{n-2} + (e_{n-1} - e_{n-2}) \\ &= e_{n+1} + e_{n-2}. \end{aligned}$$

It follows that
$$e_n = 2e_{n-1} + e_{n-4}.$$

The recurrence relations for $\{c_n\}$ and $\{d_n\}$ are much more complicated. We have $c_n - d_n = c_{n-1} - d_{n-1} - c_{n-2} + d_{n-2} - c_{n-3} + d_{n-3}$ by subtracting (5) from (4). This is equivalent to

$$c_n - c_{n-1} + c_{n-2} + c_{n-3} = d_n - d_{n-1} + d_{n-2} + d_{n-3}. \tag{6}$$

Subtracting $c_{n-1} = c_{n-2} + d_{n-3} + e_{n-3} + d_{n-4}$ from (4), we have

$$\begin{aligned} & d_n - c_{n-1} \\ &= (d_{n-1} - c_{n-2}) + (c_{n-2} - d_{n-3}) + (c_{n-4} + d_{n-4}) + (c_{n-3} - d_{n-4}) \\ &= d_{n-1} - d_{n-3} + c_{n-4} + c_{n-3}. \end{aligned}$$

This is equivalent to

$$d_n - d_{n-1} + d_{n-3} = c_{n-1} + c_{n-3} + c_{n-4}. \tag{7}$$

From (6) and (7), $d_{n-2} = (c_n - c_{n-1} + c_{n-2} + c_{n-3}) - (c_{n-1} + c_{n-3} + c_{n-4})$. This is equivalent to

$$d_n = c_{n+2} - 2c_{n+1} + c_n - c_{n-2}.$$

Substituting this into (7), we have

$$\begin{aligned} & c_{n-1} + c_{n-3} + c_{n-4} \\ &= d_n - d_{n-1} + d_{n-3} \\ &= (c_{n+2} - 2c_{n+1} + c_n - c_{n-2}) - (c_{n+1} - 2c_n + c_{n-1} - c_{n-3}) \\ & \quad + (c_{n-1} - 2c_{n-2} + c_{n-3} - c_{n-5}). \end{aligned}$$

This is equivalent to

$$c_n = 3c_{n-1} - 3c_{n-2} + c_{n-3} + 3c_{n-4} - c_{n-5} + c_{n-6} + c_{n-7}.$$

By symmetry, we also have

$$d_n = 3d_{n-1} - 3d_{n-2} + d_{n-3} + 3d_{n-4} - d_{n-5} + d_{n-6} + d_{n-7}.$$

Problem 2b.
There is a Rook on the square in the middle of the leftmost column of a $3 \times n$ chessboard. It may pass through each square at most once, and may not pass through all four squares which form a 2×2 subboard. What is the number of different Rook paths which ends on any square of the rightmost column?

By symmetry, we may assume that the destination is the square in the middle of the rightmost column, while the starting square is
(e) at the bottom left corner;
(e) at the top left corner;
(f) in the middle of the leftmost column.
Note that the first two cases of Problem 2b are equivalent to the last case of Problem 2a. Hence both are also labeled (e). We only have to deal with the last case of Problem 2b. Call the numbers of such paths f_n. Below are some initial data.

n	1	2	3	4	5	6	7	8	9	10
e_n	1	2	4	8	17	36	76	160	337	710
f_n	1	1	3	7	15	31	65	137	289	609

The first move of the Rook may be to the right, and the number of ways of continuing from there is f_{n-1}. Suppose the first move is upward or downward. Then the next two moves must be to the right. The number of ways of continuing from either position is e_{n-2}.

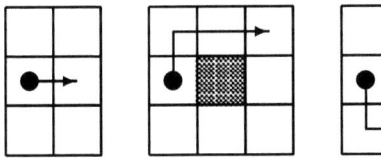

Figure 1.3

It follows that
$$f_n = f_{n-1} + 2e_{n-2}.$$
This is equivalent to $e_{n-2} = \frac{1}{2}(f_n - f_{n-1})$. Since $e_n = 2e_{n-1} - e_{n-4}$, we have $\frac{1}{2}(f_{n+2} - f_{n+1}) = f_{n+1} - f_n - \frac{1}{2}(f_{n-2} - f_{n-3})$. This is equivalent to
$$f_n = 3f_{n-1} - 2f_{n-2} + f_{n-4} - f_{n-5}.$$

Actually, $\{f_n\}$ satisfies a simpler recurrence relation, in fact, the same one satisfied by $\{e_n\}$. We prove this by mathematical induction. The basis can easily be verified from the initial data. Suppose $f_n = 2f_{n-1} + f_{n-4}$ holds for some $n \geq 4$. Then

$$\begin{aligned} f_{n+1} &= 3f_n - 2f_{n-1} + f_{n-3} - f_{n-4} \\ &= 2f_n + (2f_{n-1} + f_{n-4}) - 2f_{n-1} + f_{n-3} - f_{n-4} \\ &= 2f_n + f_{n-3}. \end{aligned}$$

This completes the inductive argument.

Exercises

1. Without worrying about the minimum number of crossing required, prove by mathematical induction on n that the crossing is possible for any positive number n of thieves.

2. While in the *Martian Garden*, Commander Spock considered the following approach. In the mth test for $1 \leq m \leq 29$, answer the mth question True and all other questions False. Show that he could pass the quiz on the 30th test.

3. There is a Queen on the square at the bottom left corner of a $2 \times n$ chessboard. It may pass through each square at most once, and may not pass through all four squares which form a 2×2 subboard. The number of different Queen paths which ends on the top right corner is denoted by a_n, and the number of different Queen paths which ends on the bottom right corner is denoted by b_n. Find a recurrence relation for each of a_n and b_n.

Bibliography

[1] Daniel Chiu and Kelvin Shih, Rook paths, Mathematics Competitions **28** (2015) #1 57–63.

[2] Yen-Kang Fu and Te-Cheng Liu, The Martian Citizenship Quiz, Mathematics Competition **29** #2 67–70.

[3] Bob Henderson and Dick Hess, private communications.

[4] International Mathematics Tournament of the Towns A-Level Senior Problem 7 in Fall 2008.

[5] International Mathematics Tournament of the Towns O-Level Junior Problem 5 and Senior Problem 3 in Spring 2014.

Solution to Exercises

1. For $n = 1$ or 2, the whole party can cross together. For $n = 3$, the crossing can be accomplished in the following five steps.

   (1) Alibaba and the thieves ranked 1 and 2 go to the far shore.
   (2) The thieves ranked 1 and 2 come back to the near shore.
   (3) The thieves ranked 2 and 3 go to the far shore.
   (4) Alibaba and the thief ranked 2 come back to the near shore.
   (5) Alibaba and the thieves ranked 1 and 2 go to the far shore.

   Suppose the crossing is possible for $n = 1, 2, \ldots, k$ for some $k \geq 3$. Then the crossing for $n = k+1$ can be accomplished in the following five steps.

   (1) Alibaba and the thieves ranked 1 to k go to the far shore.
   (2) The thieves ranked $k-1$ and k come back to the near shore.
   (3) The thieves ranked k and $k+1$ go to the far shore.
   (4) Alibaba and the thieves ranked 1 to $k-2$ come back to the near shore.
   (5) Alibaba and the thieves ranked 1 to $k-1$ go to the far shore.

   Note that Step (1), Step (4) and Step (5) are possible by the induction hypothesis, with $n = k$, $k-2$ and $k-1$ respectively. This is why we need to include the case $k = 3$ as part of the basis.

2. Let k_m be the number of correct answers in the mth test. We consider two cases.
   Case 1. $k_1 = k_2 = \cdots = k_{29}$.
   This means that the answers to the first 29 questions are the same. Hence the answer key must be one of TT...TT, TT...TF, FF...FT and FF...FF. These correspond to 1, 2, 28 and 29 being the common value of $k_1 = k_2 = \cdots = k_{29}$. Spock can pass the quiz on the 30th test.
   Case 2. $k_1, k_2, \ldots, k_{29}$ are not all the same.
   These numbers differ from one another by 0 or 2. By symmetry, we may assume that $a_1 - a_2 = 2$. Then the correct answer to question 1 is True and that to question 2 is False. For $3 \leq m \leq 29$, the correct answer to the mth question is True if $a_1 - a_m = 0$, and False if $a_1 - a_m = 2$. Finally, from the value of a_1 and the number of correct answers in the first 29 questions, Spock will know the correct answer to the 30th question.

3. The initial data are $a_1 = b_1 = 1$ and $a_2 = b_2 = 3$. For $n \geq 3$, the diagram below shows the only combinations of the first few moves.

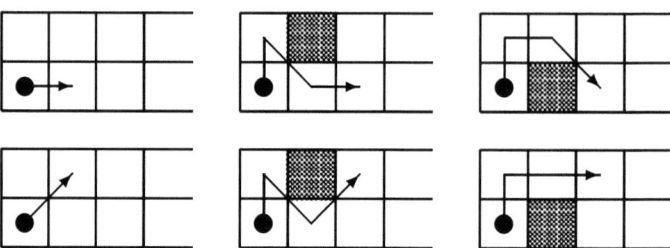

Figure 1.4

This yields $a_n = a_{n-1} + 2a_{n-2} + b_{n-1} + 2b_{n-2}$. Similarly, we also have $b_n = b_{n-1} + 2b_{n-2} + a_{n-1} + 2a_{n-2}$. It follows that $a_n = b_n$ for all $n \geq 1$, and we have $a_n = 2a_{n-1} + 4a_{n-2}$ as well as $b_n = 2b_{n-1} + 4b_{n-2}$.

Chapter Two
A Sample Minicourse
TESSELLATIONS

Section 1. Platonic and Archimedean Tilings

A **tessellation** or **tiling** of the plane is a dissection of the plane into regions called **tiles**. This definition is far too general to be of any interest. Thus we will impose additional conditions as we go along. The following two properties are basic to all tessellations under discussion.

- Each tile is a convex polygon.
- When two tiles meet, the common part is a complete edge in each tile. Such a tessellation is said to be **edge-to-edge**.

Figure 2.1 shows three familiar tessellations.

Figure 2.1

These three tessellations share several additional properties.

- Each tile is a regular polygon.
- All tiles are identical. Such a tessellation is said to be **monohedral**.
- All vertices are surrounded by tiles in the same way. Such a tessellation is said to be **monomorphic**.

Remark:
The usual name for what we call a monomorphic tessellation is a *uniform* tessellation. Later, we will come across tessellations in which a vertex may be surrounded by tiles in one of two ways. The usual name for such a tessellation is a *2-uniform* tessellation. Like *dual-monopoly*, this name is an **oxymoron**. We avoid it, and use **dimorphic** instead.

A monohedral tessellation with regular polygons as tiles is necessarily monomorphic. Such a tessellation is called a **Platonic** tiling.

To form a tessellation, the tiles must be able to cover a point completely and without overlap. The first step is to determine which combinations of regular polygons have this property.

The sum of the exterior angles of a regular n-gon is $360°$. Hence each of its interior angles is $(\frac{1}{2} - \frac{1}{n})360°$. We consider four cases.

Case 1. The point is covered by 3 non-overlapping regular polygons. Let $a \leq b \leq c$ be the numbers of sides of the three regular polygons. Then $\frac{1}{2} - \frac{1}{a} + \frac{1}{2} - \frac{1}{b} + \frac{1}{2} - \frac{1}{c} = 1$. This may be rewritten as $\frac{1}{a} + \frac{1}{b} + \frac{1}{c} = \frac{1}{2}$. We consider four subcases.

Subcase 1(a). $a = 3$.
We have $2bc + 6b + 6c = 3bc$ so that $c = \frac{6b}{b-6}$. Hence $b > 6$. When $b = 7$, $c = 42$. When $b = 8$, $c = 24$. When $b = 9$, $c = 18$. When $b = 10$, $c = 15$. When $b = 11$, c is not integral. When $b = 12$, $c = 12$. There are no further values since $b \leq c$.

Subcase 1(b). $a = 4$.
We have $bc + 4b + 4c = 2bc$ so that $c = \frac{4b}{b-4}$. Hence $b > 5$. When $b = 5$, $c = 20$. When $b = 6$, $c = 12$. When $b = 7$, c is not integral. When $b = 8$, $c = 8$. There are no further values since $b \leq c$.

Subcase 1(c). $a = 5$.
We have $2bc + 10b + 10c = 5bc$ so that $c = \frac{10b}{3b-10}$. Note that $b \geq a = 5$. When $b = 5$, $c = 10$. When $b = 6$, c is not integral. When $b \geq 7$, $c < b$. Thus there are no further values.

Subcase 1(d). $a \geq 6$.
Suppose $a = 6$. Then $bc + 6b + 6c = 3bc$ so that $c = \frac{3b}{b-3}$. Note that $b \geq a = 6$. When $b = 6$, $c = 6$. There are no further values since $a \leq b \leq c$.

Case 2. The point is covered by 4 non-overlapping regular polygons. Let $a \leq b \leq c \leq d$ be the numbers of sides of the four regular polygons. Then $\frac{1}{a} + \frac{1}{b} + \frac{1}{c} + \frac{1}{d} = 1$. We consider four subcases.

Subcase 2(a). $a = b = 3$.
We have $2cd + 3c + 3d = 3cd$ so that $d = \frac{3c}{c-3}$. Hence $c > 3$. When $c = 4$, $d = 12$, When $c = 5$, d is not integral. When $c = 6$, $d = 6$. There are no further values since $c \leq d$.

Subcase 2(b). $a = 3$ and $b = 4$.
We have $7cd + 12c + 12d = 12cd$ so that $d = \frac{12c}{5c-12}$. Note that $c \geq b = 4$. When $c = 4$, $d = 6$. When $c \geq 5$, $d < c$. Thus there are no further values.

Subcase 2(c). $a = 3$ and $b \geq 5$.
Since $\frac{1}{3} + \frac{1}{5} + \frac{1}{5} + \frac{1}{5} < 1$, there are no values.

Subcase 2(d). $a = 4$.
Suppose $b = 4$. Then $cd + 2c + 2d = 2$ so that $d = \frac{2c}{c-2}$. Note that $c \geq b = 4$. When $c = 4$, $d = 4$. There are no further values since $a \leq b \leq c \leq d$.

Case 3. The point is covered by 5 non-overlapping regular polygons.
Let $a \le b \le c \le d \le e$ be the numbers of sides of the five regular polygons.
Then
$$\frac{1}{a}+\frac{1}{b}+\frac{1}{c}+\frac{1}{d}+\frac{1}{e}=\frac{3}{2}.$$
We consider two subcases.
Subcase 3(a). $a = b = c = 3$.
We have $2de + 2d + 2e = 3de$ so that $e = \frac{2d}{d-2}$. When $d = 3$, $e = 6$. When $d = 4$, $e = 4$. There are no further values since $d \le e$.
Subcase 3(b). $a = b = 3$, $c \ge 4$.
Since $\frac{1}{3}+\frac{1}{3}+\frac{1}{4}+\frac{1}{4}+\frac{1}{4} < \frac{3}{2}$, there are no values.
Case 4. The point is covered by 6 or more non-overlapping regular polygons.
Suppose we have 6 regular polygons of sides $a \le b \le c \le d \le e \le f$. Then
$$\frac{1}{a}+\frac{1}{b}+\frac{1}{c}+\frac{1}{d}+\frac{1}{e}+\frac{1}{f}=2.$$
The only values are $a = b = c = d = e = f = 3$. Thus there are no values if we have 7 or more regular polygons.

In summary, there are seventeen combinations, namely, (3,7,42), (3,8,24), (3,9,18), (3,10,15), (3,12,12), (4,5,20), (4,6,12), (4,8,8), (5,5,10), (6,6,6), (3,3,4,12), (3,3,6,6), (3,4,4,6), (4,4,4,4), (3,3,3,3,6), (3,3,3,4,4) as well as (3,3,3,3,3,3). We get four more because (3,3,4,12), (3,3,6,6), (3,4,4,6) and (3,3,3,4,4) may be modified into (3,4,3,12), (3,6,3,6), (3,4,6,4) and (3,3,4,3,4) respectively.

We observe that there are only three Platonic tilings, those shown in Figure 1.

Of the other eighteen combinations, we have to reject six of them outright, namely, (3,7,42), (3,8,24), (3,9,18), (3,10,15), (4,5,20) and (5,5,10). In each combination, if we choose a polygon with an odd number of sides, then the other two polygons do not have the same number of sides, and cannot be placed alternately around the chosen polygon. We say that such a combination has a *local* problem.

If we want monomorphic tessellations, we have to reject four more combinations, namely (3,3,4,12), (3,4,3,12), (3,3,6,6) and (3,4,4,6). Consider the first two together. If we surround a dodecagon with 12 triangles, we will create points where two squares meet. The same thing happens if we surround it with 9 triangles and 3 evenly spaced squares. If we surround the dodecagon with 8 triangles and 4 evenly spaced squares, we will create points where two dodecagons meet (actually overlap). Finally, if we surround the dodecagon with 6 triangles alternately with 6 squares, we will create points where three triangles meet. It is easy to show that the other two combinations also have *global* problems.

The remaining eight combinations can all be extended to monomorphic tessellations. Six of them are dihedral and the other two are trihedral. Together, they are called **Archimedean** tilings. These are shown in Figure 2.2.

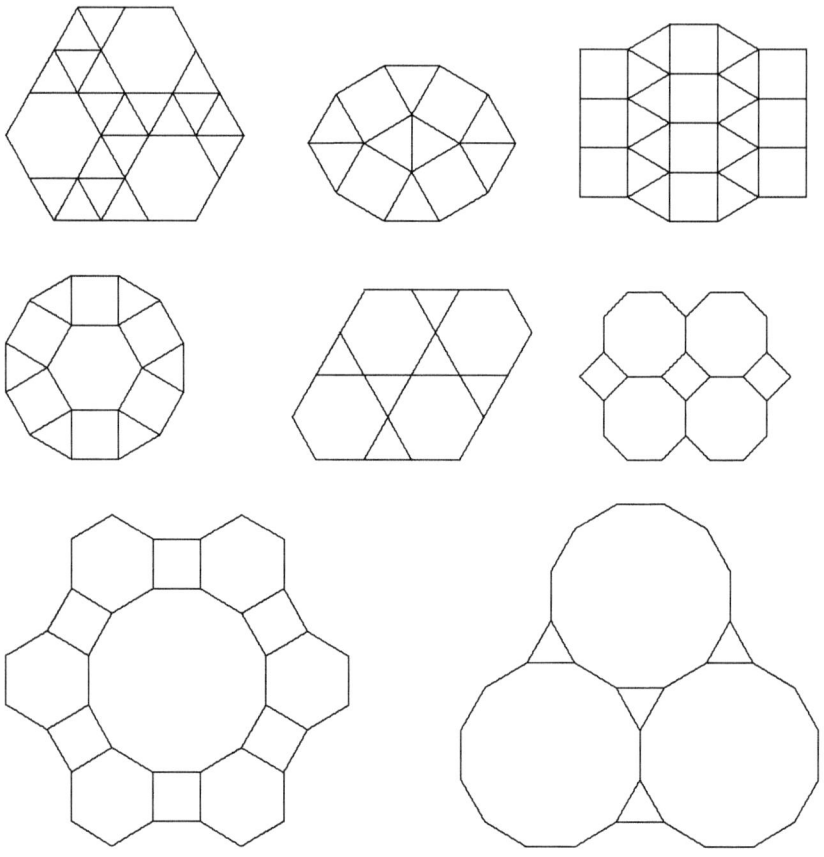

Figure 2.2

How can we justify that the Platonic and Archimedean tilings do not have global problems? To do so, we abandon local constructions of the tessellations by adding tiles to existing structures, and give global constructions of entire tessellations.

If we take two mutually perpendicular infinite sets of parallel lines which are evenly spaced at the same interval apart, we obtain the (4,4,4,4) Platonic tiling. Suppose we take two infinite sets of parallel lines which are evenly spaced at the same interval apart, but making 60° angles with one another. Then we add a third infinite set of parallel lines, also evenly spaced at the same interval apart.

If these lines pass through the points of intersection of the first two sets, we have the (3,3,3,3,3,3) Platonic tiling. If these lines pass through the points halfway between the points of intersections of the first two sets, we have the (3,6,3,6) Archimedean tiling.

The other eight tessellations may be obtained from these basic constructions by various means. The simplest case is to take alternate strips from the (4,4,4,4) and the (3,3,3,3,3,3) tessellations, which yields the (3,3,3,4,4) Archimedean tiling.

Another way is by means of combining certain existing tiles into larger tiles. Figure 2.3 shows how the (6,6,6) and the (3,3,3,3,6) tessellations may be obtained from the (3,3,3,3,3,3) tessellations. Note that the basic (3,6,3,6) tessellations may also be obtained this way.

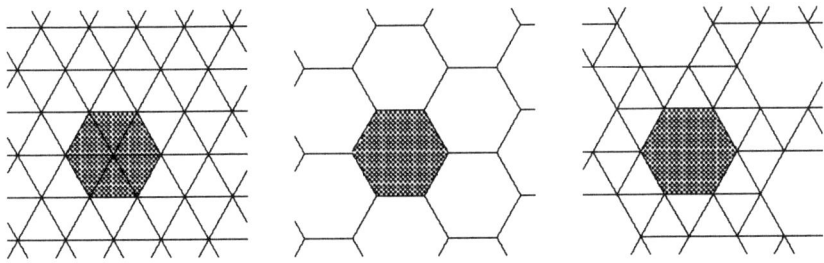

Figure 2.3

Another method is that of cutting corners and then combining them. Let us construct the (4,8,8) tessellation from the basic (4,4,4,4) tessellation. We cut each square tile into five pieces, as shown in Figure 2.4, consisting of a regular octagon at the center and four congruent right isosceles triangles at the corners. Let the edge length of the square tile be 1 and the length of the hypotenuse of the triangles be x. Then the legs of the triangles have length $\frac{x}{\sqrt{2}}$. From $\frac{x}{\sqrt{2}} + x + \frac{x}{\sqrt{2}} = 1$, we have $x = \sqrt{2} - 1 \approx 0.412$. When we merge the triangles across four square tiles, we obtain the (4,8,8) tessellations.

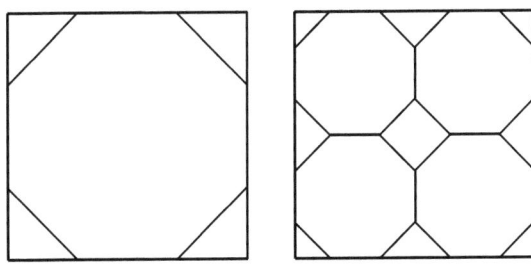

Figure 2.4

We now use the same approach to construct the (3,12,12) tessellation from the (6,6,6) tessellation. We cut each hexagonal tile into seven pieces, as shown in the diagram below on the left, consisting of a regular dodecagon at the center and six congruent isosceles triangles at the corners, with vertical angles 120°. Let the edge length of the hexagonal tile be 1 and the length of the base of the triangles be x. Then the equal sides of the triangles have length $\frac{x}{\sqrt{3}}$. From $\frac{x}{\sqrt{3}} + x + \frac{x}{\sqrt{3}} = 1$, we have $x = 2\sqrt{3} - 3 \approx 0.464$. When we merge the triangles across three hexagonal tiles, we obtain the (3,12,12) tessellation, as shown in the Figure 2.5.

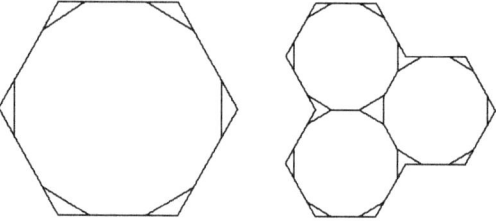

Figure 2.5

See what happens if we cut the corners all the way to the midpoints of the sides of the tile. From the (4,4,4,4) tessellation, we get it back. From the (6,6,6) tessellation, we get the basic (3,6,3,6) tessellation a third time.

The (3,4,6,4) tessellation can also be obtained from the basic (3,3,3,3,3,3) tessellation. This time, we cut each triangular tile into seven pieces, as shown in Figure 2.6, consisting of an equilateral triangle at the center, three congruent half-squares along the sides and three congruent kites at the corners, with angles 120°, 90°, 60° and 90°. Let the edge length of the triangular tile be 1 and the length of the side of the equilateral triangle be x. Then the short sides of the kite have length $\frac{x}{2}$ and the long sides $\frac{\sqrt{3}x}{2}$. From $\frac{\sqrt{3}x}{2} + x + \frac{\sqrt{3}x}{2} = 1$, we have $x = \frac{\sqrt{3}-1}{2} \approx 0.366$. When we merge across six triangular tiles the kites into regular hexagons and the half-squares into squares, we obtain the (3,4,6,4) tessellation.

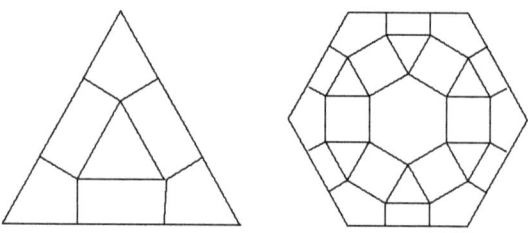

Figure 2.6

Platonic and Archimedean Tilings   33

The (4,6,12) tessellation can now be obtained from the (3,4,6,4) tessellation without cutting. Each dodecagon in the new tessellation is obtained by merging one regular hexagon, six squares and six equilateral triangles in the old tessellation, as shown in Figure 2.7.

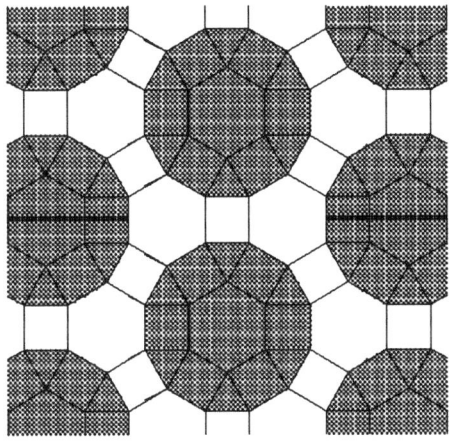

Figure 2.7

The last tessellation, namely (3,3,4,3,4), is the most difficult to get. It is obtained from the basic (4,4,4,4) tessellation with an intermediate step. We first modify the square tile by cutting out two isosceles triangles with vertical angles 150°, based on two opposite sides of the square tile, and attaching them to the other two sides. This modified non-convex tile can also tile the plane, as shown in Figure 2.8.

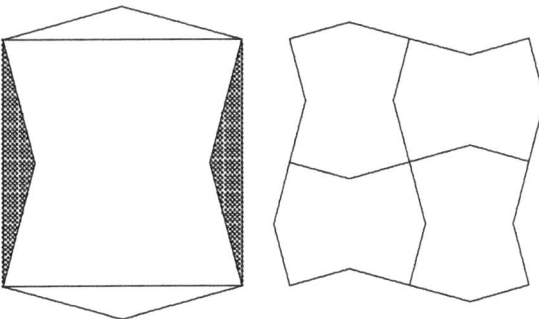

Figure 2.8

We now cut each modified tile into six pieces, as shown in Figure 2.9, consisting of two congruent equilateral triangles and four congruent right isosceles triangles. When we merge the right isosceles triangles into squares across four modified tiles, we obtain the (3,3,4,3,4) tessellation.

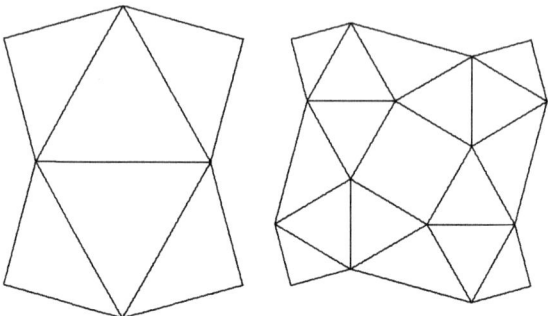

Figure 2.9

To make use of the four combinations which do not have local problems, namely, (3,3,4,12), (3,4,3,12), (3,3,6,6) and (3,4,4,6), we go to dimorphic tessellations.

The tessellation in Figure 2.10 has two kinds of vertex sequences, namely, (3,3,4,12) and (3,3,3,3,3,3).

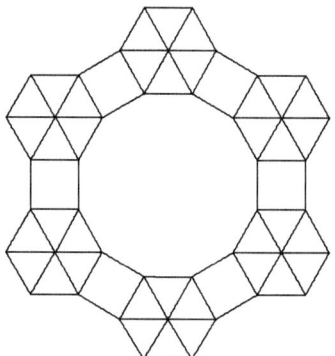

Figure 2.10

The tessellation in Figure 2.11 has two kinds of vertex sequences, namely, (3,4,3,12) and (3,12,12).

Platonic and Archimedean Tilings 35

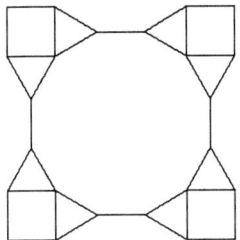

Figure 2.11

The tessellations in Figure 2.12 have two kinds of vertex sequences, one of which is (3,3,6,6). The others are (3,3,3,3,6), (3,6,3,6) and (3,3,3,3,3,3), respectively.

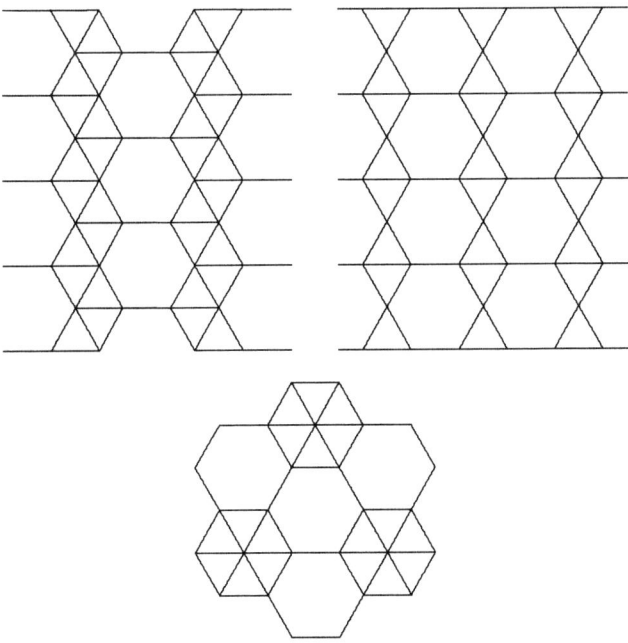

Figure 2.12

The tessellations in Figure 2.13 have two kinds of vertex sequences, one of which is (3,4,4,6). The others are (3,6,3,6), (3,6,3,6) and (3,4,6,4) respectively.

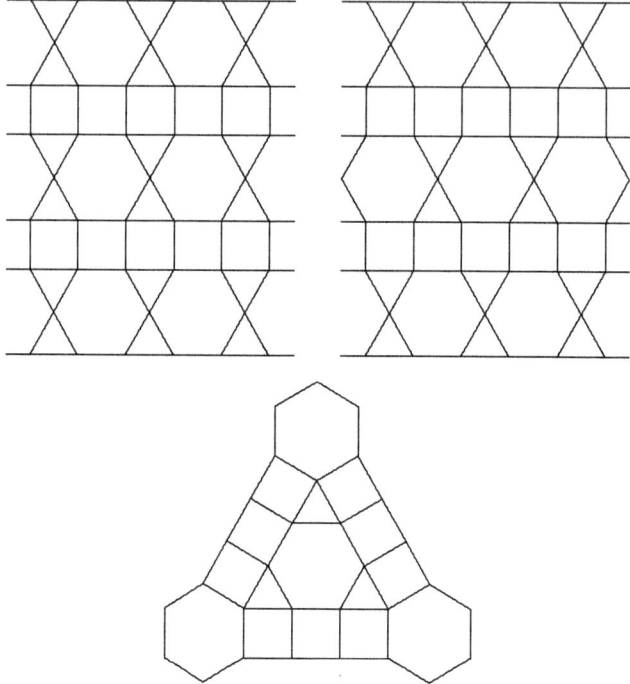

Figure 2.13

There are twenty dimorphic tessellations altogether. The proof is much more complicated, and is omitted. Four of them are obtained by putting together strips from the (4,4,4,4) and the (3,3,3,3,3,3) tessellations. The tessellations in Figure 2.14 feature the vertex sequences (3,3,3,3,3,3) and (3,3,3,4,4).

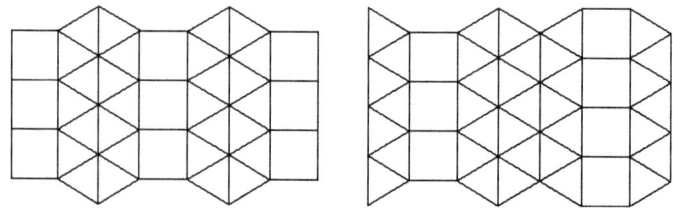

Figure 2.14

The tessellations in Figure 2.15 feature the vertex sequences (3,3,3,4,4) and (4,4,4,4).

Platonic and Archimedean Tilings

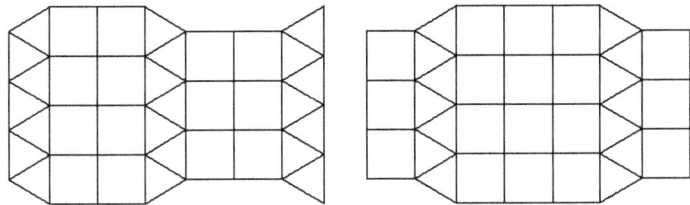

Figure 2.15

In the tessellation on the left, each (4,4,4,4) vertex is adjacent to two other (4,4,4,4) vertices as well as two (3,3,3,4,4) vertices. In the tessellation on the right, each (4,4,4,4) vertex is adjacent to three other (3,3,3,3,3,3) vertices as well as one (3,3,3,4,4) vertex. This uniformity will not be maintained if we insert a fourth column of squares between two columns of equilateral triangles, even though no new vertex sequences are introduced. For the same reason, we cannot insert a fourth column of equilateral triangles between two columns of squares in Figure 2.14.

The remaining eight dimorphic tessellations feature the following combinations of vertex sequences.

1. (3,3,3,3,3,3) and (3,3,3,3,6);
2. (3,3,3,3,3,3) and (3,3,3,3,6), a different form;
3. (3,3,3,4,4) and (3,3,4,3,4);
4. (3,3,3,4,4) and (3,3,4,3,4), a different form;
5. (3,3,3,3,3,3) and (3,3,4,3,4);
6. (3,3,4,3,4) and (3,4,6,4);
7. (3,3,3,4,4) and (3,4,6,4);
8. (3,4,6,4) and (4,6,12).

These are left as exercises.

Exercises

1. Obtain the (3,6,3,6) tessellation from
   (a) the (3,3,3,3,3,3) tessellation;
   (b) the (6,6,6) tessellation.

2. Find two dimorphic tessellations for each of the following combinations of vertex sequences.

   (a) (3,3,3,3,3,3) and (3,3,3,3,6);
   (b) (3,3,3,4,4) and (3,3,4,3,4).

3. Find a dimorphic tessellation for each of the following combinations of vertex sequences.

   (a) (3,3,3,3,3,3) and (3,3,4,3,4);
   (b) (3,3,4,3,4) and (3,4,6,4);
   (c) (3,3,3,4,4) and (3,4,6,4);
   (d) (3,4,6,4) and (4,6,12).

Section 2. From Tessellations to Rectifications

We now turn to a different kind of tiles. A **polyomino** is a figure consisting of unit squares joined edge to edge. There are 1 monomino, 1 domino, 2 trominoes, 5 tetrominoes, 12 pentominoes and 35 hexominoes. They are shown in Figure 2.16, with the pentominoes and hexominoes separated by their smaller cousins which are shaded.

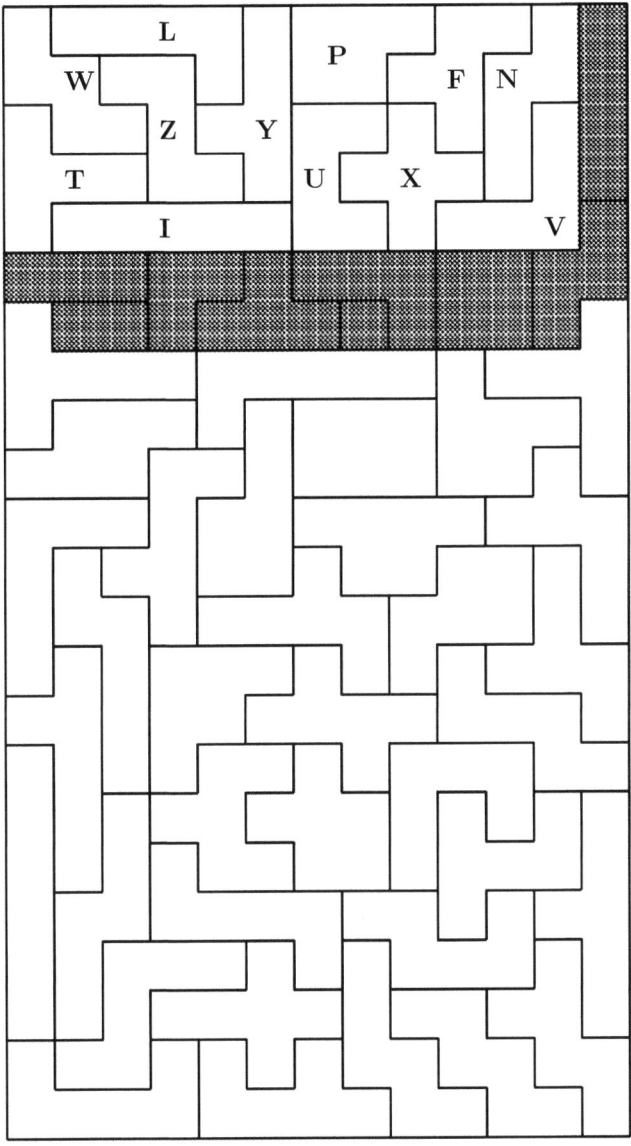

Figure 2.16

Each of the twelve pentominoes has a single-letter name based on its shape, as shown in Figure 2.16. All pentominoes tile the plane. We consider them in four groups of three.

Group 1. The I-, N- and Y-pentominoes.
We tile the plane with a decomino as shown in Figure 2.17 on the left. This decomino can be formed with two copies of the I-, N- or Y-pentomino.

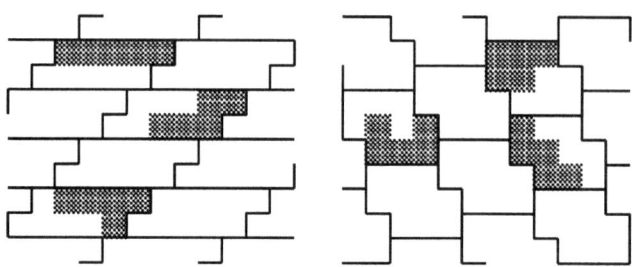

Figure 2.17

Group 2. The P-, U- and W-pentominoes.
We tile the plane with a decomino as shown in Figure 2.17 on the right. This decomino can be formed with two copies of the P-, U- or W-pentomino.

Group 3. The F-, T- and X-pentominoes.
We tile the plane with an icosomino as shown in Figure 2.18 on the left. This icosomino can be formed with four copies of the F-, T- or X-pentomino.

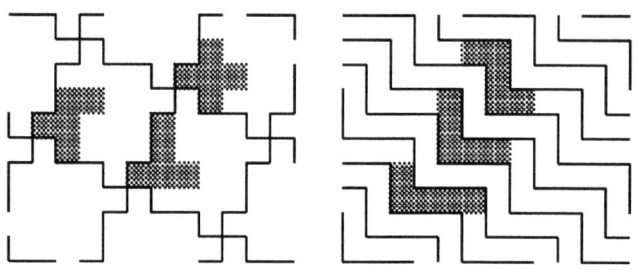

Figure 2.18

Group 4. The L-, V- and Z-pentominoes.
We tile the plane with an infinite region of width 1, as shown in Figure 2.18 on the right. This region can be tiled by copies of the L-, V- or Z-pentomino.

The I-pentomino is itself a rectangle, and copies of the L-, P- and Y-pentominoes can tile rectangles, as shown Figure 2.19.

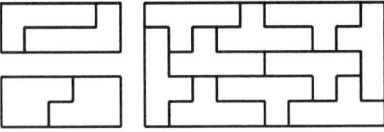

Figure 2.19

The N- and W-pentominoes cannot tile rectangles since neither can fill a finite edge. Figure 2.20 below shows that either of them can tile an infinite bent strip. Of course, anything that can tile a rectangle can tile a bent strip.

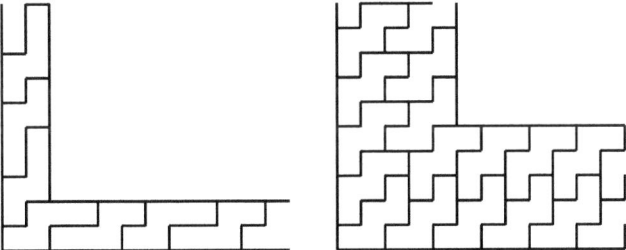

Figure 2.20

The F- and V-pentominoes cannot tile a bent strip since neither can fill a corner. Figure 2.21 shows that either of them can tile an infinite branched strip. Of course, anything that can tile a bent strip can tile a branched strip simply by putting two tiled bent strips side by side.

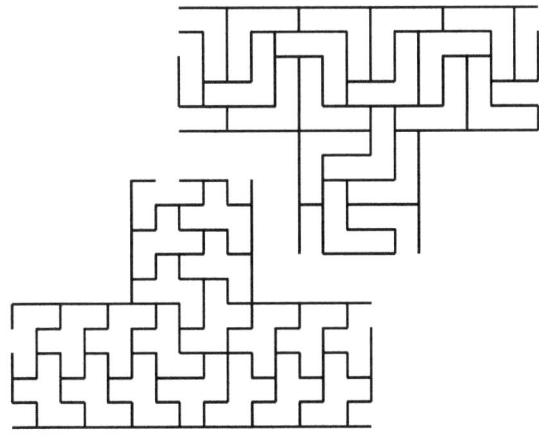

Figure 2.21

Anything that can tile a branched strip can tile an infinite crossed strip, simply by putting two tiled branched strips side by side. Figure 2.22 shows a tiling of a crossed strip with the F-pentomino that is not obtained this way.

Figure 2.22

Anything that can tile a crossed strip can tile an infinite straight strip. Essentially, we ignore three of its four arms and extend the fourth one into a straight strip. To see that this is always possible, let the height of this arm be h and the width of the polyomino in question be w. Divide this arm into infinitely many $2w \times h$ rectangles. Within each rectangle, there is at least one zig-zag path going from the top edge to the bottom edge without cutting any copy of the polyomino. This zig-zag path can take finitely many different shapes, say n of them for some positive integer n. If we take $n+1$ rectangles, the Pigeonhole Principle guarantees that there are two zig-zag lines which are identical. The tiles between these two zig-zag lines form a larger polyomino that can tile a straight strip. This is illustrated in Figure 2.23, again with the F-pentomino.

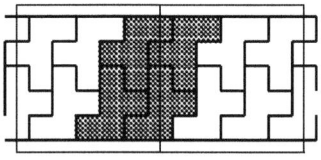

Figure 2.23

The T-, U-, X- and Z-pentominoes cannot tile an infinite strip. Figure 2.24 shows all possible positions of them against an edge of a straight strip. In each case, either the lone shaded square cannot be filled, or one of the shaded squares cannot be filled.

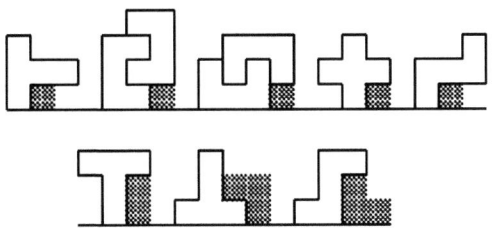

Figure 2.24

We have established the following tiling hierarchy.
Class 1. Those which can tile a rectangle.
Class 2. Those which can tile a bent strip but not a rectangle.
Class 3. Those which can tile a branched strip but not a bent strip.
Class 4. Those which can tile a crossed strip but not a branched strip.
Class 5. Those which can tile a straight strip but not a crossed strip.
Class 6. Those which can tile the plane but not a straight strip.
Class 7. Those which cannot tile the plane.

Of the twelve pentominoes, the I-, L-, P- and Y-pentominoes are in Class 1, the N- and W-pentominoes are in Class 2, the F- and V-pentominoes are in Class 3, and the T-, U-, X- and Z-pentominoes are in Class 6. There are no pentominoes in Class 4, 5 or 7.

Of the smaller polyominoes, all are in Class 1 except the S-tetromino which is in Class 2. Polyominoes in Class 1 are said to be *rectifiable*, and the process of forming rectangles is called **rectification**.

We now turn our attention to the hexominoes. We give their classifications with the tessellations, but without the justification that, except for those in class 1, they are not in an earlier class. The arguments are similar to those for the pentominoes.

Of the 35 hexominoes, 2 are rectangles themselves and 8 others are rectifiable. They are shown in Figure 2.25, with 5 of them requiring only 2 copies while the other 3 require 4, 18 and 92 copies respectively!

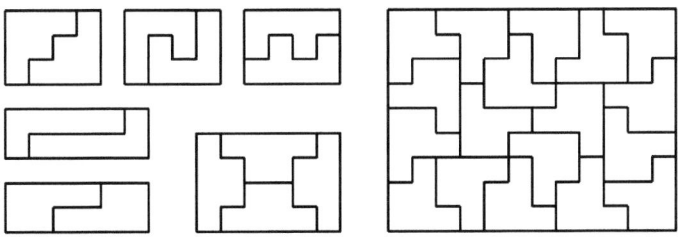

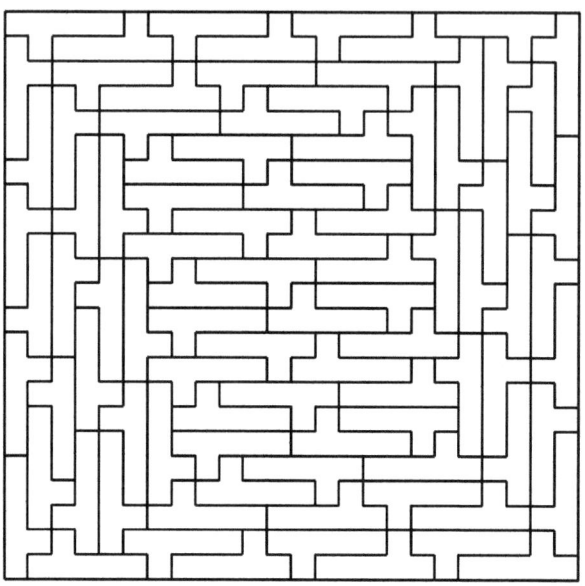

Figure 2.25

Figure 2.26 shows the 5 hexominoes in Class 2.

Figure 2.26

From Tessellations to Rectifications 45

There are no hexominoes in Class 7, and Figure 2.27 shows the 10 hexominoes which are in Class 6.

Figure 2.27

Figure 2.28 shows four hexominoes in Class 3.

Figure 2.28

The remaining 6 hexominoes may also be in Class 3, but are currently not. They are left as exercises.

Our tiling hierarchy focuses on strips. Other meaningful tiling hierarchies may be defined. The following is an example. It focuses on quadrants.
Class A. Those which can tile a rectangle.
Class B. Those which can tile a quadrant of the plane but not a rectangle.
Class C. Those which can tile a half plane but not a quadrant of the plane.
Class D. Those which can tile the plane but not a half-plane.
Class E. Those which cannot tile the plane.

There is an obvious class which may be added, namely, those which can tile three quadrants of the plane, but not a quadrant of the plane. Since anything which tiles a quadrant can tile three quadrants, this class sits below Class B. It is not clear how it compares with Classes C, D and E. Thus the inclusion of this class may make the tiling hierarchy non-linear.

Exercises

4. Find three hexominoes currently in Class 4. They may have expectations of promotion to Class 3, but this is unlikely.

5. Find three hexominoes currently in Class 5. They may have expectations of promotion to Class 4, but this is unlikely. They may even have great expectations of promotion to Class 3, and this is most unlikely.

6. In all the examples of rectification we have encountered, the number of copies of the polyomino used is always even. Find a non-rectangular polyomino an odd number of copies of which can form rectangle.

Section 3. Frieze and Wallpaper Patterns

A **Frieze pattern** is a periodic tiling of an infinite straight strip. This means that the tiling has a translation symmetry along the axis of the strip, that is, the line halfway between the edges of the strip. We first consider a few examples using the tetrominoes as tiles.

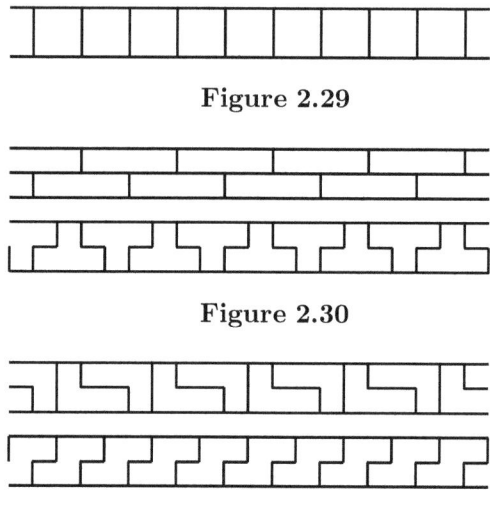

Figure 2.29

Figure 2.30

Figure 2.31

The Frieze pattern in Figure 2.29 and the bottom Frieze pattern in Figure 2.31 map into themselves after a translation of 2 units along their axes. The Frieze patterns in Figure 2.30 and the top Frieze pattern in Figure 2.31 map into themselves after a translation of 4 units along its axis.

All three Frieze patterns have another kind of symmetry in common, namely, half-turns or 180° rotations. These must be about points which lie on the axes. In the bottom Frieze pattern in Figure 2.30, the centers must be at the midpoints of edges of the tiles along the axis. In all of the other Frieze patterns, there are two kinds of half-turn centers.

There is a third symmetry shared by the Frieze patterns in Figures 2.29 and 2.30, but not by the Frieze pattern in Figure 2.31, namely, lateral reflections. These are reflections across lines perpendicular to the axis. In Figure 2.29, there are two kinds of lateral reflections, with the lines of reflection along the vertical edges, or along the perpendicular bisectors of the horizontal bases. In Figure 2.30, the lines of reflections must be along the perpendicular bisectors of the horizontal bases.

The Frieze pattern in Figure 2.29 possesses a fourth kind of symmetry, namely, a central reflection across the axis. This is not shared by the Frieze patterns in Figures 2.30 and 2.31.

Frieze and Wallpaper Patterns

A fifth kind of symmetry is shared by the Frieze patterns in Figures 2.29 and 2.30 but not by the Frieze pattern in Figure 2.31, namely, glide reflections. A glide reflection is a composition of a central reflection with a translation in either order. The Frieze patterns in Figures 2.29 and 2.30 map into themselves after a central reflection and a translation of 2 units along their axes.

We now make a simple observation.

Theorem 1.
If a Frieze pattern has central reflection symmetry, then it automatically has glide reflection symmetry.

Our next observation is a bit more involved.

Theorem 2.
Suppose a Frieze pattern has two of half-turn symmetry, lateral reflection symmetry and glide reflection symmetry. Then it will also have the third kind.

Proof:
Figure 2.32 shows four possible orientations of an infinite strip, and the symmetries which take them among one another. The composition of any two of half-turn symmetry, lateral reflection symmetry and glide reflection symmetry is the third kind, and the desired conclusion follows.

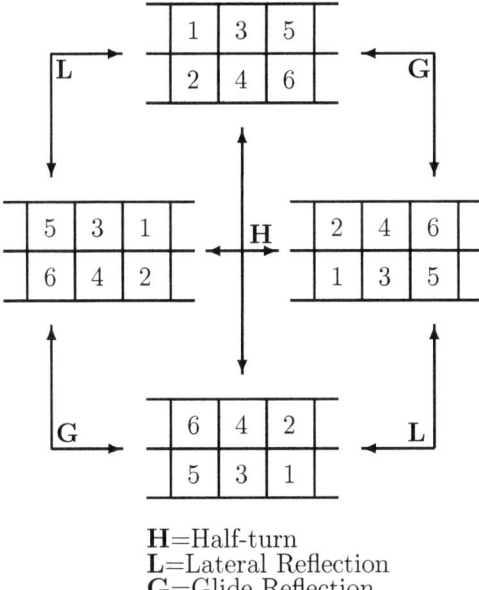

H=Half-turn
L=Lateral Reflection
G=Glide Reflection

Figure 2.32

These two theorems will help us determine the number of different Frieze patterns according to what symmetries they have.

With four kinds of optional symmetry in addition to translation, there are sixteen possible cases which are summarized in the following table.

Half -turns	Reflections			Realizable Cases?
	Lateral	Central	Glide	
Yes	Yes	Yes	Yes	Figure 2.29
Yes	Yes	Yes	No	Theorem 1/2
Yes	Yes	No	Yes	Figure 2.30
Yes	Yes	No	No	Theorem 2
Yes	No	Yes	Yes	Theorem 2
Yes	No	Yes	No	Theorem 1
Yes	No	No	Yes	Theorem 2
Yes	No	No	No	Figure 2.31
No	Yes	Yes	Yes	Theorem 2
No	Yes	Yes	No	Theorem 1
No	Yes	No	Yes	Theorem 2
No	Yes	No	No	Figure 2.33
No	No	Yes	Yes	Figure 2.34
No	No	Yes	No	Theorem 1
No	No	No	Yes	Figure 2.35
No	No	No	No	Figure 2.36

As it turns out, nine of the sixteen cases are not realizable. Either Theorem 1 or Theorem 2 is used as the justification. The remaining seven cases are realizable. Thus there are seven types of Frieze patterns. An example of each is shown in Figures 2.29 to 2.31 and 2.33 to 2.36. We use the domino in Figure 2.33 and hexominoes in Figures 2.34, 2.35 and 2.36.

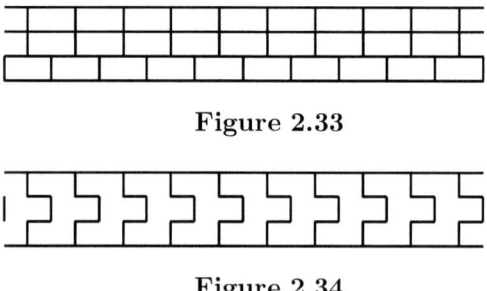

Figure 2.33

Figure 2.34

Frieze and Wallpaper Patterns

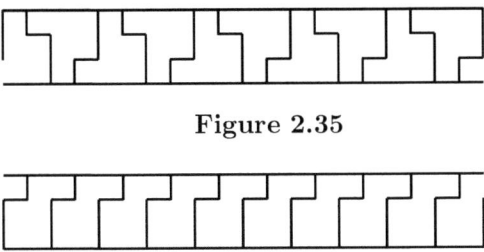

Figure 2.35

Figure 2.36

A **Wallpaper pattern** is a periodic tiling of the plane. This means that the tiling has a translation symmetry along two non-parallel axes. The terms central and lateral reflections are no longer meaningful, and rotations are not restricted to half-turns or 2-fold symmetries.

It is known that each rotation must be a 2-fold, 3-fold, 4-fold or 6-fold symmetry. The proof is omitted. Clearly, a pattern with a 4-fold symmetry also has 2-fold symmetries, and a pattern with a 6-fold symmetry also has 2-fold and 3-fold symmetries. However, a pattern may not have both 3-fold and 4-fold symmetries, as otherwise it would have a 12-fold symmetry. Thus the Wallpaper patterns may be put into five groups.

1. **Group I.** Those with no rotational symmetries.
2. **Group II.** Those with 2-fold symmetries only.
3. **Group III.** Those with 4-fold symmetries only.
4. **Group IV.** Those with 3-fold symmetries.
5. **Group V.** Those with 6-fold symmetries. There are two types.

Every reflection is automatically a glide reflection, but the converse is not true. The reflectional symmetries put the Wallpaper patterns four classes.

1. **Class A.** Those with no glide reflections.
2. **Class B.** Those with at least one glide reflection which is also a reflection, and at least one glide reflection which is not a reflection.
3. **Class C.** Those with at least one glide reflection which is also a reflection, but no glide reflections which are not reflections.
4. **Class D.** Those with at least one glide reflection which is not a reflection, but no glide reflections which are also reflections.

The following table summarizes the twenty combinations, six of which are not realizable while three contain two types of patterns each. We omit the justification for the non-realizable cases, as well as the proof that the seventeen types of Wallpaper patterns constitute a complete list.

Groups	Class A	Class B	Class C	Class D
I	2.37(c)	2.37(a)	2.37(b)	2.37(d)
II	2.38(a)	2.38(b,e)	2.38(c)	2.38(d)
III	2.39(c)	2.39(a,b)	Not Realizable	
IV	2.40(b)	2.40(a,c)		
V	2.41(a)	2.41(b)		

The realizable cases are shown in Figures 2.37 to 2.41. Centers of rotations are marked with black dots. Axes of glide reflections are drawn as dotted lines. Those which run partially or totally along edges of the tiles are marked with arrows. Polyominoes are used as tiles in Figures 2.37 to 2.39.

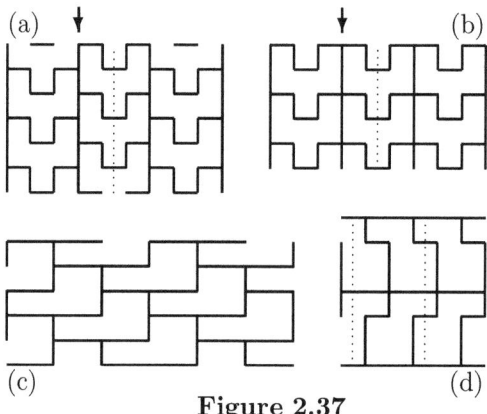

Figure 2.37

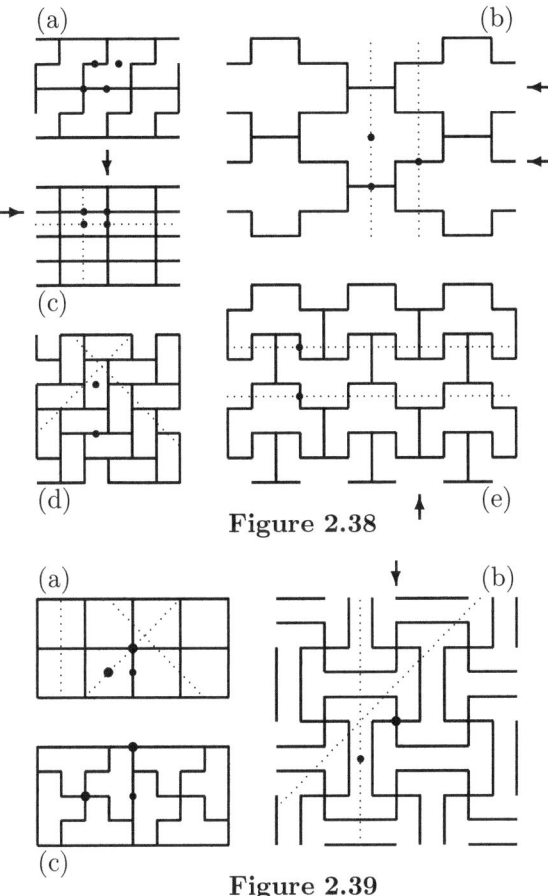

Figure 2.38

Figure 2.39

Since polyominoes do not have 3-fold or 6-fold rotational symmetries, we turn to their cousins, the polyiamonds, for use as tiles in Figures 2.40 and 2.41. A **polyiamond** is a figure consisting of unit equilateral triangles joined edge to edge. In particular, a regular hexagon is a hexiamond. Note that the tessellation in Figure 2.40(c) is dihedral.

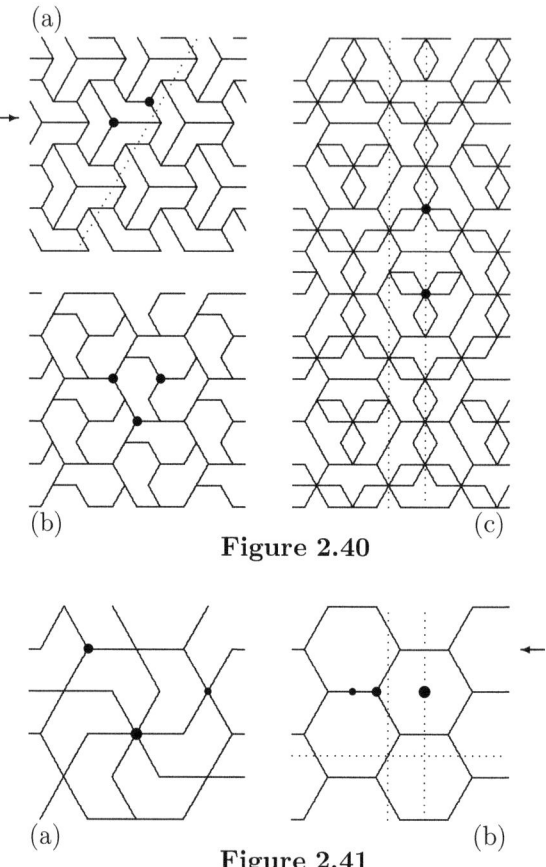

Figure 2.40

Figure 2.41

Exercises

7. Find an example of each of Types 1, 2 and 3 of Frieze patterns using hexominoes in strips of width 3.

8. Since polyiamonds do not have 4-fold rotational symmetry, the first type of Frieze patterns cannot be formed with polyiamonds. The problematic four type, which required dominoes for its construction, is also not possible with polyiamonds. Find an example of each of the other five types formed with polyiamonds

9. Do there exist monohedral tessellations of the plane which are non-periodic?

Bibliography

[1] Dahlke, Karl, The Y-hexomino has order 92, J. Comb. Theory Series A **51** (1989) #1 127–128.

[2] Gardner, Martin, Polyominoes and rectification, in *The Mathematical Magic Show*, Mathematical Association of America (1989) 172–187.

[3] Gardner, Martin, Tiling with convex polygons, in *Time Travel*, W. H. Freeman (1987) 163–176.

[4] Gardner, Martin, Tiling with polyominoes, polyiamonds and polyhexes, in *Time Travel*, W. H. Freeman (1987) 177–187.

[5] Gardner, Martin, Penrose tiling, in *Penrose Tiles to Trapdoor Ciphers*, Mathematical Association of America (1997) 1–30.

[6] Golomb, Solomon, Tiling with polyominoes, J. Comb. Theory **1** (1966) 280–296.

[7] Golomb, Solomon, Tiling with sets of polyominoes, J. Comb Theory **9** (1970) 60–71.

[8] Golomb, Solomon, *Polyominoes: Puzzles, Patterns, Problems & Packings*, Princeton University Press (1994).

[9] Grünbaum, Branko and Shephard, G. C., *Tilings and Patterns*, W. H. Freeman (1986).

[10] Marlow, T. W., Grid dissectionc, Chessics **23** (1985) 78–79.

[11] Martin, George, *Transformation Geometry*, Springer-Verlag (1982).

[12] Schattschneider, Doris, *Visions of Symmetry*, W. H. Freeman (1990).

Solution to Exercises

1. (a) The (3,6,3,6) tessellation is obtained from the (3,3,3,3,3,3) tessellation by combining tiles, as shown in Figure 2.42 on the left.

Figure 2.42

   (b) The (3,6,3,6) tessellation is obtained from the (6,6,6) tessellation by cutting corners and then combining them, as shown in Figure 2.43 on the right.

2. (a) The dimorphic tessellations featuring (3,3,3,3,3,3) and (3,3,3,3,6) are in Figure 2.43.

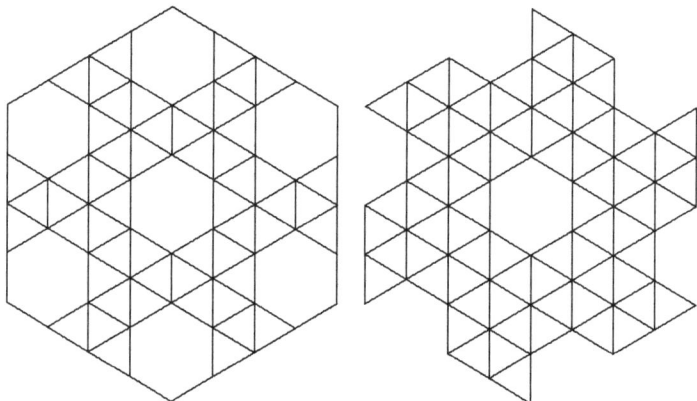

Figure 2.43

   (b) The dimorphic tessellations featuring (3,3,3,4,4) and (3,3,4,3,4) are in Figure 2.44.

Solution to Exercises

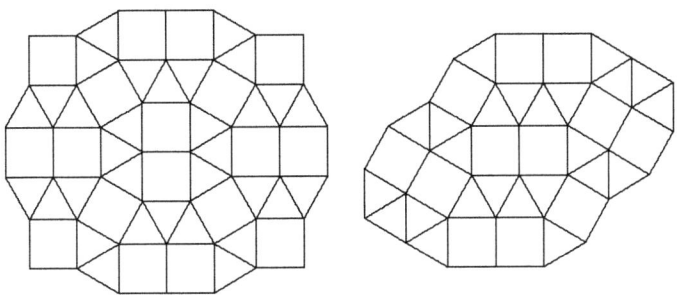

Figure 2.44

3. (a) The dimorphic tessellation featuring (3,3,3,3,3,3) and (3,3,4,3,4) is in Figure 2.45 on the left.

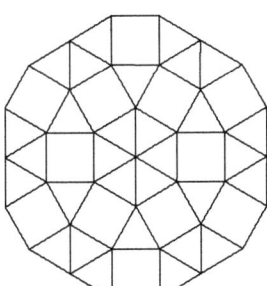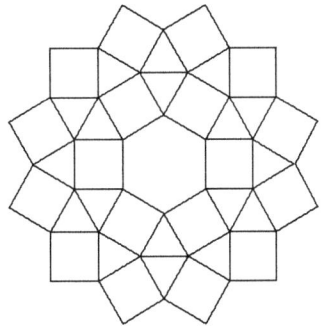

Figure 2.45

(b) The dimorphic tessellation featuring (3,3,4,3,4) and (3,4,6,4) is in Figure 2.45 on the right.

(c) The dimorphic tessellation featuring (3,3,3,4,4) and (3,4,6,4) is in Figure 2.46 on the left.

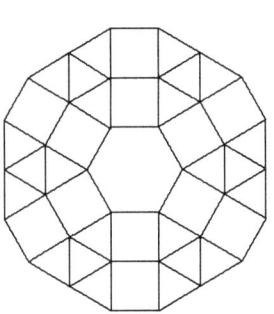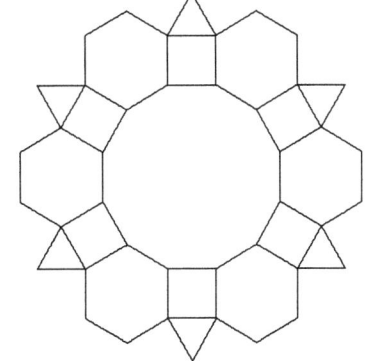

Figure 2.46

(d) The dimorphic tessellation featuring (3,4,6,4) and (4,6,12) is in Figure 2.46 on the right.

4. Figure 2.47 shows three hexominoes currently in Class 4.

Figure 2.47

Solution to Exercises

5. Figure 2.48 shows three hexominoes currently in Class 5.

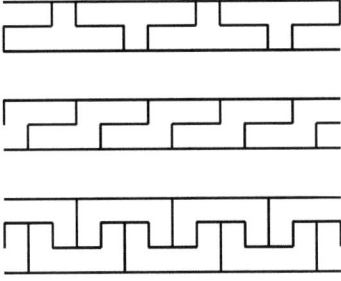

Figure 2.48

6. Figure 2.49 shows that 15 copies of the V-tromino can form a 5×9 rectangle.

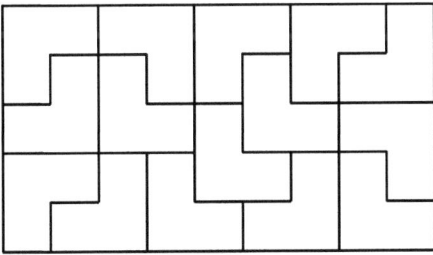

Figure 2.49

7. These are shown in Figures 2.50, 2.51 and 2.52.

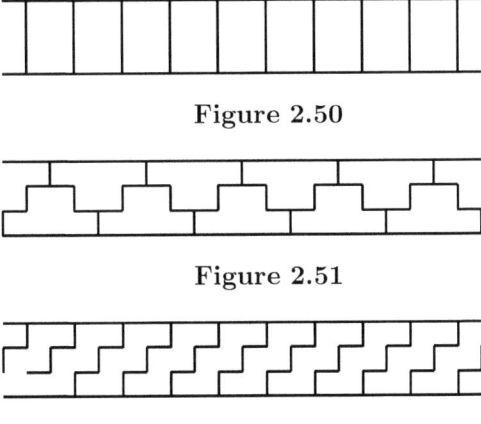

Figure 2.50

Figure 2.51

Figure 2.52

8. These are shown in Figures 2.53 (with tetriamonds), 2.54 (with noniamonds) and 2.55 (with hexiamonds).

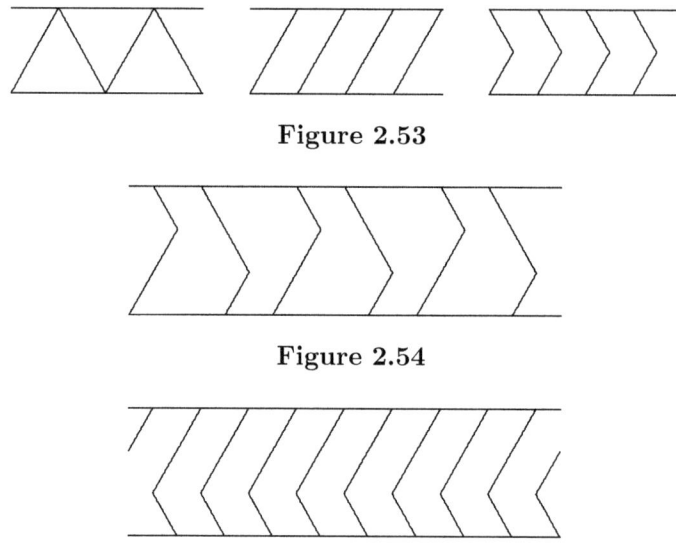

Figure 2.53

Figure 2.54

Figure 2.55

9. Monohedral non-periodic tessellations of the plane exist. For example, draw five evenly spaced rays from the origin, dividing the plane into five sectors with central angles 72°. We use a 72° − 54° − 54° triangle as our tile. It can tessellate each of the sectors. Recall that periodic tessellations can only have 2-fold, 3-fold, 4-fold and 6-fold rotational symmetries. Since our example has 5-fold rotational symmetry, it is non-periodic. Note however that our tile can also be used to form a periodic tessellation. Whether there exists a single tile which can form only non-periodic tessellations is an open problem.

PART TWO
MATHEMATICAL COMPETITIONS

The S.M.A.R.T. Circle is not intended to serve as a training ground for competitions. Nevertheless, many Circle members benefit from their experience in the Circle and perform extremely well in various competitions.

Most of our members take part in the Edmonton Junior High Mathematics Contest. This is organized by a group of dedicated teachers. It was founded in 1977 as a joint venture between Calgary and Edmonton. In the first six years, the paper was set in Calgary in odd-numbered years and in Edmonton in even-numbered years. In 1983, the two branches became separate contests. The Edmonton group was headed initially by **Bill Bober**, then by **Emily Kalwarowski** and later by **Robert Wong**.

Up until 1999, the contest paper consisted of questions with multiple choices, questions requiring answers only and problems requiring full solutions. Logistic reasons dictated that the last group had to be abandoned. The S.M.A.R.T. Circle picked up the slack, and instituted the Edmonton Junior High Mathematics Invitational in 2000. The paper consists of four problems requiring full solutions, usually one in number theory, one in algebra, one in geometry and one in combinatorics. The questions and solutions from 2000 to 2015 are given in Chapter 3.

In 1999, a new international contest was formed in Kaohsiung, Taiwan. It was called the Invitational World Youth Mathematics Intercity Competition for junior high school students. It was meant to be a junior version of the International Mathematical Olympiad. In 2008, it merged with the Elementary Mathematics International Contest to form the International Mathematics Competition. The current president, **Wen-Hsien Sun**, is the founder of our sister organization, the *Chiu Chang Mathematical Circle* of Taipei, Taiwan. For the questions and solutions from 1999 to 2013, see *An In-Depth Study of the International Mathematics Competition: Junior High School Division, 1999–2013*. It was written by W. H. Sun, H. Zheng, and H. W. Zhu, and published by Chiu Chang Mathematics Publishers, Taipei, in 2014.

The participating units are individual cities rather than countries or regions. This was done deliberately to downplay political rhetorics. Edmonton is so far the only Canadian city to participate, largely because of the S.M.A.R.T. Circle. We started in 2005, and had participated every year since 2007. The details are in the following table.

Year	Host	Leaders	Members
2005	Kaohsiung Taiwan	Robert Wong Alan Tsay	Sean Jai, Chengxi Qiu, Ray Yang, Sven Zhang
2007	Changchun China	Gilbert Lee Emily Cliff	Yuri Delanghe, Ranek Kiil, Michael Meyers, Mariya Sardarli
2008	Chiangmai Thailand		Mariya Sardarli
2009	Durban South Africa		Mariya Sardarli
2010	Incheon South Korea	Andy Liu B.-K. Chun David Rhee	Dale Chensong, Weilian Chu, Heejoo Nam, Michael Rue, Desmond Sisson, Angus Tulloch, Giavanna Valacco, Lingfeng Zhu
2011	Denpasar Indonesia	Andy Liu Sean Graves	Dale Chensong, Michael Cao, Andrew Ho, Daniel Jin, Darren Li, Ling Long Desmond Sisson, Dennis Situ, Henry Song, Angus Tulloch, Kevin Wang, Scott Wang
2012	Taipei Taiwan	Andy Liu Sean Graves Ryan Morrill Alan Tsay	Jack Chen, Richard Kang, Daniel Jin, Darren Li, Ling Long, Dennis Situ, Henry Song, Johnson Tang, Sammy Wu
2013	Burgas Bulgaria	Andy Liu Ryan Morrill Weilian Chu	Jack Chen, Brian Kehrig, Richard Kang, Dennis Situ, Johnson Tang, Longxiang Wang, Daniel Zhou, Jeffrey Zhou
2014	Daejon South Korea	Sean Graves Ryan Morrill	Jack Chen, Josh Geng, Richard Kang, Richard Mah, Mark Nie, Longxiang Wang, Poplar Wang, Daniel Zhou
2015	Changchun China	Sean Graves Ryan Morrill	Richard Kang, Mark Nie, Steven Shi, Poplar Wang

Josh Geng, Richard Kang, Brian Kehrig, Heejoo Nam, Desmond Sisson, Steven Shi and Jeffrey Zhou were from Calgary, and Angus Tulloch was from Rimbey. Outside of Alberta, Daniel Zhou came from Saskatchewan and Scott Wang from Taiwan. They participated under the Edmonton banner. Among team leaders, **Alan Tsay, Gilbert Lee, Emily Cliff, Byung-Kyu Chun, David Rhee** and **Weilian Chu** were former Circle members.

Naturally, the top members of the S.M.A.R.T. Circle have aspirations for membership on the Canadian National Team which competes in the International Mathematical Olympiad. We had contributed a number of team members somewhat disproportionate to our relatively small population. They were **Byung-Kyu Chun** in 1995–1997, the late *Robert Barrington Leigh* in 2002–2003, **David Rhee** in 2004–2006 and **Maria Sardarli** in 2011.

At about the same level as the I.M.O. is the International Mathematics Tournament of the Towns, arguably the best mathematics competition in the world. In each academic year, there is a Fall Round and a Spring Round. In each Round, there is a Junior paper and a Senior paper, and each paper has an O-Level version and an A-Level version. Unlike the International Mathematical Olympiad, students write the contest locally. Thus there is no travel cost, and any number of students may participate.

The problems are in general hard, but not just for the sake of being hard. Most have elegant ideas behind them. Our Circle members love this contest, even when they are unable to solve any of the problems. These are the kind of problems which you really want to know how to solve, and not just to get them out of the way.

The glorious history of the Tournament is now well-documented. The founding father is the great **Nikolay Konstantinov**. In recent years, the mantle had been handed over to **Sergey Dorichenko** and **Boris Frenkin**. For more details, see the following books published by the Australian Mathematics Trust, Canberra, under the leadership of **Peter Taylor**.

[1] Taylor, P. J. *International Mathematics Tournament of the Towns: 1980–1984*, AMT, Canberra, 1993.

[2] Taylor, P. J. *International Mathematics Tournament of the Towns: 1984–1989*, AMT, Canberra, 1992.

[3] Taylor, P. J. *International Mathematics Tournament of the Towns: 1989–1993*, AMT, Canberra, 1994.

[4] Storozhev, A. M. and Taylor, P. J. *International Mathematics Tournament of the Towns: 1993–1997*, AMT, Canberra, 1983.

[5] Storozhev, A. M. *International Mathematics Tournament of the Towns : 1997–2002*, AMT, Canberra, 2006.

[6] Liu, A. and Taylor, P. J. *International Mathematics Tournament of the Towns : 2002–2007*, AMT, Canberra, 2009.

Over the years, the International Mathematics Tournament of the Towns has proposed consistently outstanding problems, each with an elegant idea behind it. We collect a score of the best of them in Chapter 4. They are not necessarily the very best, but each leads naturally to further investigations. We give each a detailed development.

Top performers in the Tournament are invited to a Summer Seminar, usually held within Russia. Circle members Matthew Wong and Daniel van Vliet went in 1993 to Beloretsk. Circle member David Rhee went in 2005 to Mir Town in Belarus, along with Jerry Lo of the Chiu Chang Mathematical Circle.

The camp consists of eight days of intensive activities divided into two parts. After the Opening Ceremony on the first day, six projects, each consisting of a sequence of questions, are presented. The students are allowed the second and the third days to work on them, with whatever solutions they can come up with due at 10:00 pm of the third days. They are not expected to work on all the projects. In fact, most concentrated on just one or two of them. Progress is difficult, but rewarding. Students may work individually or in teams, and may consult accompanying teachers.

The fourth day is the transition from the first part to the second part. While solutions to the questions posed so far are presented, a follow-up sequence of questions for each project is presented. One of the next three days is for an excursion, the day dependent on weather conditions. The other two days are for continuing work on the projects. The final deadline is at 10:00 pm of the seventh day.

Most of the eighth day is free while the Jury completes the meticulous grading. In the Closing Ceremony that evening, each student receives a detailed report card on their accomplishments in the camp, along with appropriate prizes. It is a most wonderful junior research experience.

Chapter Three
Past Papers of the Edmonton
Junior High Mathematics Invitational

Section 1. Problems.

2000

1. A computer is infected by viruses. At night, each virus splits into two viruses. When this is discovered, there are already 64 of them. An anti-virus is introduced and it kills off one virus, so that 63 are left. Like a virus, each anti-virus splits into two anti-viruses at night, each of which will kill off one virus the next day. Will all the viruses be killed off eventually if nothing else is done?

2. Each year, the Easter Bunny awards children with Easter eggs. A Bunny Committee recommends 1 for Ace, 2 for Bea and 3 for Cec. The Easter Bunny is fair-minded and imposes the rule that each child should be given the same number of Easter eggs. The Bunny Committee gets around this by recommending 2 for each of Bea and Cec, and the next day, it recommends 1 for each of Ace and Cec. The Easter Bunny is forgetful, and goes along with any recommendation which follows the rule on the day.

    (a) Suppose the Bunny Committee wants to give 1 Easter egg to Ace, 3 to Bea, 6 to Cec, 8 to Dee and 9 to Eve. What is the minimum number of days needed to get around the Easter Bunny?

    (b) What is the highest total number of Easter eggs such that no matter how they are distributed, among any number of children, it will take no more than two days to get around the Easter Bunny?

3. A computer infected by an anti-virus is locked into the following program. It shows on the screen the numbers 2, 3 and 1 in succession. The next number is equal to the units digit of the sum of the last three numbers. So the fourth number is 2+3+1=6 but the fifth is 0 because 3+1+6=10, and so on.

    (a) Will this computer ever show on the screen the numbers 1, 2 and 3 in succession?

    (b) Will there be a sequence of three numbers which will be shown on the screen in succession on at least two occasions?

    (c) Will this computer ever show on the screen the numbers 2, 3 and 1 in succession a second time?

4. Figure 3.1 shows a irregular five-pointed star $ABCDE$. What is the value of $\angle EAB + \angle ABC + \angle BCD + \angle CDE + \angle DEA$?

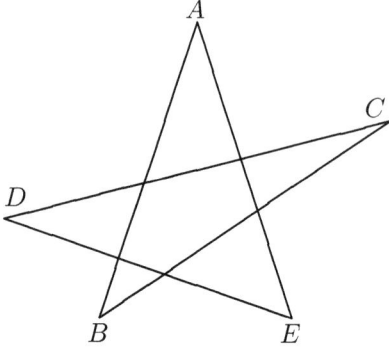

Figure 3.1

2001

1. Find 100 different positive integers such that the product of any five of them is divisible by the sum of these five numbers.

2. You are trying to cover a 5×5 chessboard with any combinations of the three kind of pieces in Figure 3.2.

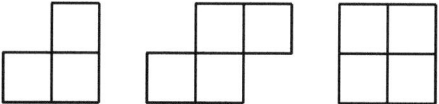

Figure 3.2

Each square of a piece covers a square of the chessboard. No overlapping within or protrusions beyond are allowed. Is the task possible? Either give such a covering or prove that no such coverings exist.

3. Let x and y be any numbers. Prove that $(1+x)^2 + (1+y)^2 + x^2 + y^2 \geq 1$.

4. $ABCD$ is a rectangle. E is any point on BC, and F is any point on CD. AE and BF intersect at G. AF and DE intersect at H. BF and DE intersect at K. Which is larger, the area of $AGKH$ or the sum of the areas of BEG, DFH and $CEKF$?

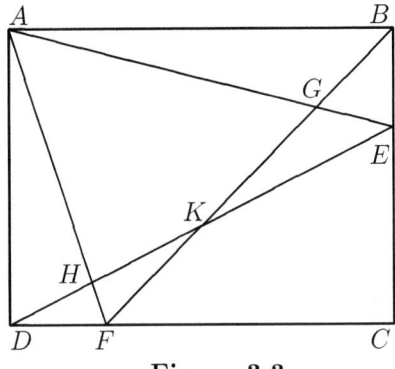

Figure 3.3

2002

1. On each planet of a star system, there is an astronomer observing the nearest planet. The number of planets is odd, and pairwise distances between them are different. Prove that at least one planet is not under observation.

2. A lottery ticket has a six-digit number. A ticket is said to be "lucky" if the sum of the first three digits of its number is equal to the sum of the last three digits. Prove that the sum of the numbers of all lucky tickets is divisible by 13.

3. In a chess tournament, each of eight participants play a game against each of the others. A win is worth 1 point, a draw $\frac{1}{2}$ point, and a loss 0 points. At the end of the tournament, each participant has a different total score, and that of the participant in second place is equal to the sum of those of the bottom four participants. What is the result of the game between the participants in third and seventh places?

4. E, F, G and H are points on the extensions of the sides. AB, BC, CD and DA of a convex quadrilateral $ABCD$, such that A, B, C and D are the midpoints of DE, AF, BG and CH, respectively. Prove that the area of $EFGH$ is five times that of $ABCD$.

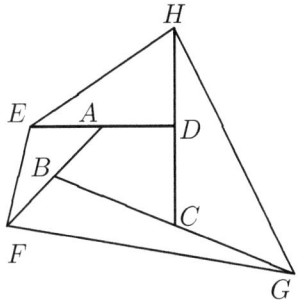

Figure 3.4

2003

1. Markers are to be placed in some squares of a 4×4 board.

   (a) Place seven markers so that if the markers on any two rows and any two columns are removed, at least one marker remains on the board.

   (b) Prove that no matter how six markers are placed on the board, then it is always possible to choose two rows and two columns so that no markers remain on the board when all markers in these rows and columns are removed.

2. Pokémon cards have become a craze in Dilbertville. There are only three kinds of cards, and only three official dealers through whom all transactions must be conducted. Dogbert will trade one dog card for one cat card and one rat card, or vice versa. Catbert will trade one cat card for two dog cards and one rat card, or vice versa. Ratbert will trade one rat card for three dog cards and one cat card, or vice versa. No other trades are allowed. Starting with one cat card, try to obtain

   (a) some rat cards, but without any dog or cat cards;
   (b) some dog cards, but without any cat or rat cards;
   (c) more cat cards, but without any dog or rat cards.

   For each task, either prove that it is impossible or find a sequence of trades which will result in the minimum number of cards of the specified type.

3. Let x and y be such that $(x + \sqrt{x^2 + 1})(y + \sqrt{y^2 + 1}) = 1$. What is the value of $x + y$?

4. $ABCDEF$ is a regular hexagon with center O. M is the midpoint of CD and N is the midpoint of OM. If the area of triangle NBC is 3, what is the area of triangle NAB?

2004

1. In the Canadian Cyber League, twelve teams played one another 16 times in a season. The numbers of games played among the teams so far were shown in Figure 3.5, with the teams identified by their Roman numerals only.

#	I	II	III	IV	V	VI	VII	VIII	IX	X	XI	XII
I	0	8	12	10	13	5	8	8	8	10	11	14
II	8	0	11	9	12	6	7	9	7	11	10	13
III	12	11	0	6	5	9	4	12	6	14	7	10
IV	10	9	6	0	7	7	2	10	4	12	5	6
V	13	12	5	7	0	10	5	13	7	15	8	11
VI	5	6	9	7	10	0	5	5	5	7	8	11
VII	8	7	4	2	5	5	0	8	2	10	3	6
VIII	8	9	12	10	13	5	8	0	8	6	11	14
IX	8	7	6	4	7	5	2	8	0	10	5	8
X	10	11	14	12	15	7	10	6	10	0	13	16
XI	11	10	7	5	8	8	3	11	5	13	0	9
XII	14	13	10	6	11	11	6	14	8	16	9	0

Figure 3.5

Determine the two teams which had already played all 16 games between them, given the following numbers of games played between certain pairs of teams.

First Team	Second Team	#
Edmonton Emails	Ottawa Outputs	4
London Laptops	Regina Robots	4
Montreal Modems	Vancouver Viruses	5
Winnipeg Websites	Vancouver Viruses	5
Calgary Computers	Winnipeg Websites	6
Hamilton Hexidecimals	Toronto Terminals	8
Halifax Hackers	Hamilton Hexidecimals	9
Edmonton Emails	Saskatoon Spams	11
Calgary Computers	London Laptops	12

2. After a spill of some toxic substances, 55 people were evacuated from their homes and spent the night in the Community center, with its floor plan shown in Figure 3.6. The 55 people were distributed among the ten rooms so that there was at least one person in each room, and no two rooms had the same number of people. To pass the time, the people in each room eavesdropped on the rooms sharing a wall with their own room, and were able to come up with an exact count of the total number of people in those rooms. Seven rooms filed the following reports: Library 25, Meeting Room 19, Office 24, Assembly Hall 36, Reception Area 27, Store Room 18 and Dining Room 10. What should the missing reports be from the Play Room, the Closets and the Bath Room?

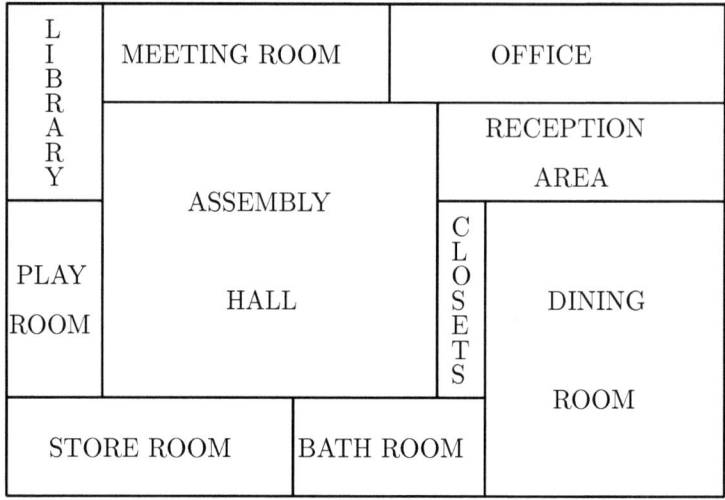

Figure 3.6

3. Let a, b and c be numbers such that $bc+ca+ab = 1$ and $\frac{1}{a}+\frac{1}{b}+\frac{1}{c} = 1$.

   (a) Explain why $abc = 1$.
   (b) Explain why $1+a+ca = a(1+c+bc)$, $1+b+ab = b(1+a+ca)$ and $1+c+bc = c(1+b+ab)$.
   (c) Determine the value of $\frac{1}{1+a+ca} + \frac{1}{1+b+ab} + \frac{1}{1+c+bc}$.

4. EF, FD and DE are the tangents to a circle at the points A, B and C respectively, such that the circle is inside triangle DEF. Let $\angle DBC = \alpha$, $\angle ECA = \beta$ and $\angle FAB = \gamma$.

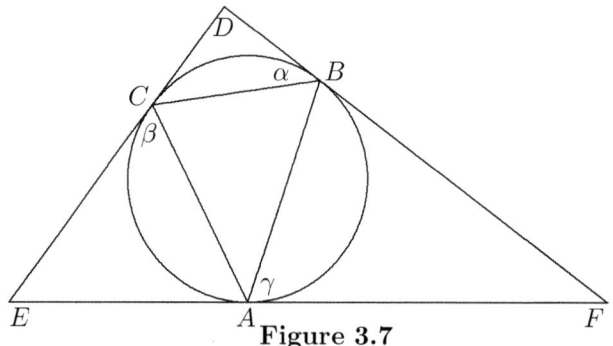

Figure 3.7

   (a) Explain why $\angle BDC = 180° - 2\alpha$, $\angle CEA = 180° - 2\beta$ and $\angle AFB = 180° - 2\gamma$.
   (b) Determine the value of $\alpha + \beta + \gamma$.
   (c) Explain why $\angle CAB = \alpha$, $\angle ABC = \beta$ and $\angle BCA = \gamma$.

2005

1. Three people are accused of being terrorists and are subjected to random justice. In a public trial, each will have a spot marked on his forehead. The spot is either black or white. Each can see the colors of the spots of his two co-defendants, but not that of his own. Each may either pass or guess the color of his spot. If all three pass, they will be found guilty. If at least one of them guesses wrong, all three will be found guilty. They will only be free if at least one of them guesses, and all guesses are correct. They may not confer with one another once the trial has begun, but may plot strategies beforehand. For instance, they may designate one of them as the guesser and have the other two pass. Then their chances of being free is $\frac{1}{2}$. Find a strategy which improves their chances to $\frac{3}{4}$.

2. A square box contains one square of area 17 square centimeters and eight squares each of area 68 square centimeters, as shown in Figure 3.8.

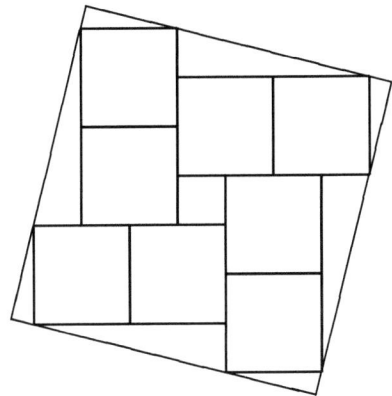

Figure 3.8

   (a) The empty spaces consist of four large triangles and four small ones. Explain why all eight triangles are similar to one another.
   (b) What is the area of each small triangle?
   (c) What is the length of a side of the box?

3. In triangle ABC, $BC = a$, $CA = b$ and $AB = c$. D is a point on BC such that AD is perpendicular to BC.

   (a) Let $CD = x$ and $AD = y$. Applying Pythagoras' Theorem to triangle CAD, we have $y^2 = b^2 - x^2$. Apply Pythagoras' Theorem to triangle BAD and obtain a similar expression for y^2 in terms of a, c and x.

(b) Equate these two expressions and find an expression for x in terms of a, b and c.

(c) The expression $y^2 = b^2 - x^2$ is a difference of two squares, and may be factored into $(b+x)(b-x)$. Substitute the expression for x and use further factoring to show that

$$4a^2y^2 = (a+b+c)(a+b-c)(b+c-a)(c+a-b).$$

(d) Let $s = \frac{a+b+c}{2}$ be the semiperimeter of triangle ABC. Explain why $\frac{a+b-c}{2} = s - c$ and find similar expressions for $\frac{b+c-a}{2}$ and $\frac{c+a-b}{2}$ which involve s.

(e) The area of triangle ABC is given by $\frac{ay}{2}$. Explain why it is also equal to

$$\sqrt{s(s-a)(s-b)(s-c)}.$$

4. A positive integer n is the sum of the squares of its smallest four divisors that are positive integers.

(a) Explain why n must be even.

(b) What are the possible values of the remainder when the square of an integer is divided by 4?

(c) Explain why n is not divisible by 4.

(d) Explain why the third smallest divisor d of n must be odd.

(e) Explain why the fourth smallest divisor of n is $2d$.

(f) Explain why $d = 5$.

(g) What is the value of n?

2006

1. There are eight machines numbered 2 to 9. If the positive integer m is inputted into machine n, the machine will consider all digits of the input m and all digits of the product mn, and output the smallest of all these digits. Which machine would you use, and which positive integer would you use as input, in order to obtain the largest possible output?

2. (a) Let x and y be positive numbers such that $x + y = 2k$ for some constant k. Clearly, if $x = y = k$, then $xy = k^2$. Prove that $xy < k^2$ in all other cases.

(b) You have a fence of length 60 meters. You may use it to form three sides of a rectangular garden, the fourth side of which is part of a straight wall of length 60 meters.

Naturally, you want the area of your garden to be as large as possible. How long should you make the side of the garden perpendicular to the wall, and what is the area of this optimal garden?

3. Each member of a social club has a piece of juicy gossip, and is eager to spread it among the other members. Club members communicate only by three-way conference-calls. During such a call, all pieces of gossip known to each participant are shared with the other two participants.

   (a) If there are nine club members, show how everybody can know everything using only six calls.

   (b) Prove that if there are eight club members, not everybody can know everything using only five calls.

4. ABC is a triangle with $\angle C = 90°$, $BC = a$, $CA = b$, $AB = c$ and inradius r.

   (a) By considering lengths, prove that $r = \frac{a+b-c}{2}$.

   (b) By considering areas, prove that $r = \frac{ab}{a+b+c}$.

   (c) Deduce from (a) and (b) a well-known result about right triangles.

2007

1. In a group dinner, 30 people ordered chicken Kiev while 125 people ordered minced chicken. Each order cost an integral number of dollars. The total bill for the chicken Kiev was higher than that for the minced chicken, but this would have been reversed had there been an extra order of minced chicken. Tony and Cindy had chicken Kiev while their daughter Cherry had minced chicken. Would 100 dollars be enough to pay for their shares of the dinner?

2. A positive integer is the product of a prime number and the square of a different prime number. The sum of four different positive divisors of this number is equal to 100. What are the possible values of this number?

3. In triangle ABC, E is the midpoint of CA and F is the midpoint of AB. BE and CF intersect at the point G, and the extension of AG intersects BC at D. Denote by $[T]$ the area of triangle T. Let $[FAG] = x$ and $[AGE] = y$.

   (a) Express $[FBG]$ in terms of x and $[GCE]$ in terms of y.

   (b) Express $[BDG]$ in terms of x and then in terms of y.

   (c) Use (b) to explain why D is the midpoint of BC.

4. Cherry and her brother Eddy are dividing three cakes of equal size. Cherry cuts the first cake into two pieces, not necessarily of equal size. In fact, the smaller piece may have size 0. Eddy may decide to choose first or second. Whoever chooses first takes the larger of the two pieces, unless they are of equal size. This process is repeated for the second cake and the third cake. However, Eddy must let Cherry choose first at least once.

   (a) What is the largest amount of cake Cherry can guarantee for herself?
   (b) How can she get the amount of cake in (a)?
   (c) Why can she not get more than the amount of cake in (a)?

2008

1. There are $n + 1$ lily-pads in a row, numbered from 0 to n. A frog starts on lily-pad 0, and wishes to reach lily-pad n by a sequence of alternating left and right hops. Each hop may be of length 2, 3 or 4. The frog may not land on the same lily-pad more than once. What is the minimum number $n > 1$ for which the task is impossible?

2. When two integers a and b, each between 10 and 99 inclusive, are written one after another, a four-digit number is formed. How many such numbers are divisible by the product ab?

3. An alien spaceship is going to land on an 8×8 patch of land in an attempt to conquer earth. The earthlings place mines in some of the 64 squares. The alien spaceship is equipped with a mine-sweeper which will clear all mines in any 4 rows and any 4 columns. If at least one mine remains after the mine-sweeping, the alien spaceship will be blown up upon landing, and earth is saved. What is the minimum number of mines the earthlings have to place, and where are they to be placed, in order to save their planet?

4. In triangle ABC, $AB = AC$ and $\angle CAB = 90°$. H, D and K are points on BC such that $BH = HD = DK = KC$. E is a point on CA such that $CE = 3EA$, and F is a point on AB such that $BF = 3FA$. Determine $\angle FHE + \angle FDE + \angle FKE$.

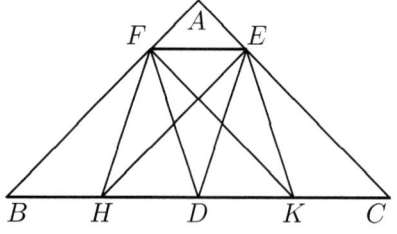

Figure 3.9

2009

1. (a) Find twelve different even numbers between 1 and 49 so that the sum of any two of them is not 50.
   (b) Prove that given any thirteen even numbers between 1 and 49, you can find among them a pair with the sum exactly 50.

2. Consider all the twelve-digit numbers formed by using each of the digits 1, 2 and 3 exactly four times.
   (a) How many of these numbers are prime?
   (b) How many of these numbers are squares of integers?

3. ABC is a triangle with $AB = AC = a$ and P is a point on the side BC. D is a point on AB such that PD is perpendicular to AB and E is a point on AC such that PE is perpendicular to AC.
   (a) Prove that the area of triangle ABC is equal to $\frac{a(x+y)}{2}$, where $x = PD$ and $y = PE$.
   (b) If P moves on the edge BC, explain why the sum $PD + PE$ does not change.

4. Let a and b be two consecutive integers and let c be their product. Prove that there exists an integer d such that $a^2 + b^2 + c^2 = d^2$.

2010

1. Animal House has six rooms in a row on each of two floors. Four dogs, four cats and four rats live there, one in each room. At night, the light is left on if the animal in the room is
   (1) a cat with a rat upstairs or downstairs;
   (2) a rat immediately between two cats on the same floor;
   (3) a dog immediately between a cat and a rat on the same floor.
   Otherwise, the light is turned off and the animal goes to sleep. One night, seven lights are left on, as shown by the light circles in the diagram below. Determine for each room what animal is in it.

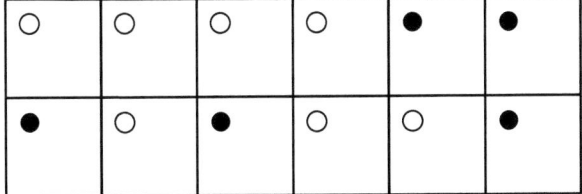

Figure 3.10

2. Each of Aaron, Betty and Cyrus had an integral number of dollars. They wished to buy a horse which cost a positive integral number of dollars. Aaron said to Betty and Cyrus, "If each of you give me half your money, I will have just enough to buy the horse." Betty said to Cyrus and Aaron, "If each of you give me one third of your money, I will have just enough to buy the horse." Cyrus said to Aaron and Betty, "If each of you give me one quarter of your money, I will have just enough to buy the horse." Assuming that they had coins worth half a dollar, one third of a dollar or one quarter of a dollar, what was the minimum cost of the horse?

3. A *table* number is a number which appears in the multiplication table. For instance, 25 is a table number because $25 = 5 \times 5$ but $52 = 4 \times 13$ is not a table number. Arrange the digits 1, 2, 3, 4, 5, 6, 7, 8 and 9 in a row so that each of the eight pairs of adjacent digits is a two-digit table number.

4. In triangle ABC, $AB = AC$ and $\angle CAB = 20°$. P is the point on AC such that $AP = BC$. We wish to determine $\angle ABP$. Let G be the point inside ABC such that BCG is an equilateral triangle. Extend BG to meet AC at E.

   (a) Explain why $EA = EB$.
   (b) Explain why $EP = EG$.
   (c) Explain why $GP = GC$.
   (d) Determine $\angle ABP$.

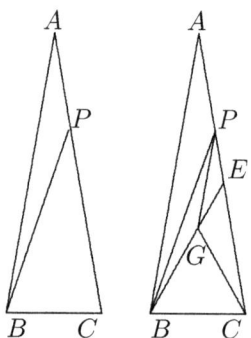

Figure 3.11

2011

1. Without finding their actual values, determine which of the following two numbers is larger:

$$\frac{1+1+2+3+5+8+13+21+34+55+89}{3+3+7+11+19+31+51+83+135+219+355}$$

or

$$\frac{2+2+3+4+6+9+14+22+35+56+90}{4+4+8+12+20+32+52+84+136+220+356}?$$

2. The positions of some positive integers in a row are numbered consecutively, starting with a 1 in position 0. Each number except for those in the first or the last position is 1 more than the average of its two neighbors.

    (a) If the number in position 1 is n, calculate numbers in the next four positions in terms of n.

    (b) If the number in position 1 is n, determine the number in position k in terms of n and k.

    (c) If the row is as long as possible, prove that the number in the last position is 1.

3. (a) Compute the product of $x+y$ and $x+y$.

    (b) Compute the product of $x+y$ and $x-y$.

    (c) Express $x^2 - 6xy + 9y^2 - 4z^2$ as a product of two factors neither of which is ± 1.

4. The following are three basic inequalities in elementary geometry. Use the first two to prove the third.

    The Triangle Inequality.
    In any triangle ABC, $AB + BC > CA$.

    The SaS Inequality.
    In triangles ABC and DEF, if $AB = DE$, $BC = EF$ but $CA > FD$, then $\angle ABC > \angle DEF$.

    The Exterior Angle Inequality.
    If D is a point on the extension of the side BC of triangle ABC, then $\angle ACD > \angle CAB$.

2012

1. Without using a calculator, determine

    (a) the smallest positive integer n such that $n\sqrt{2} > 30$;

    (b) the smallest positive integer n such that $n^{n+1} > 30^{20}$.

2. Let x, y and z be numbers such that $x+y+z=1$, $yz+zx+xy=2$ and $xyz=3$.
   (a) Evaluate $x^2+y^2+z^2$.
   (b) Evaluate $x^3+y^3+z^3$.

3. A and B are two points on a circle with center O and M is the midpoint of the segment AB. Consider the following statement about a straight line.
   (1) It passes through O.
   (2) It passes through M.
   (3) It is perpendicular to AB.
   Prove that if two of these statements are true, then the third one is also true.

4. A club has nine members. Each club meeting is attended by exactly six of them. They sit around a table. However, if two members have occupied adjacent seats before, they refuse to sit next to each other again. What is the maximum number of meetings this club can have?

2013

1. (a) Prove that $3^{2012}+3^{2011}+3^{2010}+\cdots+3^2+3+1=\frac{3^{2013}-1}{2}$.
   (b) Find all positive integer powers of 3 each of which is 1 more than some positive integer power of 2.

2. Lana and Derek belong to a sports club with eight members, including themselves. The club has only one ping-pong table, and all members form a single line at random. The first two members in the line play each other. The loser goes out, and is replaced by the next member in the line. The session is over when the member last in line has played a game. The members are evenly matched so that both players in a game have the same chance of winning. What is the probability that Lana and Derek get to play each other during the session?

3. A, B and C are points on a circle with center O and PQ is a line passing through C, as shown in the diagram below. Prove that if one of the following three statements is true, then the other two are also true.
   (1) PQ is perpendicular to OC.
   (2) $\angle PCB = \angle BAC$.
   (3) $\angle QCA = \angle ABC$.

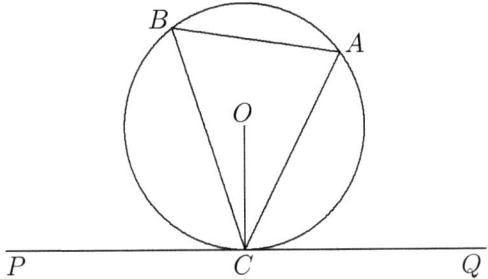

Figure 3.12

4. (a) Prove that for all integers $n \geq 2$, $\dfrac{3n+1}{3n-2} < \left(\dfrac{2n}{2n-1}\right)^2 < \dfrac{n}{n-1}$.

   (b) Determine the integer part of $\dfrac{2 \times 4 \times 6 \times 8 \times 10 \times 12 \times 14 \times 16}{1 \times 3 \times 5 \times 7 \times 9 \times 11 \times 13 \times 15}$ without computing its value.

2014

1. Determine the number of ways of expressing 2014 as a sum of powers of 2, if each distinct power may appear at most three times.

2. (a) Using only the fact that the square of a real number is greater than or equal to 0, prove that $x + \frac{1}{x} \geq 2$ for any positive real number x.

   (b) Let x, y and z be positive numbers such that $xyz(x+y+z) = 1$. When $(x+y)(y+z)$ attains its minimum value, determine the maximum value of y.

3. A vertical cross-section of a chain of mountains is bounded above by a polygonal line consisting of segments with slope either 1 or -1. The two endpoints have altitude 0 while all other vertices have positive altitudes. Two mountaineers from opposite endpoints move towards each other by going over the top of the mountains at the same constant speed. They must be at the same altitude all the time. However, they may change directions at any time.

   (a) If the chain of mountains is as shown in Figure 3.13, where will the two mountaineers meet?

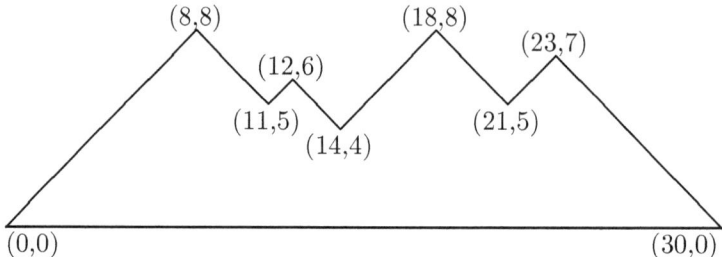

Figure 3.13

(b) If the chain of mountains is as shown in Figure 3.14, where will the two mountaineers meet?

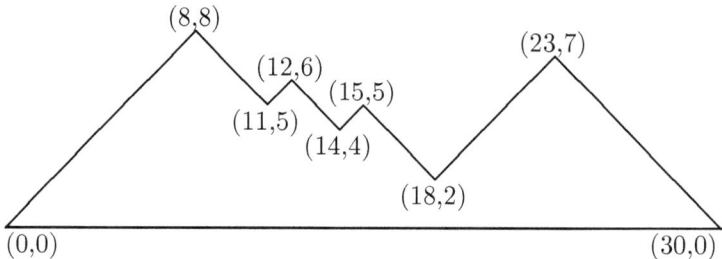

Figure 3.8

(c) Prove that for every chain of mountains, the two mountaineers can always meet.

4. $ABCD$ is a convex quadrilateral. K, L, M and N are the respective midpoints of AB, BC, CD and DA, as shown in Figure 3.15 on the left.

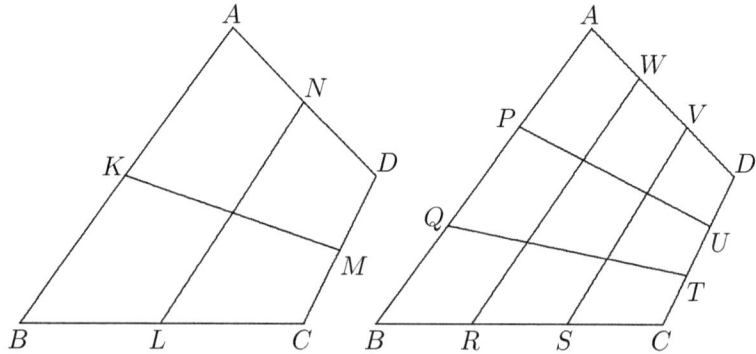

Figure 3.15

(P, Q), (R, S), (T, U) and (V, W) are the respective points of trisections of AB, BC, CD and DA, as shown in Figure 3.15 on the right.

(a) Prove that the segments KM and LN bisect each other.

(b) KM and LN divide $ABCD$ into a 2×2 skewed chessboard. Prove that the total area of any two of the four quadrilaterals not in the same row and not in the same column is equal to half the area of $ABCD$.

(c) Prove that the segments PU, QT, RW and SV trisect one another.

(d) PU, QT, RW and SV divide $ABCD$ into a 3×3 skewed chessboard. Prove that the total area of any three of the nine quadrilaterals with no two in the same row and no two in the same column is equal to one third the area of $ABCD$.

2015.

1. Find the number of positive integers n less than 10000 such that $2^n - n^2$ is divisible by 7.

2. A small movie theater has 30 seats; their numbers are indicated on the tickets. The first 21 people, including Edmond, who come to a show take 21 seats without paying attention to the seat numbers. But the remaining 9 ticket holders are sticklers. If any of them finds her or his assigned seat occupied, the person sitting there is evicted; that person then looks for his or her proper seat and evicts the usurper; and so forth. This migration ends with the spectator whose assigned seat is unoccupied. Find the probability that Edmond will not have to change his seat?

3. Find all possible values of a such that $(x-a)(x-10)+1 = (x-b)(x-c)$ for all x and some integers b and c.

4. A unit circle is one with radius 1. For a given non-negative integer n, we wish to find a line and a circle of any radius, such that exactly n unit circles can be tangent to both. For what values of n is this possible?

Section 2. Solutions

2000

1. Denote the number of viruses on day n by v_n, and the number of anti-viruses on day n by a_n. On night n, the number of anti-viruses doubles to $2a_n$, and this does not change during the day. It follows that $a_{n+1} = 2a_n$. On night n, the number of viruses also doubles, to $2v_n$, but the next day, each of the a_{n+1} anti-viruses will kill off one of them. Hence $v_{n+1} = 2v_n - a_{n+1} = 2v_n - 2a_n$. We have

$$\frac{v_{n+1}}{a_{n+1}} = \frac{2v_n - 2a_n}{2a_n} = \frac{v_n}{a_n} - 1.$$

So the ratio $\frac{v_n}{a_n}$ decreases by 1 each day. Since $\frac{v_0}{a_0} = 64$, $\frac{v_{64}}{a_{64}} = 0$, and all the viruses will be killed off by the 64th day.

2. (a) If we give 1 egg per child on the first day, we give away 5 eggs. There is no point in giving 2 eggs per child. If we give 3 each, we have to leave out Ace for now and give away 12 eggs. If we give 6 each, the total is 18. If we give 8 each, the total is 16. Finally, if we give 9 each, only Eve will get any. So it appears that the best strategy is to give 6 eggs to each of Cec, Dee and Eve on the first day. Now Ace needs 1, Bea 3, Dee 2 more and Eve 3 more. On the second day, we can either give 3 eggs each to Bea and Eve, or 2 eggs each to Bea, Dee and Eve. On the third day, the first method will leave Ace wanting 1 egg and Dee wanting 2, whereas the second method will have Ace, Bea and Eve each wanting 1. Clearly, the second way is better, and the whole thing could be done in 3 days. Clearly, 1 day is not enough. Let us leave out Dee and Eve and just concentrate on Ace, Bea and Cec. No matter what we do on the first day, we will have at least two children wanting different numbers of eggs on the second day, and the task must extend to a third day.

   (b) Suppose the total number of eggs is 6. If there are less than three children, we can certainly get around the Easter Bunny in at most two day, by awarding only one child per day. Hence we may assume that there are at least three children. The only ways to partition 6 into three or more positive integers are 1+1+1+1+1+1, 2+2+2, 1+1+1+1+2, 1+1+1+3, 1+1+2+2, 1+1+4 and 1+2+3. The first two cases can be handled in 1 day. The next three cases can be handled in 2 days. The last day can also be handled in 2 days since on the first day, we can give 1 egg to each child except the one who is getting 2 eggs. The argument in (a) shows that the total number of eggs cannot be 7, if Ace, Bea and Cec are to get 1, 2 and 4 eggs respectively.

3. (a) We start with an even number 2. This is followed by two odd numbers, 3 and 1. The fourth and fifth numbers must be even while the sixth and seventh numbers must be odd. Since the even-odd-odd pattern recurs at the fifth to seventh numbers, we will continue to get two even numbers followed by two odd numbers indefinitely. Since the sequence 1, 2 and 3 has one even number between two odd ones, it cannot appear.

(b) Each block of three successive numbers must be one of $(0,0,0)$, $(0,0,1), \ldots, (9,9,9)$. Since there are 10 choices for each number, the number of different blocks is $10 \times 10 \times 10 = 1000$. Since this computer keeps on churning out numbers, it will have shown 1001 blocks at some point. Since we only have 1000 different blocks, one of them must reappear.

(c) Let the numbers be $x_1 = 2$, $x_2 = 3$, $x_3 = 1$, x_4, ... In (b), we have proved that some block must reappear. Let its first appearance be (x_n, x_{n+1}, x_{n+2}) and its second appearance be $(x_{n+k}, x_{n+k+1}, x_{n+k+2})$ for some positive integer k. If $n = 1$, there is nothing further to prove. Suppose $n > 1$. Since x_{n+2} in uniquely determined by x_{n-1}, x_n and x_{n+1}, x_{n-1} is also uniquely determined by x_n, x_{n+1} and x_{n+2}. It follows that $x_{n-1} = x_{n_k-1}$. Continuing this argument, we can show eventually that the block $(x_1, x_2, x_3) = (2, 3, 1)$ must reappear.

4. Let BC and DE intersect at F, as shown in Figure 3.16. Then $\angle EFB = \angle DFC$ and $180° = \angle CBE + \angle BED + \angle EFB$. We also have $\angle DEA + \angle EAB + \angle ABC + \angle CBE + \angle BED = 180°$ as well as $\angle BCD + \angle CDE + \angle DFC = 180°$. Adding these four equations and cancelling terms which appear on both sides, we are left with

$$\angle EAB + \angle ABC + \angle BCD + \angle CDE + \angle DEA = 180°.$$

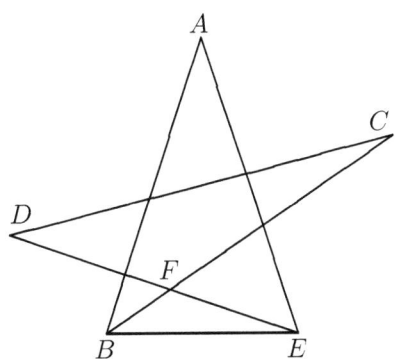

Figure 3.16

2001

1. Let our numbers be M, $2M$, $3M$, ..., $100M$ for some suitably chosen positive integer M. Let aM, bM, cM, dM and eM be any five of these numbers. Then their product is $P = abcdeM^5$ while their sum is $S = (a+b+c+d+e)M$. Since

$$\begin{aligned} 15 &= 1+2+3+4+5 \\ &\leq a+b+c+d+e \\ &\leq 96+97+98+99+100 \\ &= 490, \end{aligned}$$

we can choose $M = 490 \times 489 \times 488 \times \cdots \times 17 \times 16 \times 15$. Then M is a multiple of $a+b+c+d+e$, so that P is indeed divisible by S.

2. Suppose we use x, y and z copies of the three shapes respectively. Then $3x + 4(y+z) = 25$. We can have $x = 7$ and $y+z = 1$, or $x = 3$ and $y+z = 4$. So the total number of pieces used in any successful covering is 8 or 7. In Figure 3.17, nine of the squares are shaded. Each of the three shapes can cover at most one of them. Hence at least 9 pieces are needed. This is a contradiction. Hence the desired covering cannot exist.

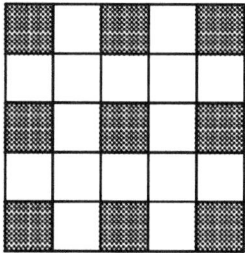

Figure 3.17

3. We have

$$\begin{aligned} & (1+x)^2 + (1+y)^2 + x^2 + y^2 \\ &= 1 + 2x + x^2 + 1 + 2y + y^2 + x^2 + y^2 \\ &= (1 + x^2 + y^2 + 2x + 2y + 2xy) + (x^2 + y^2 - 2xy) + 1 \\ &= (1 + x + y)^2 + (x - y)^2 + 1 \\ &\geq 1. \end{aligned}$$

Equality holds if and only if both $1+x+y$ and $x-y$ are 0. In other words, the minimum value 1 of the given expression is attained if and only if $x = y = -\frac{1}{2}$.

4. To $AGKH$, we add the triangles ABG and FHK to make it up to triangle ABF, as shown in Figure 3.18. Its area is $\frac{1}{2}AB$ times the distance between AB and DC.

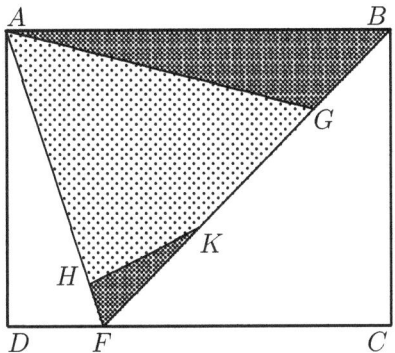

Figure 3.18

We now add the same two triangles ABG and FHK to the team of BEG, DFH and $CEKF$. They turn into two triangles ABE and CDE, as shown in Figure 3.19. The area of the first is AB times $\frac{1}{2}BE$ and the area of the second is CD times $\frac{1}{2}CE$. Since $AB = CD$ and $BE + CE = BC$, their total area is the same as the area of triangle ABF. So both teams have equal total area after reinforcement, and since the reinforcement are the same for each team, they must have equal total area originally.

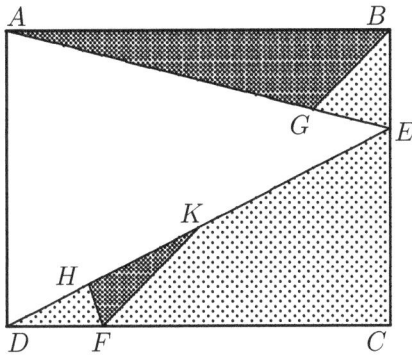

Figure 3.19

2002

1. Consider the two planets closest to each other. It is obvious that the astronomers on these planets are watching each other. If any other astronomer watches either of these two planets, we will not have enough astronomers to watch every planet.

If nobody else watches these two planets, then we can apply the same argument to the remaining planets by considering the two which are closest to each other. Since the number of planets is odd, the last planet left will not be watched by any astronomer.

2. If the lucky ticket has the number n, then the ticket with the number $999999 - n$ is also lucky, and these two tickets are distinct. The sum of their numbers $999999 = 999 \times 1001$ is divisible by 13. This is because $1001 = 7 \times 11 \times 13$. Hence the sum of the numbers of all lucky tickets is also divisible by 13.

3. The bottom four participants played six games among themselves, and therefore the sum of their scores was no less than 6 points even if they lost every game against the top four. Hence the score of the runner-up was at least 6 points. If the winner scored 7 points, then the runner-up lost the game against the winner, and finished with no more than 6 points. If the winner scored 6.5 points, again the runner-up could have no more than 6 points. Hence the runner-up had exactly 6 points, and the bottom four participants had exactly 6 points among themselves, meaning that they lost every game against the top four participants. In particular, the game between the participants in third and seventh places was won by the player in the third place.

4. Denote the area of polygon P by $[P]$. Since $EA = AD$ and $HD = DC$, $[HEA] = [HAD] = [CAD]$. Similarly, $[FGC] = [FCB] = [ACB]$. Hence $[HED] + [FGB] = 2[ABCD]$. In the same way, we can prove that $[EFA] + [GHC] = 2[ABCD]$, so that $[EFGH] = 5[ABCD]$.

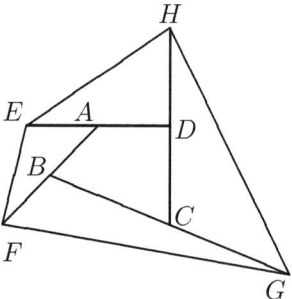

Figure 3.20

2003

1. (a) Clearly, we should conduct transactions with each trader in one direction only, getting more rat cards. Suppose we trade x times with Ratbert, y times with Dogbert and z times with Catbert.

Counting dog cards, we have $-3x - y + 2z = 0$. Counting cat cards, we have $1 - x + y - z = 0$. Addition yields $1 - 4x + z = 0$ or $z = 4x - 1$. Substituting back into either equation yields $y = 5x - 2$. Thus the smallest solution is $(x, y, z) = (1, 3, 3)$, resulting in only 7 rat cards. A possible sequence of trades is shown below:

Number of Cards	After Trade with XXX-bert					
	Cat	Dog	Dog	Cat	Cat	Dog	Rat
Rat	1	2	3	4	5	6	7
Dog	2	1	0	2	4	3	0
Cat	0	1	2	1	0	1	0

(b) Let s be the sum of the numbers of rat and cat cards. In any trade, the value of s changes by 0 or 2. Since $s = 1$ initially, we can never have only dog cards, because that means $s = 0$.

(c) This case is similar to (a) except that we must begin trading with Catbert in the wrong direction, and thereafter trading only to get more cat cards. We use the same notations as before. Counting rat cards, we have $1 - x + y - z = 0$. Counting dog cards, we have $2 + 3x - y - 2z = 0$. This time, we have $x = \frac{3(z-1)}{2}$ and $y = \frac{5(z-1)}{2}$. We must have $z > 1$ as otherwise we will end up with 1 cat card. Thus the smallest solution is $(x, y, z) = (3, 5, 3)$, resulting in only 11 cat cards. Any 11 trades with this distribution will work, after the initial one with Catbert.

2. (a) Consider the arrangement of seven markers shown in Figure 3.21. To remove the marker at the bottom right corner requires the crossing out of a row or column without removing another marker. By symmetry, we may assume that a row is crossed out. No matter which two columns are removed, we will be left with two markers in the third column, and we cannot remove both by crossing out another row.

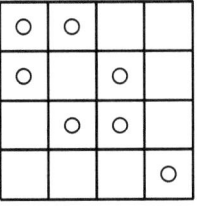

Figure 3.21

(b) If there are at most six markers, consider the two columns which contain between them the fewest markers. If the total is at least three, then one of these two columns must contain at least two markers, whereas one of the other two columns must contain at most one, contradicting the minimality assumption of our pair. Hence the total is at most two. Cross out the other two columns, and we have at most two markers left. We can now cross out at most two rows in order to remove any remaining markers.

3. Let $x+\sqrt{x^2+1} = A$. Then $\sqrt{x^2+1} = A-x$. This yields $1 = A^2 - 2Ax$ so that $x = \frac{A^2-1}{2A}$. Similarly, $y = \frac{B^2-1}{2B}$ if we let $B = y + \sqrt{y^2+1}$. Now $x + y = \frac{A^2B - B + AB^2 - A}{2AB} = \frac{B - B + A - A}{2} = 0$ since $AB = 1$.

4. Extend OM to cut the extensions of BA and BC at Q and R respectively. Then $QA = AB = BC = CR$ and $OQ = OR = 2OM = 4ON$. Hence we have $RN : NQ = 3 : 5$, so that $[BRN] : [BNQ] = 3 : 5$. Since $BC = CR$ and $BA = AQ$, we have $[NBC] = \frac{1}{2}[BNR]$ and $[NAB] = \frac{1}{2}[BNQ]$. Thus $[NBC] : [NAB] = 3 : 5$. Since $[NBC] = 3$, we have $[NAB] = 5$.

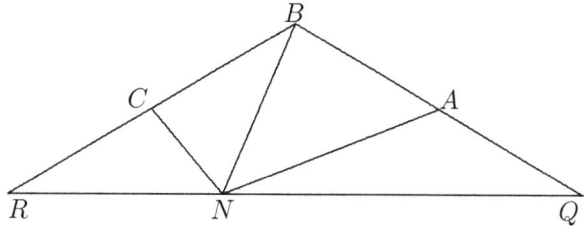

Figure 3.22

2004

1. Edmonton had played Ottawa 4 times and Saskatoon 11 times, and must therefore be team III. Hence Ottawa was team VII and Saskatoon team II. London had played Regina 4 times and Calgary 12 times, and must therefore be team IV. Hence Regina was team IX and Calgary team X. Since Calgary had played Winnipeg 6 times, Winnipeg was team VIII. Since Winnipeg had played Vancouver 5 times, Vancouver was team VI. Since Vancouver had played Montreal 5 times, Montreal was team I. Finally, Hamilton had played Toronto 8 times and Halifax 9 times, and must be team XI. Hence Toronto was team V and Halifax team XII. It follows that the two teams which had already played all the games between them were Calgary and Halifax.

Solutions

2. Only the Dining Room was not adjacent to the Assembly Hall. Hence the total number of people in these two rooms was $55 - 36 = 19$. Hence there were 9 in one and 10 in the other. Now the total number of people in the Meeting Room and the Play Room was $25 - 10 = 15$ or $25 - 9 = 16$. The most we could have was 7 in one and 8 in the other, so that there were 10 people in the Assembly Hall and 9 in the Dining Room. The total number of people in the Meeting Room and the Reception Area was $24 - 10 = 14$. Hence the number of people in the Meeting Room must be 8, that in the Reception Area 6 and that in the Play Room 7. Moreover, the number of people in the Bath Room must be $18 - 10 - 7 = 1$, and the number of people in the Closets must be $10 - 6 - 1 = 3$. The total number of people in the Office and the Library was $19 - 10 = 9$, so that there were 4 in one and 5 in the other. Hence the number of people in the Store Room must be 2. Finally, the number of people in the Office was $27 - 10 - 9 - 3 = 5$, and the number of people in the Library 4. Thus the missing reports were Play Room $4 + 10 + 2 = 16$, the Bath Room $2 + 10 + 3 + 9 = 24$ and the Closets $1 + 9 + 6 + 10 = 26$.

3. (a) We have $abc = abc(\frac{1}{a} + \frac{1}{b} + \frac{1}{c}) = bc + ca + ab = 1$.

   (b) We have
   $$\begin{aligned} 1 + a + ca &= a + ca + abc = a(1 + c + bc), \\ 1 + b + ab &= b + ab + abc = b(1 + a + ca), \\ 1 + c + bc &= c + bc + abc = c(1 + b + ab). \end{aligned}$$

   (c) We have
   $$\begin{aligned} &\frac{1}{1+a+ca} + \frac{1}{1+b+ab} + \frac{1}{1+c+bc} \\ &= \frac{1}{1+a+ca} + \frac{1}{b(1+a+ca)} + \frac{1}{c(1+b+ab)} \\ &= \frac{1}{1+a+ca} + \frac{1}{b(1+c+ca)} + \frac{1}{bc(1+a+ca)} \\ &= \frac{1}{1+a+ca}\left(1 + \frac{1}{b} + \frac{1}{bc}\right) \\ &= \frac{1}{a(1+c+bc)} \cdot \frac{1+c+bc}{bc} \\ &= \frac{1}{abc} \\ &= 1. \end{aligned}$$

4. (a) Since DB and DC are tangents to the same circle from the same point, they are equal in length. Hence DBC is an isosceles triangle so that $\angle DCB = \angle DBC = \alpha$. Now
$$\angle BDC = 180° - \angle DBC - \angle DCB = 180° - 2\alpha.$$
Similarly, $\angle CEA = 180° - 2\beta$ and $\angle AFB = 180° - 2\gamma$.

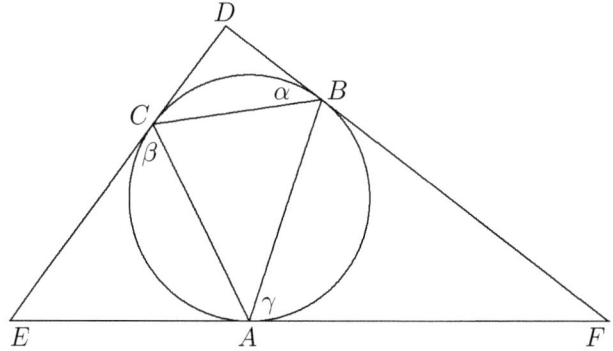

Figure 3.23

(b) We have
$$\begin{aligned}180° &= \angle BDC + \angle CEA + \angle AFB \\ &= 180° - 2\alpha + 180° - 2\beta + 180° - 2\gamma \\ &= 540° - 2(\alpha + \beta + \gamma).\end{aligned}$$
Hence $2(\alpha+\beta+\gamma) = 540° - 180° = 360°$ so that $\alpha+\beta+\gamma = 180°$.

(c) We have $\angle CAB = 180° - \angle EAC - \angle FAB = 180° - \beta - \gamma = \alpha$. Similarly, $\angle ABC = \beta$ and $\angle BCA = \gamma$.

2005

1. The strategy is for each defendant to pass if he sees two spots of different colors, and guesses the other color when he sees two spots of the same color. Since each spot is either black or white, there are $2 \times 2 \times 2 = 8$ different scenarios. Of the two scenarios where all three spots are of the same color, all three defendants will guess and each will be wrong. Of the other six scenarios where two spots are of one color and the third of the other color, only one defendant will guess and he will be correct. Hence the chances of their being free is $\frac{3}{4}$.

2. (a) Clearly, the four large triangles are congruent to one another, as are the four small ones. Label the vertices of the large triangle and the small one along the bottom edge as shown in Figure 3.24. Then $\angle ABC = 90° = \angle CED$. Since AB is parallel to CD, $\angle BAC = \angle ECD$. Hence the triangles are similar to each other.

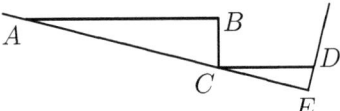

Figure 3.24

(b) The side length of each large square is $2\sqrt{17}$ centimeters. Hence $AB = 4\sqrt{17}$ centimeters and $BC = \sqrt{17}$ centimeters. It follows that the area of triangle ABC is $\frac{1}{2}AB \cdot BC = 34$ square centimeters. By Pythagoras' Theorem, $AC = 17$ centimeters. Since $CD = 2\sqrt{17}$ centimeters, the linear ratio of the similar triangles ABC and CED is $\sqrt{17} : 2$. Hence triangle CED has area 8 square centimeters.

(c) The total area of the box is $17 + 4 \times (68 \times 2 + 34 + 8) = 729$ square centimeters. Hence its side length is $\sqrt{729} = 27$ centimeters.

3. (a) If n is odd, then all its divisors are odd, and so are their squares. However, as a sum of four odd numbers, n would have been even. This contradiction shows that n must be even.

(b) An even integer has the form $2k$ for an arbitrary integer k. When $(2k)^2 = 4k^2$ is divided by 4, the remainder will be 0. An odd integer has the form $2k+1$ for an arbitrary integer k. When $(2k+1)^2 = 4k(k+1)+1$ is divided by 4, the remainder will be 1.

(c) Clearly, the smallest divisor of n is 1, and by (a), the second smallest one is 2. Since n is even, it is the sum of two odd squares and two even squares. When divided by 4, it will leave a remainder of 2. Thus n is not divisible by 4.

(d) Suppose the third smallest divisor of n is an even number $d = 2m$. Then m is also a divisor of n. Since it is smaller than d, it must have already appeared. However, if $m = 1$, then $2m = 2$ and 2 is the second smallest divisor of n. Hence $m = 2$ and $2m = 4$, but by (c), 4 is not a divisor of n. It follows that d is odd.

(e) Since n is even by (a), the fourth smallest divisor must be an even number 2ℓ, where $\ell = 1, 2$ or d. Since we have already ruled out 1 and 2, this divisor is $2d$.

(f) Now $n = 1^2 + 2^2 + d^2 + (2d)^2 = 5(1 + d^2)$. Hence 5 is a divisor of n. Since 4 does not divide n, we must have either $d = 5$ or $2d = 5$. Clearly, the second alternative is not possible.

(g) Finally, $n = 5(1 + 5^2) = 130 = 1^2 + 2^2 + 5^2 + 10^2$.

4. (a) We have $BD = BC - CD = a - x$. Hence
$$y^2 = AD^2 = AB^2 - BD^2 = c^2 - (a-x)^2.$$

(b) From $b^2 - x^2 = c^2 - a^2 + 2ax - x^2$, we have $a^2 + b^2 - c^2 = 2ax$ so that $x = \frac{a^2+b^2-c^2}{2a}$.

(c) We have
$$\begin{aligned} y^2 &= (b+x)(b-x) \\ &= \left(b + \frac{a^2+b^2-c^2}{2a}\right)\left(b - \frac{a^2+b^2-c^2}{2a}\right) \\ &= \frac{a^2 + 2ab + b^2 - c^2}{2a} \cdot \frac{c^2 - a^2 + 2ab - b^2}{2a}. \end{aligned}$$

It follows that
$$\begin{aligned} 4a^2y^2 &= ((a+b)^2 - c^2)(c^2 - (a-b)^2) \\ &= (a+b+c)(a+b-c)(b+c-a)(c+a-b). \end{aligned}$$

(d) We have $a + b - c = (a+b+c) - 2c = 2s - 2c$. It follows that $\frac{a+b-c}{2} = s - c$. Similarly, we have $\frac{b+c-a}{2} = \frac{(a+b+c)-2a}{2} = s - a$ and $\frac{c+a-b}{2} = \frac{(a+b+c)-2b}{2} = s - b$.

(e) The area of triangle ABC is given by
$$\begin{aligned} \frac{ay}{2} &= \sqrt{\frac{4a^2y^2}{16}} \\ &= \sqrt{\frac{a+b+c}{2} \cdot \frac{a+b-c}{2} \cdot \frac{b+c-a}{2} \cdot \frac{c+a-b}{2}} \\ &= \sqrt{s(s-a)(s-b)(s-c)}. \end{aligned}$$

2006

1. The largest possible output is 6. This may be obtained by putting 98 into machine 7 or 97 into machine 8. We now prove that no larger output is possible. Consider any input m into machines 2 to 6. The first digit of $2m, 3m, 4m, 5m$ or $6m$ is at most 5 if there is carrying to this digit. Otherwise, the first digit of m is at most 4. Hence we cannot get an output greater than 5 from machines 2 to 6. The same is true for machine 9 since the sum of the last digits of any input m and $9m$ is 10. Suppose the output from machine 7 is 7 or more. Then the first digit of $7m$ is at least 7, but then the first digit of m must be 1. This is a contradiction. Suppose the output from machine 8 is 7 or more. Then the last digit of $8m$ must be 8, but then the last digit of m is either 1 or 6. This is also a contradiction. It follows that the largest possible output is 6.

2. (a) We may assume by symmetry that $x > k > y$. Since $x + y = 2k$, we must have $x = k + d$ and $y = k - d$ for some number d where $0 < d < y$. Then $xy = (k+d)(k-d) = k^2 - d^2 < k^2$ since $d^2 > 0$.

   (b) Let the length of the side of the garden perpendicular to the wall be x meters. Then the length of the side of the garden parallel to the wall is $60 - 2x$ meters. Thus the area in square meters of the garden is $x(60 - 2x) = 2x(30 - x)$. Note that $x + (30 - x) = 30$ is constant. By (a), the largest possible value of $x(30 - x)$ occurs when $x = 15$. Clearly, at $x = 15$, $2x(30 - x)$ also takes its maximum value, which is $2(15)(30 - 15) = 450$.

3. Let I be the incenter of triangle ABC. Let D, E and F be the points of tangency of the incircle with the sides BC, CA and AB respectively. Denote the area of triangle T by $[T]$.

   (a) Since $\angle IDC = \angle DCE = \angle CEI = 90°$, $CDIE$ is a rectangle. Since $ID = IE$, $CDIE$ is a square, so that $CD = CE = r$. Note that $AE = AF$ because they are tangents from A to the incircle. Similarly, $BF = BD$. It follows that

   $$\begin{aligned} 2r &= CD + CE = CD + (BD - BF) + CE + (AE - AF) \\ &= (CD + BD) + (CE + AE) - (AF + BF) = a + b - c, \end{aligned}$$

   so that $r = \frac{a+b-c}{2}$.

   (b) We have $\frac{ab}{2} = [ABC] = [IBC] + [ICA] + [IAB] = \frac{r(a+b+c)}{2}$. Hence $r = \frac{ab}{a+b+c}$.

   (c) From $\frac{a+b-c}{2} = \frac{ab}{a+b+c}$, we have $(a+b)^2 - c^2 = 2ab$, which simplifies to $a^2 + b^2 = c^2$. This is Pythagoras' Theorem.

4. (a) Let the nine club members be A_1, A_2, A_3, B_1, B_2, B_3, C_1, C_2 and C_3. In the first three calls, A_1 calls A_2 and A_3, B_1 calls B_2 and B_3, and C_1 calls C_2 and C_3. In the last three calls, A_1 calls B_1 and C_1, A_2 calls B_2 and C_2, and A_3 calls B_3 and C_3. Now everybody knows everything.

   (b) Consider the moment after the third call has been made. If not everyone has participated, clearly nobody knows everything at this point. If everyone has participated, then each has made exactly one call except one who has made two calls. Still, nobody knows everything at this point. It follows that each of the eight members must participate in at least one more call. This means that three further calls are required, so that five calls are not sufficient.

2007

1. Let each order of chicken Kiev cost k dollars and each order of minced chicken m dollars. Then $125m < 30k < 126m$. The inequality on the left simplifies to $25m < 6k$ or $25m + 1 \le 6k$, while the one on the right simplifies to $5k < 21m$ or $5k \le 21m - 1$. It follows that we have $125m + 5 \le 30k \le 126m - 6$ or $11 \le m$. Now $5k \ge 21m - 1 \ge 230$ so that $k \ge 46$. Hence the family's share was $2k + m \ge 103$, and 100 dollars would not be enough.

2. Let $n = pq^2$ where p and q are distinct primes. We shall prove that the only solutions are $n = 63, 76, 261$ and 2107. We consider five cases:
   Case 1. $q = 2$.
   The divisors are 1, 2, 4, p, $2p$ and $4p$. If three of the four divisors which sum to 100 are 1, 2 and 4, the remaining one must be 93, which cannot be any of p, $2p$ and $4p$. If three of the four divisors which sum to 100 are p, $2p$ and $4p$, then the fourth must be 2 so that $100 - 2 = 98$ is divisible by 7. However, we cannot have $p = 14$. The only alternative is to take two of 1, 2 and 4 and two of p, $2p$ and $4p$. Examining $100 - 1 - 2 = 97$, $100 - 2 - 4 = 94$ and $100 - 1 - 4 = 95$, we must have $95 = p + 4p$ so that $p = 19$. This yields $n = 76$.
   Case 2. $q = 3$.
   The divisors are 1, 3, 9, p, $3p$ and $9p$. If three of the four divisors which sum to 100 are 1, 3 and 9, the remaining one must be 87. We must have $87 = 3p$ so that $p = 29$. This yields $n = 261$. If three of the four divisors which sum to 100 are p, $3p$ and $9p$, then the fourth must be 9 so that $100 - 9 = 91$ is divisible by 13. Hence $p = 7$ and this yields $n = 63$. The only alternative is to take two of 1, 3 and 9 and two of p, $3p$ and $9p$. Examining $100 - 1 - 3 = 96$, $100 - 3 - 9 = 88$ and $100 - 1 - 9 = 90$. The only possibilities are $96 = p + 3p$, $88 = p + 3p$ and $90 = p + 9p$. However, we cannot have any of $p = 24, 22$ and 10.
   Case 3. $q = 5$.
   The divisors are 1, 5, 25, p, $5p$ and $25p$. If three of the four divisors which sum to 100 are 1, 5 and 25, the remaining one must be 69, which cannot be any of p, $5p$ and $25p$. If three of the four divisors which sum to 100 are p, $5p$ and $25p$, no fourth divisor will work. The only alternative is to take two of 1, 5 and 25 and two of p, $5p$ and $25p$. Examining $100 - 1 - 5 = 94$, $100 - 5 - 25 = 70$ and $100 - 1 - 25 = 74$. However, none of them is divisible by 6, 30 or 26. Thus there are no solutions in this case.

Case 4. $q = 7$.
The divisors are 1, 7, 49, p, $7p$ and $49p$. If three of the four divisors which sum to 100 are 1, 7 and 49, the remaining one must be 43 and $p = 43$ is the only possibility. This yields $n = 2107$. If one of the four divisors which sum to 100 is $49p$, then we must have $p = 2$ but $2p$ is already 98. The only alternative is to take two of 1, 7 and 49 along with p and $7p$. Hence $8p = 100 - 1 - 7$, $100 - 1 - 49$ or $100 - 7 - 49$. However, none of 92, 50 and 44 is divisible by 8.

Case 5. $q \geq 11$.
Here $pq^2 \geq 121p > 121 > 100$. Hence we must have

$$100 = 1 + p + q + pq = (1+p)(1+q).$$

None of $100 = 4 = 5 \times 20 = 10 \times 10$ leads to prime values for p and q. Thus there are no solutions in this case.

3. (a) Since $AF = BF$, $[FBG] = [FAG] = x$. Since $AE = CE$, $[GCE] = [AGE] = y$.

   (b) Since $BF = AF$, $[BFC] = [AFC] = x + 2y$. It follows that $[BCG] = [BFC] - [FBG] = 2y$. Since $CE = AE$, we have $[BEC] = [BEA] = 2x+y$. Hence $[BCG] = [BEC] - [GCE] = 2x$.

   (c) From (b), we have $2x = [BCG] = 2y$. Now $BD = CD$ since

$$\frac{BD}{CD} = \frac{[BAD]}{[CAD]} = \frac{[BGD]}{[CGD]} = \frac{[BAD]-[BGD]}{[CAD]-[CGD]} = \frac{[BAG]}{[CAG]} = \frac{2x}{2y} = 1.$$

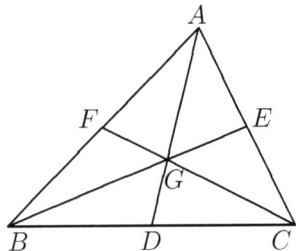

Figure 3.25

4. (a) Cherry can guarantee herself of getting $\frac{13}{8}$ cakes, leaving $\frac{11}{8}$ cakes for Eddy.

   (b) Cherry cuts the first cake into two pieces of respective sizes $\frac{5}{8}$ and $\frac{3}{8}$. If Eddy lets her choose first, she will cut the other two cakes in halves, so that she will get $\frac{5}{8} + \frac{1}{2} + \frac{1}{2} = \frac{13}{8}$. Suppose Eddy takes the larger piece. Cherry then cuts the second cake into two pieces of respective sizes $\frac{3}{4}$ and $\frac{1}{4}$. If Eddy lets her choose first, she will cut the third cake in halves, so that she will get $\frac{3}{8} + \frac{3}{4} + \frac{1}{2} = \frac{13}{8}$.

Suppose Eddy takes the larger piece again. Cherry then cuts the third cake into two pieces of respective sizes 1 and 0. Now she takes the larger piece and gets $\frac{3}{8} + \frac{1}{4} + 1 = \frac{13}{8}$ once again.

(c) We now prove that Cherry cannot get more than $\frac{13}{8}$ cakes. Suppose she cuts the first cake into two pieces of respective sizes x and $1-x$, with $x \geq \frac{1}{2}$. We consider two cases:
Case 1. $x < \frac{5}{8}$.
Eddy will let her choose first. The best Cherry can do from here on is to cut the other two cakes in halves. She will then get $x + \frac{1}{2} + \frac{1}{2} < \frac{13}{8}$.
Case 2. $x \geq \frac{5}{8}$.
Eddy will take the larger piece. Suppose Cherry now cuts the second cake into two pieces of respective sizes y and $1-y$, with $y \geq \frac{1}{2}$. If $y \geq \frac{3}{4}$, Eddy will take the larger piece. Cherry will then have $1 - x + 1 - y + 1 \leq 3 - (x+y) \leq \frac{13}{8}$. If $y < \frac{3}{4}$, Eddy will let her choose first. The best Cherry can do from here on is to cut the third cake in halves. She will then get $1 - x + y + \frac{1}{2} < \frac{13}{8}$.

2008

1. We look at all possible paths, taking each one as far as it will go. Together, they show that lily-pad 10 cannot be visited. On the other hand, the boldfaced entries show that lily-pads 2 to 9 can be visited.

   (a) 0, **2**.
   (b) 0, **3**, 1, **4**, 2, **5** or **6**.
   (c) 0, 3, 1, 5, 2, 4.
   (d) 0, 3, 1, 5, 2, 6, 4, **7** or **8**.
   (e) 0, 4, 1, 3.
   (f) 0, 4, 1, 5, 2, 6, 3, 7.
   (g) 0, 4, 2, 5, 1, 3.
   (h) 0, 4, 2, 5, 3, 6 or 7.
   (i) 0, 4, 2, 6, 3, 5, 1.
   (j) 0, 4, 2, 6, 3, 7, 5, 8 or **9**.

2. Let a and b be the two two-digit numbers, with a placed before b to form the four-digit number $100a + b$. The key observation is that a must divide b. It follows easily from the fact that $100a + b$ is divisible by a, and $100a$ is certainly divisible by a. Hence $b = na$ for some positive integer n. Since both a and b are two-digit numbers, $n \leq 9$. Now $100a + b = a(100 + n)$ is divisible by ab. This means that b is a divisor of $100 + n$. Now 101, 103, 107 and 109 are prime numbers and have no two-digit divisors. If $n = 8$, then $b \leq 54$ and a will not be a two-digit number. If $n = 6$, we must have $a = b = 53$, but $5353 = 53 \times 101$ is not divisible by 53^2. If $n = 5$, then $b = 35$ or 15 since it is 5 times a, but then a will not be a two-digit number. From $n = 4$, we get the answer 1352, and from $n = 2$, we get the answer 1734. Thus there are only two such numbers.

3. The minimum number is 13. We first show that this can be attained. Place a mine in each square on the main diagonal from (1,1) to (8,8). Place 5 more mines in (1,2), (2,3), (3,4), (4,5) and (5,1). By symmetry, we may assume that the mine-sweeper uses more rows than columns to clear the mines in (6,6), (7,7) and (8,8). If it uses 3 rows, then it uses 4 columns to clear the remaining 10 mines. However, this will leave 2 mines in different rows, and the mine-sweeper can clear only 1 more row. If it uses 2 rows, then it uses 3 columns to clear the remaining 10 mines. However, this will leave 4 mines in at least 3 different rows, and the mine-sweeper can clear only 2 more rows. We now show that 12 mines will not be enough. The mine-sweeper will clear the 4 rows containing the largest total number of mines. This total is at least 8, as otherwise a chosen row has at most 1 mine while an unchosen one has at least 2. After the row clearance, there will remain at most 4 mines, and the column clearance can take care of them.

4. Imbed triangle ABC in a square grid as shown in Figure 3.26. Then we have

$$\angle FHE + \angle FDE + \angle FKE = \angle BFH + \angle HFD + \angle DFK$$
$$= \angle BFK$$
$$= 90°.$$

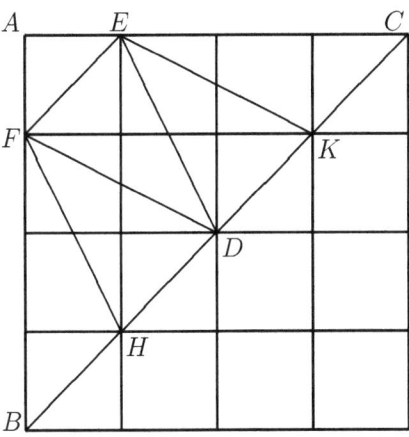

Figure 3.26

2009

1. (a) We can take the numbers 2, 4, 6, 8, 10, 12, 14, 16, 18, 20, 22 and 24. The sum of any two of them is less than 50.

(b) The twenty-four even numbers between 1 and 49 can be combined into twelve pairs with sum 50, namely, 2+28, 4+46, 6+44, 8+42, 10+40, 12+38, 14+36, 16+34, 18+32, 20+30, 22+28 and 24+26. If we choose at random thirteen numbers from these twelve pairs, we must take both numbers of some pair, and the sum of these two numbers will be 50.

2. The sum of the twelve digits of such a number is $(1+2+3) \times 4 = 24$, which is divisible by 3. Hence the number itself is also divisible by 3.

   (a) Since such a number is clearly greater than 3, it cannot be prime.

   (b) Note that 24 is not divisible by 9, and neither is the number itself. A number which is divisible by 3 but not by 9 cannot be the square of an integer.

3. (a) The area of triangle BAP is given by $\frac{ax}{2}$ and the area of triangle CAP is given by $\frac{ay}{2}$. Hence the area of triangle ABC is given by $\frac{a(x+y)}{2}$.

   (b) Since the length a of $AB = AC$ and the area $\frac{a(x+y)}{2}$ are both constant, so is $x + y$, which is $PD + PE$.

4. We have $a = n$, $b = n + 1$ and $c = n(n + 1)$ for some positive integer n. Hence

$$\begin{aligned} d^2 &= a^2 + b^2 + c^2 \\ &= n^2 + (n+1)^2 + (n^2+n)^2 \\ &= n^4 + 2n^3 + 3n^2 + 2n + 1 \\ &= (n^2 + n + 1)^2. \end{aligned}$$

It follows that $d = n^2 + n + 1$.

2010

1. There are no table numbers which start with 9. Hence the digit 9 is in the last place. The only table number which ends with 9 is $7 \times 7 = 49$. Hence the digit 4 is in the eighth place. The only table number which starts with 8 is $9 \times 9 = 81$ and the only table number which starts with 7 is $9 \times 8 = 72$. Since 9 is already in the last place, 8 must be followed by 1 and 7 by 2. Now the only table number which ends in 7 is $9 \times 3 = 27$, but 7 cannot follow 2 and then be followed by 2. Hence 7 is in the first place and 2 is in the second place. Now $6 \times 3 = 18$, $7 \times 4 = 28$ and $8 \times 6 = 48$ are table numbers. However, 8 cannot follow 1 since it is to be followed by 1, and 8 cannot follow 4 since 9 follows 4. Hence 8 is in the third place and 1 is in the fourth place.

Neither 13 nor 34 is a table number. Hence 3 is in the sixth place. Since 53 is not a table number, 5 must be in the seventh place, leaving the fifth place for 6. Now all of 72, 28, 81, 16, 63, 35, 54 and 49 are table numbers. Hence the arrangement is 728163549.

2. Room 1 on the second floor has only one neighbor on the same floor, and must be occupied by a cat. Room 1 on the first floor must be occupied by a rat. Consider Room 2 on the second floor.
Case 1. It is occupied by a dog.
Then Room 3 on the second floor must be occupied by a rat, but this rat should not leave its light on as it cannot be between two cats.
Case 2. It is occupied by a cat.
Then Room 2 on the first floor must be occupied by a rat, but this rat should not leave its light on as it cannot be between two cats.
Case 3. It is occupied by a rat.
Then Room 3 on the second floor must be occupied by a cat, and Room 3 on the first floor must be occupied by a rat. Hence Room 2 on the first floor can only be occupied by a cat. If a rat is in Room 4, it must be on the second floor, with the last cat in Room 5 on the second floor. Then the remaining rooms are occupied dogs, but the dogs in Rooms 4 and 5 on the first floor should not leave their lights on. Neither Room 4 can be occupied by a cat without a rat in the other Room 4. Hence both Room 4 are occupied by dogs. These dogs leave their lights on because there is a rat in Room 5 on the second floor, and a cat in Room 5 on the first floor. Both Room 6 are occupied by dogs. The unique solution is shown in Figure 3.27.

○ Cat	○ Rat	○ Cat	○ Dog	● Rat	● Dog
● Rat	○ Cat	● Rat	○ Dog	○ Cat	● Dog

Figure 3.27

3. (a) Since $AB = AC$, $\angle ABC = \angle ACB$. Since $\angle CAB = 20°$ and the sum of these three angles is 180°, $\angle ABC = 80°$. Since BCG is an equilateral triangle, $\angle EBC = 60°$. Hence
$$\angle EBA = \angle ABC - \angle EBC = 20° = \angle EAB.$$
It follows that $EA = EB$.
(b) Since BCG is equilateral, $GB = BC = AP$. Hence
$$EP = EA - AP = EB - BG = EG.$$

(c) Since $EP = EG$, $\angle EPG = \angle EGP$. The sum of these two angles is equal to $\angle PAB + \angle GBA$. Hence $\angle EPG = 20°$. On the other hand, $\angle ECG = \angle ACB - \angle GCB = 20°$ also. Hence $GP = GC$.

(d) We have $GB = GC = GP$ so that $\angle GBP = \angle GPB$. The sum of these two angles is equal to $\angle EGP = 20°$. Hence $\angle GBP = 10°$ and $\angle ABP = \angle ABE - \angle GBP = 10°$.

4. Let x, y and z be the number of dollars possessed by Aaron, Betty and Cyrus. Let the cost of the horse be t dollars. After clearing fractions, we have

$$2x + y + z = 2t; \qquad (1)$$
$$x + 3y + z = 3t; \qquad (2)$$
$$x + y + 4z = 4t. \qquad (3)$$

If we add twice equation (1) to equation (2), we obtain

$$5x + 5y + 3z = 7t. \qquad (4)$$

If we add three times equation (4) to twice equation (3), we obtain

$$17x + 17y + 17z = 29t. \qquad (5)$$

Since 17 divides $29t$ but 17 and 29 have no common factors greater than 1, 17 must divide t, so that t cannot be less than 17. We now show that the minimum value is indeed 17. Taking $t = 17$, equation (5) becomes

$$x + y + z = 29. \qquad (6)$$

Subtracting equation (6) from equation (1), we have $x = 2t - 29 = 5$. Subtracting equation (6) from equation (2), we have $2y = 3t - 29 = 22$ so that $y = 11$. Subtracting equation (6) from equation (3), we have $3z = 4t - 29 = 39$ so that $z = 13$. Taking half the money of Betty and Cyrus, Aaron will have $5 + \frac{11+13}{2} = 17$ dollars. Taking one third of the money of Cyrus and Aaron, Betty will have $11 + \frac{13+5}{3} = 17$ dollars. Taking one quarter of the money of Aaron and Betty, Cyrus will have $13 + \frac{5+11}{4} = 17$ dollars. Everything checks out, so that 17 is indeed the minimum value of t.

2011

1. Let the first fraction be $\frac{a}{b}$ where $0 < a < b$. Then the second fraction is $\frac{a+11}{b+11}$. We have

$$\frac{a+11}{b+11} - \frac{a}{b} = \frac{(a+11)b - a(b+11)}{b(b+11)} = \frac{11(b-a)}{b(b+11)} > 0$$

because $b - a > 0$ and $b + 11 > b > 0$. It follows that the second fraction is larger.

2. (a) We have
$$(x+y)(x+y) = x(x+y) + y(x+y)$$
$$= x^2 + xy + yx + y^2$$
$$= x^2 + 2xy + y^2.$$

(b) We have
$$(x+y)(x-y) = x(x-y) + y(x-y)$$
$$= x^2 - xy + yx - y^2$$
$$= x^2 - y^2.$$

(c) We have
$$x^2 - 6xy + 9y^2 - 4z^2 = (x-3y)^2 - (2z)^2$$
$$= (x-3y+2z)(x-3y-2z).$$

3. (a) Let the number in position 2 be a. Then $n = \frac{1+a}{2} + 1$ so that $a = 2n - 3$. Let the number in position 3 be b. Then we have $2n - 3 = \frac{b+n}{2} + 1$, so that $b = 3n - 8$. Let the number in position 4 be c. Then $3n - 8 = \frac{c+2n-3}{2} + 1$, so that $c = 4n - 15$. Finally, let the number in position 5 be d. Then $4n - 15 = \frac{d+3n-8}{2} + 1$, so that $d = 5n - 24$.

(b) The pattern in (a) suggests that the number in position k is $kn - (k^2 - 1)$. We have verified this for $k \leq 5$. Let x be the number in position $k+1$. Then $kn - (k^2 - 1) = \frac{x+(k-1)n-((k-1)^2-1)}{2} + 1$, so that $x = 2(kn-k^2) - (k-1)n + (k^2 - 2k) = (k+1)n - ((k+1)^2 - 1)$, which completes the proof by mathematical induction on k.

(c) From (b), the number in position n is $n^2 - (n^2 - 1) = 1$. This is as long as the row can get as the next number will be negative.

4. To draw the desired conclusion $\angle ACD > \angle CAB$, we have to apply the SaS Inequality to triangles ACD and CAB. We already have $AC = CA$. Since the Exterior Angle Inequality does not depend on exactly where D is on the extension of BC, we can choose D so that $CD = AB$.

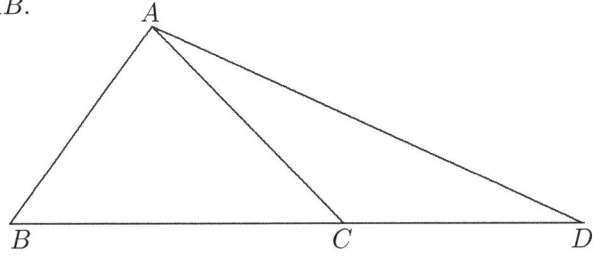

Figure 3.28

All we need now is to prove that $AD > BC$. This must come from the Triangle Inequality. When this is applied to triangle BAD, we have $AB + AD > BD = BC + CD$. Since $AB = CD$, we have $AD > BC$ as desired.

2012

1. (a) Since $882 = (21\sqrt{2})^2 < 900 = 30^2 < 968 = (22\sqrt{2})^2$, the smallest value of n is 22.

   (b) From (a), $(21\sqrt{2})^{20} < 30^{20} < (22\sqrt{2})^{20}$. Since $21^2 < 2^{10}$, we have $(21\sqrt{2})^{20} = 21^{20}2^{10} > 21^{22}$. On the other hand, since $22^3 > 2^{10}$, $(22\sqrt{2})^{20} = 22^{20}2^{10} < 22^{23}$. Hence $21^{22} < 30^{20} < 22^{23}$. It follows that the smallest value of n is also 22.

2. (a) We have $(x+y+z)^2 = x^2 + y^2 + z^2 + 2yz + 2zx + 2xy$. Hence
$$\begin{aligned} x^2 + y^2 + z^2 &= (x+y+z)^2 - 2(xy+yz+zx) \\ &= 1^2 - 2(2) \\ &= -3. \end{aligned}$$

   (b) We have
$$\begin{aligned} (x+y+z)^3 &= x^3 + y^3 + z^3 + 3y^2z + 3z^2x + 3x^2y \\ &\quad + 3yz^2 + 3zx^2 + 3xy^2 + 6xyz \end{aligned}$$

   and
$$\begin{aligned} & (x+y+z)(yz+zx+xy) \\ &= y^2z + z^2x + x^2y + yz^2 + zx^2 + xy^2 + 3xyz. \end{aligned}$$

   Hence
$$\begin{aligned} & x^3 + y^3 + z^3 \\ &= (x+y+z)^3 - 3(x+y+z)(yz+zx+xy) + 3xyz \\ &= 4. \end{aligned}$$

3. First, we assume that (1) and (2) are true, and prove that (3) is also true. In Figure 3.29 on the left, $OM = OM$. We have $OA = OB$ as both are radii of the circle. Moreover, it is given that $AM = BM$. Hence the three sides of OAM are equal to the corresponding sides of OBM, so that the two triangles are congruent. It follows that $\angle OMA = \angle OMB$. Because their sum is $180°$, each is $90°$ so that OM is indeed perpendicular to AB.

Next, we assume that (1) and (3) are true, and prove that (2) is also true. In Figure 3.29 on the left, let N be the foot of perpendicular from O to AB, so that $\angle ONA = 90° = \angle ONB$. We have $ON = ON$ and $OA = OB$ as before, so that triangles ONA and ONB are congruent. It follows that $AN = BN$, which means that N coincides with M, and the line indeed passes through M.

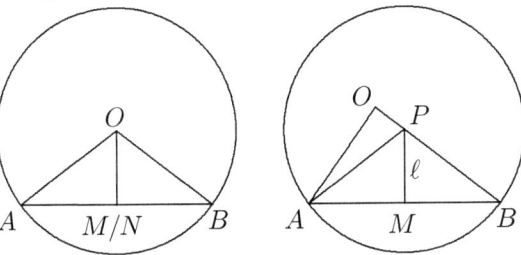

Figure 3.29

Finally, we assume that (2) and (3) are true, and prove that (1) is also true. Suppose to the contrary that O does not lie on the perpendicular bisector ℓ of AB. We may assume that O is on the same side of ℓ as A, as shown in Figure 3.29 on the right. This means that the segment OB must intersect ℓ, at some point which we denote by P. As before, we can prove that triangles PMA and PMB are congruent, so that $PA = PB$. Now $OA < OP + PA = OP + PB = OB$ by the Triangle Inequality. This contradicts the given condition that O is the center of the circle.

4. Since everyone at a meeting sits between two others, each member can attend at most 4 meetings. Since exactly 6 members attends a meeting, the number of meetings is at most $4 \times 9 \div 6 = 6$. We now construct a set of 6 meetings. Arrange the 9 members in a 3×3 configuration as shown below.

$$\begin{array}{ccc} 1 & 2 & 3 \\ 4 & 5 & 6 \\ 7 & 8 & 9 \end{array}$$

In the first meeting, members in the first row and the second row occupy alternate seats round the table. The second meeting is attended by members in the second and the third rows, and the third meeting is attended by members in the third and the first rows, with alternating seating as shown in the diagram below.

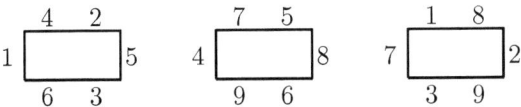

Figure 3.30

The last three meetings operate on columns instead of rows, as shown in the diagram below.

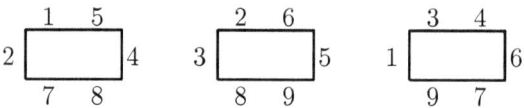

Figure 3.31

2013

1. (a) Let $S = 3^{2012} + 3^{2011} + 3^{2010} + \cdots + 3^2 + 3 + 1$. Then
$$3S = 3^{2013} + 3^{2012} + 3^{2011} + \cdots + 3^3 + 3^2 + 3.$$
Subtraction yields $2S = 3^{2013} - 1$ so that $S = \frac{2^{2013}-1}{2}$.

   (b) Let $3^m - 1 = 2^n$ for some positive integers m and n. Consider first when m is odd, so that $m = 2k+1$. By (a),
$$2^n = 3^{2k+1} - 1 = 2(3^{2k} + 3^{2k-1} + \cdots + 3 + 1).$$

   The second factor is the sum of an odd number of odd integers, and is therefore odd. Since the only positive integer power of 2 that is odd is 1, we have $2^n = 2 \times 1$. It follows that $n = 1$ and $m = 1$, so that the power of 3 is 3 itself. Consider now when m is even, so that $m = 2k$. Then $2^n = 3^{2k-1} - 1 = (3^k+1)(3^k-1)$. The two factors on the right side are consecutive even numbers both of which are powers of 2. So they must be 2 and 4 respectively. It follows that $n = 3$ and $m = 2$, so that the power of 3 is 9.

2. Let the positions in the line of Lana and Derek be m and n. We consider the following cases.
   Case 1. $\{m,n\} = \{1,2\}$.
   In this case, they will meet for sure as they each other in the first game.
   Case 2. $\{m,n\} = \{1,3\}, \{2,3\}, \{3,4\}, \{4,5\}, \{5,6\}, \{6,7\}$ or $\{7,8\}$.
   In this case, they will meet if whichever of them plays first wins a game, with probability $\frac{1}{2}$.
   Case 3. $\{m,n\} = \{1,4\}, \{2,4\}, \{3,5\}, \{4,6\}, \{5,7\}$ or $\{6,8\}$.
   In this case, they will meet if whichever of them plays first wins 2 games, with probability $\frac{1}{4}$.
   Case 4. $\{m,n\} = \{1,5\}, \{2,5\}, \{3,6\}, \{4,7\}$ or $\{5,8\}$.
   In this case, they will meet if whichever of them plays first wins 3 games, with probability $\frac{1}{8}$.

Case 5. $\{m,n\} = \{1,6\}, \{2,6\}, \{3,7\}$ or $\{4,8\}$.
In this case, they will meet if whichever of them plays first wins 4 games, with probability $\frac{1}{16}$.
Case 6. $\{m,n\} = \{1,7\}, \{2,7\}$ or $\{3,8\}$.
In this case, they will meet if whichever of them plays first wins 5 games, with probability $\frac{1}{32}$.
Case 7. $\{m,n\} = \{1,8\} \{2,8\}$.
In this case, they will meet if whichever of them plays first wins 6 games, with probability $\frac{1}{64}$.
It follows that the overall probability is $\frac{S}{1+7+6+5+4+3+2}$, where

$$S = 1 \times 1 + 7 \times \frac{1}{2} + 6 \times \frac{1}{4} + 5 \times \frac{1}{8} + 4 \times \frac{1}{16} + 3 \times \frac{1}{32} + 2 \times \frac{1}{64}.$$

Then

$$2S = 1 \times 2 + 7 \times 1 + 6 \times \frac{1}{2} + 5 \times \frac{1}{4} + 4 \times \frac{1}{8} + 3 \times \frac{1}{16} + 2 \times \frac{1}{32}.$$

Subtraction yields

$$S = 8 - \frac{1}{2} - \frac{1}{4} - \frac{1}{8} - \frac{1}{16} - \frac{1}{32} - \frac{2}{64} = 7.$$

Hence the probability is $\frac{7}{28} = \frac{1}{4}$.

3. Exend the radius CO to intersect the circle again at D. Then we have $\angle DAC = 90° = \angle DBC$, $\angle ABD = \angle ACD$ and $\angle BAD = \angle BCD$.

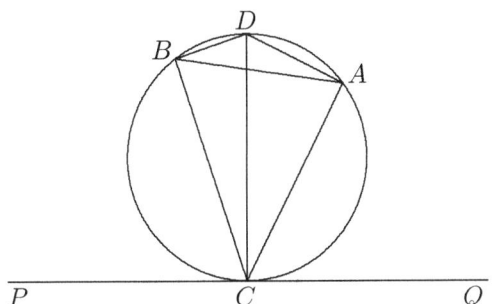

Figure 3.32

We first assume that (1) is true, that is, $\angle PCD = 90°$. Then

$$\angle PCB = 90° - \angle BCD = 90° - \angle BAD = \angle BAC,$$

which is (2). We can prove (3) in an analogous manner. We now assume that (2) is true, or $\angle PCB = \angle BAC$. Then

$$\angle PCD = \angle PCB + \angle BCD = \angle BAC + \angle BAD = \angle DAC = 90°.$$

This is (1), and (3) follows from (1). Finally, the argument is analogous if we assume that (3) is true.

4. (a) The upper bound is equivalent to $n(2n-1)^2 > (n-1)4n^2$. The left side simplifies to $4n^3 - 4n^2 + n$ while the right side simplifies to $4n^3 - 4n^2$. The left side is greater than the right side since $n > 0$. The lower bound is equivalent to $4n^2(3n+1) > (2n-1)^2(3n-2)$. The left side simpifies to $12n^3 - 8n^2$ while the right side simplifies to $12n^3 - 8n^2 - n + 1$. The left side is greater than the right side since for $n \geq 2$, $-n + 1 < 0$.

(b) By (a), $\sqrt{\frac{7}{4}} < \frac{4}{3} < \sqrt{2}$, $\sqrt{\frac{10}{7}} < \frac{6}{5} < \sqrt{\frac{3}{2}}$, $\sqrt{\frac{13}{10}} < \frac{8}{7} < \sqrt{\frac{4}{3}}$, $\sqrt{\frac{16}{13}} < \frac{10}{9} < \sqrt{\frac{5}{4}}$, $\sqrt{\frac{19}{16}} < \frac{12}{11} < \sqrt{\frac{6}{5}}$, $\sqrt{\frac{22}{19}} < \frac{14}{13} < \sqrt{\frac{7}{6}}$ and $\sqrt{\frac{25}{22}} < \frac{16}{15} < \sqrt{\frac{8}{7}}$. Multiplying these together, we have

$$\sqrt{\frac{7}{4}}\sqrt{\frac{10}{7}}\sqrt{\frac{13}{10}}\sqrt{\frac{16}{13}}\sqrt{\frac{19}{16}}\sqrt{\frac{22}{19}}\sqrt{\frac{25}{22}} < \frac{4}{3} \times \frac{6}{5} \times \frac{8}{7} \times \frac{10}{9} \times \frac{12}{11} \times \frac{14}{13} \times \frac{16}{15}$$

and

$$\frac{4}{3} \times \frac{6}{5} \times \frac{8}{7} \times \frac{10}{9} \times \frac{12}{11} \times \frac{14}{13} \times \frac{16}{15} < \sqrt{2}\sqrt{\frac{3}{2}}\sqrt{\frac{4}{3}}\sqrt{\frac{5}{4}}\sqrt{\frac{6}{5}}\sqrt{\frac{7}{6}}\sqrt{\frac{8}{7}}.$$

This simplifies to

$$\frac{5}{2} < \frac{4 \times 6 \times 8 \times 10 \times 12 \times 14 \times 16}{3 \times 5 \times 7 \times 9 \times 11 \times 13 \times 15} < \sqrt{8} < 3.$$

Multiplying by 2, we have

$$5 < \frac{2 \times 4 \times 6 \times 8 \times 10 \times 12 \times 14 \times 16}{1 \times 3 \times 5 \times 7 \times 9 \times 11 \times 13 \times 15} < 6.$$

Hence the desired integer part is 5.

2014

1. Let the number of expressions for n be a_n. We first prove two auxiliary results.
   Lemma 1. $a_{2n+1} = a_{2n}$.
   Proof:
   Every expression for $2n+1$ must contain at least one 1. The exclusion of one 1 yields an expression for $2n$. On the other hand, every expression for $2n$ can contain at most two 1s. The inclusion of one 1 yields an expression for $2n + 1$. This one-to-one correspondence establishes the desired result.

Lemma 2. $a_{2n} = a_n + a_{n-1}$.

Proof:
Every expression for $2n$ contains either no 1s or two 1s. In the former case, dividing each term by 2 yields an expression for n. In the latter case, dividing each term by 2 after the exclusion of the two 1s yields an expression for $n-1$. These two processes are clearly reversible, and the one-to-one correspondences yield the desired result.

We now prove by mathematical induction that $a_{2n} = a_{2n+1} = n+1$. In particular, $a_{2014} = 1008$. Note that $a_0 = a_1 = 1$. Suppose the result holds up to $a_{2n-2} = a_{2n-1} = n$. By the Lemmas, we have $a_{2n} = a_{2n+1} = a_n + a_{n-1}$. If $n = 2k$, then

$$a_n + a_{n-1} = a_{2k} + a_{2k-1} = k + 1 + k = 2k + 1 = n + 1.$$

If $n = 2k+1$, then

$$a_n + a_{n-1} = a_{2k+1} + a_{2k} = k + 1 + k + 1 = 2k + 2 = n + 1.$$

This completes the inductive argument.

Alternative Solution:
Let the number of expressions for n be a_n. Let

$$A(x) = a_0 + a_1 x + a_2 x^2 + a_3 x^3 + a_4 x^4 + \cdots.$$

This is called a generating function for the sequence $\{a_n\}$. We have

$$\begin{aligned} A(x) &= (a_0 + a_2 x^2 + a_4 x^4 + \cdots) + x(a_1 + a_3 x^2 + a_5 x^4 + \cdots) \\ &= (1+x)(a_0 + (a_0 + a_1)x^2 + (a_2 + a_3)x^2 + \cdots) \\ &= (1+x)((a_0 + a_1 x^2 + a_2 x^4 + \cdots) \\ &\quad + x^2(a_0 + a_1 x^2 + a_2 x^4 + \cdots)) \\ &= (1+x)(1+x^2)(a_0 + a_1(x^2) + a_2(x^2)^2 + \cdots) \\ &= (1 + x + x^2 + x^3) A(x^2). \end{aligned}$$

There is another way to interpret this formula. Take for example $n = 6$. We have $a_6 = 4$ because there are four expressions, namely, 4+2, 4+1+1, 2+2+2 and 2+2+1+1. Two of the three expressions do not use 1s. Dividing all terms by 2, we obtain two expressions for 3, namely 2+1 and 1+1+1. As for the other two expressions, we must first remove the two 1s to reduce 6 to 4, and then divide by 2 as before to obtain two expressions for 2, namely, 2 and 1+1. Hence $a_6 = a_3(x^2)^3 + x^2(a_2(x^2)^2)$, with $a_3 = 2$ and $a_2 = 2$.

It follows that $A(x^2)$ generates the numbers of expressions without 1s, $xA(x^2)$ generates the numbers of expressions with one 1, $x^2A(x^2)$ generates the numbers of expressions with two 1s and $x^3A(x^2)$ generates the numbers of expressions with three 1s. Since these are all the options, we have $A(x) = (1 + x + x^2 + x^3)A(x^2)$. Now

$$\begin{aligned} A(x) &= (1+x+x^2+x^3)A(x^2) \\ &= \frac{1-x^4}{1-x} A(x^2) \\ &= \frac{1-x^4}{1-x} \cdot \frac{1-x^8}{1-x^2} A(x^4) \\ &= \frac{1-x^4}{1-x} \cdot \frac{1-x^8}{1-x^2} \cdot \frac{1-x^{16}}{1-x^4} A(x^8) \\ &= \cdots \\ &= \frac{1}{(1-x)(1-x^2)} \\ &= \frac{C_1}{1-x} + \frac{C_2}{(1-x)^2} + \frac{C_3}{1+x}. \end{aligned}$$

Clearing fractions, we have $1 = C_1(1-x)(1+x) + C_2(1+x) + C_3(1-x)^2$. Setting $x = -1$, $1 = 4C_3$ so that $C_3 = \frac{1}{4}$. Setting $x = 1$, $1 = 2C_2$ so that $C_2 = \frac{1}{2}$. Setting $x = 0$, $1 = C_1 + C_2 + C_3$ so that $C_1 = \frac{1}{4}$. Note that

$$\begin{aligned} \frac{1}{1-x} &= 1 + x + x^2 + x^3 + \cdots, \\ \frac{1}{(1-x)^2} &= (1 + x + x^2 + x^3 + \cdots)^2 \\ &= 1 + 2x + 3x^2 + 4x^3 + \cdots, \\ \frac{1}{1+x} &= \frac{1}{1-(-x)} \\ &= 1 - x + x^2 - x^3 + \cdots. \end{aligned}$$

Hence $a_n = \frac{n+1}{2} + \frac{1}{4}(1 + (-1)^n) = \frac{2n+3+(-1)^n}{4} = \lfloor \frac{n+2}{2} \rfloor$. In particular, $a_{2014} = 1008$.

2. (a) We have $x + \frac{1}{x} - 2 = (\sqrt{x} - \frac{1}{\sqrt{x}})^2 \geq 0$. Hence $x + \frac{1}{x} \geq 2$.

(b) We have $(x+y)(y+z) = xz + y(x+y+z) = xz + \frac{1}{xz} \geq 2$, with equality if and only if $xz = 1$. Then $x + z - 2 = (\sqrt{x} - \sqrt{z})^2 \geq 0$, with equality if and only if $x = z = 1$. When $(x+y)(y+z) = 2$, $1 = y(y+x+z) \leq y(y+2)$. Hence $y \geq \frac{-2+\sqrt{2^2-4(-1)}}{2} = \sqrt{2} - 1$.

3. (a) The two mountaineers will meet at (14,4). The chart below gives their respective progress.

(0,0)	(7,7)	(5,5)	(8,8)	(11,5)	(12,6)
(30,0)	(23,7)	(21,5)	(18,8)	(15,5)	(16,6)

(b) The two mountaineers will meet at (8,8). The chart below gives their respective progress.

(0,0)	(7,7)	(2,2)	(5,5)	(4,4)	(6,6)	(5,5)
(30,0)	(23,7)	(18,2)	(15,5)	(14,4)	(12,6)	(11,5)

(c) Let the vertices be $P_0, P_1, \ldots, P_n$. The two mountaineers are in state (k, ℓ) if one of them is moving between P_{k-1} and P_k and the other is moving between $P_{\ell-1}$ and P_ℓ. Initially, they are in state $(1, n)$. If they reach state $(n, 1)$ eventually, then this means they must have passed each other at some point, and have therefore met. Two states (i, j) and (k, ℓ) are said to be neighbors if the mountaineers can move directly from one state to the other. Every state has exactly two neighbors except for $(1, n)$ and $(n, 1)$. Each of these two states has exactly one neighbor. We shall orchestrate the movement of the two mountaineers so that they will be at the same altitude all the time, and no state will occur more than once. Since there are only finitely many states, the mountaineers must end up in state $(n, 1)$. At the start, the mountaineers move towards each other. They will continue to move in the same direction until exactly one of them reaches a vertex. This mountaineer will continue on while the other mountaineer turns back. It is clear that they will be at the same altitude all the time. Now $(n, 1)$ can occur at most once since movement is over once this state is reached. Suppose some state occurs more than once. Let (k, ℓ) be the first state to occur a second time. Now (k, ℓ) cannot be $(1, n)$ as otherwise the only neighbor of $(1, n)$ must occur a second time before that. Hence (k, ℓ) has two neighbors. We may assume that when (k, ℓ) occurs for the first time, it comes after (i, j) and before (m, n). The state immediately after (m, n) cannot be (k, ℓ) again because the two mountaineers cannot both turn back simultaneously. When (k, ℓ) occurs a second time, either (i, j) or (m, n) must occur a second time before that. This is a contradiction.

4. (a) Since $\frac{AK}{AB} = \frac{1}{2} = \frac{AN}{AD}$, triangles AKN and ABD are similar. Hence KN is parallel to BD, and $KN = \frac{1}{2}BD$. Similarly, LM is parallel to BD and $LM = \frac{1}{2}BD$. Let KM intersect LN at O. Then triangles OKN and OML are congruent, so that KM and LN bisect each other at O.

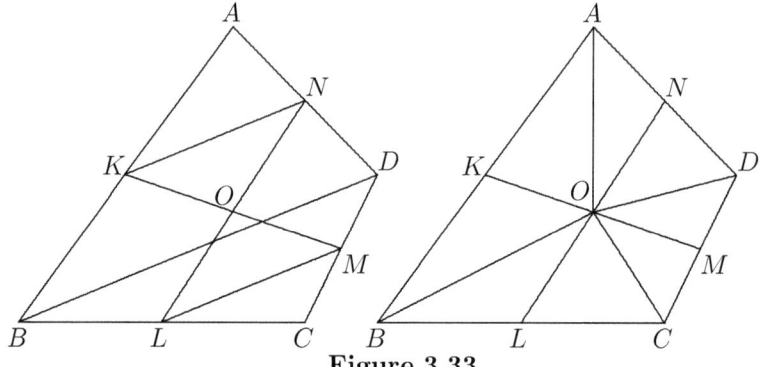

Figure 3.33

(b) We use the notation $[P]$ to denote the area of the polygon P. Since $AK = KB$, $[AOK] = [KOB]$. Similarly, $[BOL] = [LOC]$, $[COM] = [MOD]$ and $[DON] = [NOA]$. It follows that

$$[AKON] + [CMOL] = [AOK] + [NOA] + [LOC] + [COM]$$
$$= [KOB] + [DON] + [BOL] + [MOD]$$
$$= [BLOK] + [DNOM].$$

Since $[AKON] + [CMOL] + [BLOK] + [DNOM] = [ABCD]$, each of the two sums is equal to $\frac{1}{2}[ABCD]$.

(c) Since $\frac{AP}{AB} = \frac{1}{3} = \frac{AW}{AD}$, triangles APW and ABD are similar. Hence PW is parallel to BD, and $PW = \frac{1}{3}BD$. Similarly, RU is parallel to BD and $RU = \frac{2}{3}BD$. Let KM intersect LN at E. Then triangles EPW and ERU are similar with $\frac{PW}{RU} = \frac{1}{2}$. It follows that PU and RW trisect each other at E. Similarly, the other three points of intersections are also points of trisections.

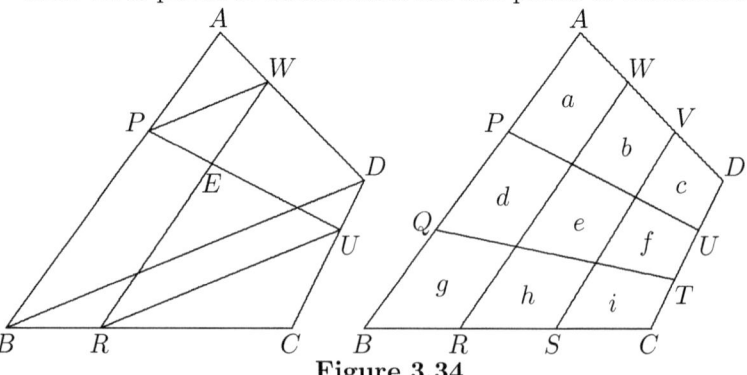

Figure 3.34

Solutions

(d) Let the area of the nine quadrilaterals be a, b, c, d, e, f, g, h and i, as shown in the diagram above on the right. By (b) and (c), we have $a + e = b + d$, $b + f = c + e$, $d + h = e + g$ and $e + i = f + h$. It follows that

$$a + e + i = b + d + f + h - e = b + f + g.$$

Also, $b + f + g = c + e + d + h - e = c + d + h$. Since

$$(a + e + i) + (b + f + g) + (c + d + h) = [ABCD],$$

each sum is $\frac{1}{3}[ABCD]$. Similarly,

$$a + f + h = b + d + i = c + e + g = \frac{1}{3}[ABCD].$$

2015

1. When successive powers of 2 are divided by 7, the remainders come in a cycle of length 3, namely, 2, 4 and 1. When successive squares are divided by 7, the remainders come in a cycle of length 7, namely 1, 4, 2, 2, 4, 1 and 0. The following table shows the remainders for both 2^n and n^2 in a cycle of length $3 \times 7 = 21$.

n	1	2	3	4	5	6	7	8	9	10
2^n	2	4	1	2	4	1	2	4	1	2
n^2	1	4	2	2	4	1	0	1	4	2
$2^n - n^2$	1	0	6	0	0	0	2	3	4	0

11	12	13	14	15	16	17	18	19	20	21
4	1	2	4	1	2	4	1	2	4	1
2	4	1	0	1	4	2	2	4	1	0
2	4	1	4	0	5	2	6	5	3	1

There are 6 such values in each cycle of 21, and the number of cycles from 1 to 9996 is $9996 \div 21 = 476$. Of the remaining three numbers 9997, 9998 and 9999, only 9998, being the second number in a new cycle, has the desired property. It follows that the total number is $6 \times 476 + 1 = 2857$.

2. Consider any distribution of seats among the first 21 people to arrive. We may assume that Edmond is the last of them. At that moment, there are 10 empty seats. We can associate a sequence of seats with each of the 9 sticklers, starting with his or her assigned seat. If that seat is unoccupied, the sequence terminates.

Otherwise, the next term is the assigned seat of the evicted person. Eventually, the sequence ends at an empty seat. Now each sequence necessarily ends in a different seat. Hence 9 of the 10 empty seats are ends of such sequences. Thus there is only one empty seat which Edmond can take without being evicted, and it follows that the desired probability is $\frac{1}{10}$.

3. We have $x^2 - (a+10)x + 10a = x^2 - (b+c)x + bc$. Since this is true for all x, we must have $a + 10 = b + c$ and $10a + 1 = bc$. Hence a must also be an integer. Let $b = a + t$ for some integer t. Then $c = 10 - t$. We have $10a + 1 = (a+t)(10-t) = 10a + (10-a)t - t^2$. Hence $1 = (10-a)t - t^2$. Since the right side is divisible by t, so is the left side, which is 1. Hence $t = \pm 1$. If $t = 1$, we have $10 - a = 2$ so that $a = 12$. If $t = -1$, we have $a - 10 = 2$ so that $a = 8$.

4. We consider three cases.
   Case 1. The line and the circle are disjoint.
   If the circle is too far away from the line, there are no such unit circles. If its nearest point to the line is exactly 1 away, there is exactly one such unit circle. These two situations are shown in Figure 3.34.

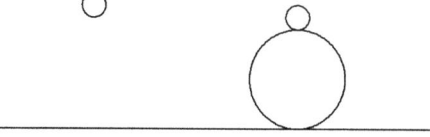

Figure 3.34

If the circle is closer to the line than that, there are always 2 solutions outside the circle, one on either side of it, as shown in Figure 3.35.

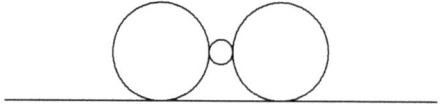

Figure 3.35

In addition, we may also have 1 or 2 such unit circles which contain the circle, as shown in Figure 3.36.

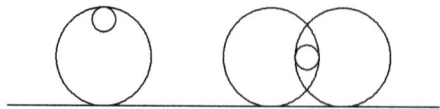

Figure 3.36

Case 2. The line and the circle are tangent.

There are always 4 such unit circles, two tangent to both the line and the circle at their point of tangency, and one on either side outside the circle and on the same side of the line as the circle.

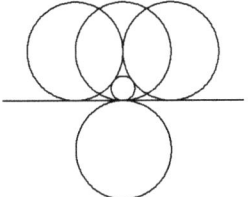

Figure 3.37

Case 3. The line and the circle intersect.

There are always 4 solutions outside the circle, two on each side of the line, as shown in Figure 3.38.

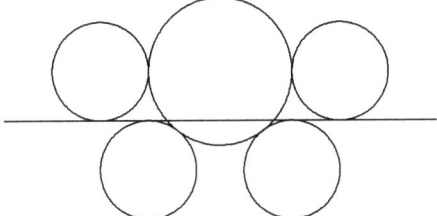

Figure 3.38

In addition, we may also have 1 to 4 such unit circles inside the circle, as shown in Figures 3.39 and 3.40.

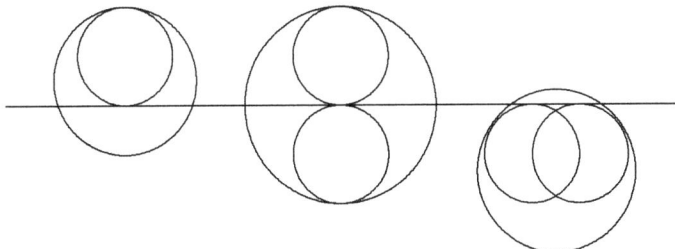

Figure 3.39

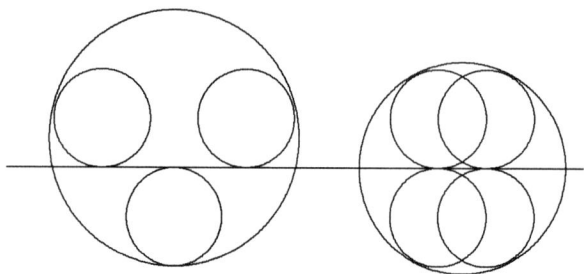

Figure 3.40

In summary, the value of n may be any number from 0 to 8.

Chapter Four
International Mathematics
TOURNAMENT OF THE TOWNS
Selected Problems

Section 1. A Problem on Area

This is based on Problem 4 in both the Junior and Senior A-Level papers of Spring 1980. We use the notation $[P]$ to denote the area of the polygon P.

Let $ABCD$ be a convex quadrilateral. Let A_1, $A-2$, ..., A_{n-1} be points on AB such that $AA_1 = A_1A_2 = \cdots = A_{n-1}B$. Let B_1, B_2, ..., B_{n-1} on BC, C_1, C_2, ..., C_{n-1} on CD and D_1, D_2, ..., D_{n-1} on DA be similarly defined. Joining corresponding points on opposite sides divides $ABCD$ into n^2 smaller quadrilaterals in an $n \times n$ configuration. For any n of these quadrilaterals with no two in the same row and no two in the same column, prove that their total area is equal to $\frac{1}{n}$ that of $ABCD$.

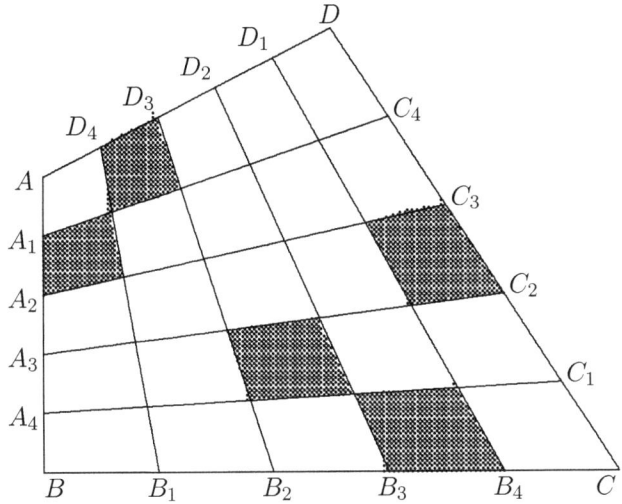

Figure 4.1

Figure 4.1 illustrates the case $n = 5$. We first consider the simple case $n = 2$.

Question 1.
Let $ABCD$ be a convex quadrilateral. Let A_1, B_1, C_1 and D_1 be the respective midpoints of AB, BC, CD and DA. A_1C_1 intersects B_1D_1 at O. Prove that $[AA_1OD_1] + [CC_1OB_1] = \frac{1}{2}[ABCD]$.

Solution:
Since $BB_1 = B_1C$ and O is equidistant from BB_1 and B_1C, we have $[OBB_1] = [OB_1C]$. Similarly, $[OCC_1] = [OC_1D]$, $[ODD_1] = [OD_1A]$ and $[OAA_1] = [OA_1B]$. The desired result follows from addition.

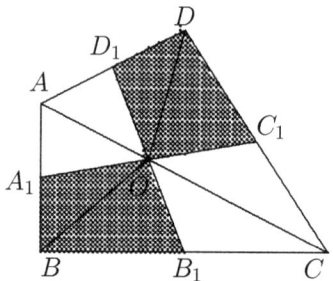

Figure 4.2

Problem 1.
Let $ABCD$ be a convex quadrilateral. Prove that $[A_1B_1C_1D_1] = \frac{1}{2}[ABCD]$, where A_1, B_1, C_1 and D_1 are the respective midpoints of AB, BC, CD and DA.

Question 2.
Let A_1 and A_2 be the points of trisection of AB, and let B_1, B_2, C_1, C_2, D_1 and D_2 be similarly defined. Join A_1C_2, A_2C_1, B_1D_2 and B_2D_1, as shown in Figure 4.3 on the left. Prove that A_1C_2, A_2C_1, B_1D_2 and B_2D_1 divide one another into three segments of equal length.

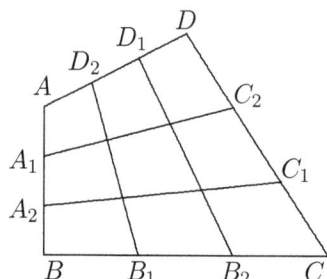 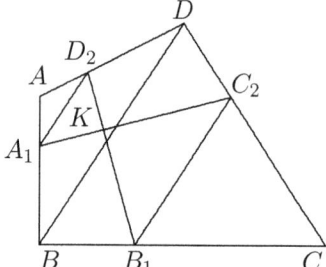

Figure 4.3

Solution:
Let K be the point of intersection of A_1C_2 and B_1D_2, as shown in Figure 4.3 on the right. We have $\frac{AA_1}{AB} = \frac{1}{3} = \frac{AD_2}{AD}$. Hence triangles A_1AD_1 and BAD are similarly, so that A_1D_1 is parallel to BD. Similarly, B_1C_2 is also parallel to BD, and hence to A_1D_1. It follows that triangles KA_1D_2 and KC_2B_1 are similar. Now $\frac{B_1C_2}{BD} = \frac{2}{3}$ while $\frac{D_2A_1}{DB} = \frac{1}{3}$. Hence $B_1C_2 = 2D_2A_1$. It follows that $KB_1 = 2KD_2$ and $KC_2 = 2KA_1$, so that K is a point of trisection of both A_1C_2 and B_1D_2. The desired result follows.

Problem 2.

Let A_1, A_2 and A_3 be the points of quadrisection of AB, and let B_1, B_2, B_3, C_1, C_2, C_3, D_1, D_2 and D_3 be similarly defined. Join A_1C_3, A_2C_2, A_3C_1, B_1D_3, B_2D_2 and B_3D_1, as shown in Figure 4.4. Prove that A_1C_3, A_2C_2, A_3C_1, B_1D_3, B_2D_2 and B_3D_1 divide one another into four segments of equal length.

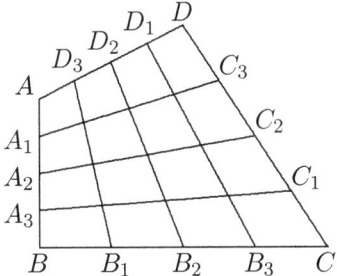

Figure 4.4

The following result is a generalization of Question 2 and Problem 2.

Question 3.

Let $ABCD$ be a convex quadrilateral and let E, F, G and H be points on AB, BC, CD and DA respectively such that we have $\frac{AE}{EB} = \frac{DG}{GC} = \alpha$ and $\frac{BF}{FC} = \frac{AH}{HD} = \beta$. Let EG and FH intersect at K. Prove that $\frac{HK}{KF} = \alpha$ and $\frac{EK}{KG} = \beta$.

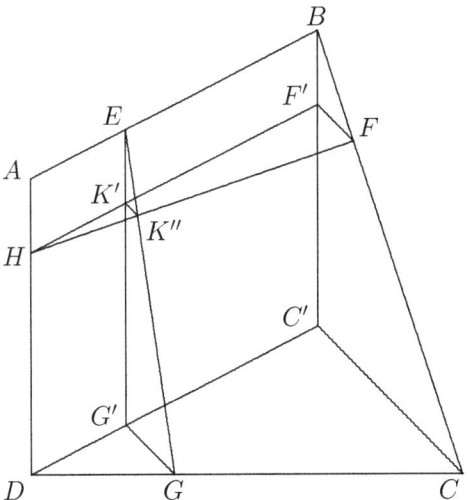

Figure 4.5

Solution:
The result is trivial if $ABCD$ is a parallelogram. If not, complete the parallelogram $ABC'D$. Let FF', GG' and $K'K''$ be parallel to CC', with F' on BC', G' on $C'D$, K' at the intersection of EG' and $F'H$, and K'' on EG. Then $\frac{FF'}{CC'} = \frac{\beta}{\beta+1}$, $\frac{GG'}{CC'} = \frac{\alpha}{\alpha+1}$ and $\frac{K'K''}{GG'} = \frac{\beta}{\beta+1}$. It follows that $\frac{K'K''}{FF'} = \frac{K'K''}{GG'} \cdot \frac{GG'}{CC'} \cdot \frac{CC'}{FF'}$, which reduces to $\frac{\alpha}{\alpha+1}$. It follows that K'' also lies on HF, so that it coincides with K. The desired conclusion follows.

Problem 3.
Let A_1 and A_2 be the points of trisection of AB, and let B_1, B_2, C_1, C_2, D_1 and D_2 be similarly defined. Join A_1C_2, A_2C_1, B_1D_2 and B_2D_1, and let them intersect one another at K, L, M and N, as shown in Figure 4.6. Prove that $[AA_1KD_2] + [KLMN] + [MB_2CC_1] = \frac{1}{3}[ABCD]$.

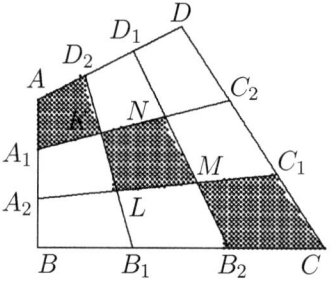

Figure 4.6

Problem 4.
Figure 4.7 shows a convex quadrilateral divided into 16 small quadrilaterals by lines joining quadrisection points on opposite sides. The area of each is marked with a lower case letter. Prove that the following twenty-four sums all have equal values.

$$\begin{array}{llll}
a+f+k+p & b+e+k+p & c+e+j+p & d+e+j+o \\
a+f+\ell+o & b+e+\ell+o & c+e+\ell+n & d+e+k+n \\
a+g+j+p & b+g+i+p & c+f+i+p & d+f+i+o \\
a+g+\ell+n & b+g+\ell+m & c+f+\ell+m & d+f+k+m \\
a+h+j+o & b+h+i+o & c+h+i+n & d+g+i+n \\
a+h+k+n & b+h+k+m & c+h+j+m & d+g+j+m
\end{array}$$

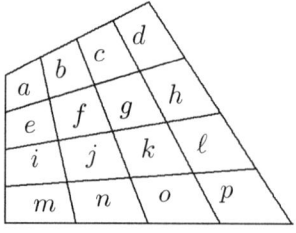

Figure 4.7

A Problem on Area

Question 4.
Each of two opposite sides of a convex quadrilateral is divided into m equal parts while each of the other two sides is divided into n equal parts. Joining corresponding points on opposite sides divides the quadrilateral into mn smaller quadrilaterals in an $m \times n$ configuration. Prove that the total area of the two smaller quadrilaterals at one pair of opposite corners is equal to the total area of the two smaller quadrilaterals at the other pair of opposite corners.

Solution:
Let $a_{i,j}$ be the area of the smaller quadrilateral in row i and column j. By Question 3, the dividing lines divide themselves into equal segments. By Question 1, we have $a_{1,j} + a_{2,j+1} = a_{1,j+1} + a_{2,j}$ for $1 \le j \le n-1$. Adding these, we have $a_{1,1} + a{2,n} = a_{1,n} + a_{2,1}$. Similarly, for $1 \le i \le m-1$, $a_{i,1} + a_{i+1,n} = a_{i,n} + a_{i+1,1}$. Adding these, we have $a_{1,1} + a_{m,n} = a_{1,n} + a_{m,1}$.

We are now in a position to tackle the original problem. Let $a_{i,j}$ be the area of the smaller quadrilateral in row i and column j. Consider the sums $a_{1,1} + a_{2,2} + \cdots + a_{n,n}$, $a_{1,2} + a_{2,3} + \cdots + a_{n,1}$, $\cdots$, $a_{1,n} + a_{2,1} + \cdots + a_{n,n-1}$. We claim that each of them have the same value. Since the sum of these sums is $[ABCD]$, each is equal to $\frac{1}{n}[ABCD]$.

Note that the first sum is the total area of the n smaller quadrilaterals on the main diagonal. We shall show that each of the other sums is equal to the first one. Suppose a sum contains a term $a_{i,j}$ with $i \ne j$. Since it has a term in every column, let the term in column i be $a_{k,i}$. By Question 4, $a_{i,j} + a_{k,i} = a_{i,i} + a_{k,j}$. We replace $a_{i,j}$ and $a_{k,i}$ in the sum with $a_{i,i}$ and $a_{k,j}$. The new sum still has a term in each row and each column, but now it has one more term on the main diagonal than before. Repeating this substitution process, we can convert it eventually to the first sum. This justifies the claim.

Section 2. A Problem on Communication

This is based on Problem 4 in both the Junior and Senior A-Level papers of Spring 1981.

Each of n girls has a different juicy piece of gossip, and is eager to share it with all the other girls. In each round, some or all of the girls engage in phone conversation in pairs, telling each other all the pieces they have heard. What is the minimum number of rounds required in order for every girl to have heard all the pieces of gossip?

Let $f(n)$ denote the minimum number of rounds required for n girls to share their pieces of gossip. Clearly, $f(1) = 0$ and $f(2) = 1$. Let there be three girls A, B and C. In each round, someone has to sit out. we may assume that A call B in the first round. C will have to be involved in the second round. By symmetry, we may assume that C calls A. Now both A and C know everything, but B has to make a call to either of them in the third round. Hence $f(3) = 3$.

Paradoxically, $f(4) = 2$, so that the function $f(n)$ is not an increasing function. Let the girls be A, B, C and D. In the first round, we may have A calling B and C calling D. If A calls C and B calls D in the second round, then everybody knows everything. Hence $f(4) = 2$ since one round is clearly not sufficient. The idea used in this special case yields the following general result.

Question 1.
Prove that $f(2m) \leq f(m) + 1$ for all positive integer m.

Solution:
Let the girls be $A_1, A_2, \ldots, A_m, B_1, B_2, \ldots, B_m$. In the first round, let A_i call B_i for $1 \leq i \leq 2^k$. Then all $2m$ pieces of gossip are known among $A_1, A_2, \ldots, A_m$ as well as among $B_1, B_2, \ldots, B_m$. Another $f(m)$ rounds are sufficient to complete the exchange of information.

Note that individual knowledge cannot increase by more than a factor of 2 per round. Since each girl knows 1 piece of gossip initially, she knows at most 2^k pieces after k rounds. Hence $f(2^k) \geq k$ for all non-negative integers k. On the other hand, by Question 1,

$$f(2^k) \geq f(2^{k-1}) + 1 \geq \cdots \geq f(2) + (k-1) = k.$$

It follows that $f(2^k) = k$ for all non-negative integers k.

Problem 1.

(a) Prove that $f(5) = 4$.

(b) Prove that $f(7) = 4$.

A Problem on Communication

We now turn our attention to positive integers n which are not powers of 2. Let k be the positive integer such that $2^k < n < 2^{k+1}$. Then $f(n) \geq k+1$. We distinguish between two cases according to the parity of n. The case where n is odd is easier. Let $n = 2^k + 2m + 1$ for some positive integers k and m with $m < 2^{k-1}$. After k rounds, nobody knows everything. In the $(k+1)$-st round, someone has to sit out, and this girl must make another call, so that $k+2$ rounds are necessary.

We now show that $k+2$ rounds are sufficient. Let the girls be $A_1, A_2, \ldots, A_{2^k}, B_1, B_2, \ldots, B_{2m-1}$. In the first round, let B_i call A_i for $1 \leq i \leq 2m-1$. Then all n pieces of gossip are known among $A_1, A_2, \ldots, A_{2^k}$. In the next k rounds, they complete the exchange of information among themselves. For $1 \leq i \leq 2m-1$, A_i returns the call of B_i in the last round and pass on all information. It follows that $f(n) = k+2$.

Problem 2.

(a) Prove that $f(10) = 4$.

(b) Prove that $f(12) = 4$.

The case where n is even but not a power of 2 is harder. Let $n = 2^k + 2m$ for some positive integers k and m with $m < 2^{k-1}$. Then $f(n) \geq k+1$. As it turns out, we have equality. Let the girls be $A_1, A_2, \ldots, A_{2^{k-1}+m}, B_1, B_2, \ldots, B_{2^{k-1}+m}$. We always have A_i calling B_i in the first round, so that the two pieces of gossip known initially to them may be denoted by g_i, $1 \leq i \leq 2^{k-1} + m$.

Question 2.

(a) Prove that $f(6) = 3$.

(b) Prove that $f(14) = 4$.

Solution:

(a) In the second round, let A_1 call B_2, A_2 call B_3, and so on. In the third round, the calls in the first round are repeated. By symmetry, we only need to follow the progress of A_1. In the second ground, she learns g_2 from B_2 while B_1 learns g_3 from A_3. In the third round, A_1 learns g_3 from B_1. Hence she knows everything after the third round. It follows that $f(6) = 3$.

(b) The process and its result are summarized in the chart below, listing the calls and the newly acquired information.

Girls	Round 1		Round 2		Round 3		Round 4	
A_1	B_1	g_1	B_2	g_2	B_4	g_3, g_4	B_1	g_5, g_6, g_7
A_2	B_2	g_2	B_3	g_3	B_5	g_4, g_5	B_2	g_6, g_7, g_1
A_3	B_3	g_3	B_4	g_4	B_6	g_5, g_6	B_3	g_7, g_1, g_2
A_4	B_4	g_4	B_5	g_5	B_7	g_6, g_7	B_4	g_1, g_2, g_3
A_5	B_5	g_5	B_6	g_6	B_1	g_7, g_1	B_5	g_2, g_3, g_4
A_6	B_6	g_6	B_7	g_7	B_2	g_1, g_2	B_6	g_3, g_4, g_5
A_7	B_7	g_7	B_1	g_1	B_3	g_2, g_3	B_7	g_4, g_5, g_6
B_1	A_1	g_1	A_7	g_7	A_5	g_5, g_6	A_1	g_2, g_3, g_4
B_2	A_2	g_2	A_1	g_1	A_6	g_6, g_7	A_2	g_3, g_4, g_5
B_3	A_3	g_3	A_2	g_2	A_7	g_7, g_1	A_3	g_4, g_5, g_6
B_4	A_4	g_4	A_3	g_3	A_1	g_1, g_2	A_4	g_5, g_6, g_7
B_5	A_5	g_5	A_4	g_4	A_2	g_2, g_3	A_5	g_6, g_7, g_1
B_6	A_6	g_6	A_5	g_5	A_3	g_3, g_4	A_6	g_7, g_1, g_2
B_7	A_7	g_7	A_6	g_6	A_4	g_4, g_5	A_7	g_1, g_2, g_3

We generalize the algorithm in Question 2 as follows. In the t-th round for $2 \le t \le k$, let A_1 call $B_{2^{t-1}}$, A_2 call $B_{2^{t-1}+1}$, and so on. We claim that in this round, A_1 will learn $g_{2^{t-2}+1}, g_{2^{t-2}+2}, \ldots, g_{2^{t-1}}$. We use mathematical induction on t. For $t = 2$, since A_1 calls B_2 in the second round, she will learn g_2. Suppose the result holds for some t where $2 \le t < k$. Then after the t-th round, A_1 learns $g_{2^{t-2}+1}, g_{2^{t-2}+2}, \ldots, g_{2^{t-1}}$. At the same time, $A_{2^{t-2}+1}$ learns $g_{2^{t-1}+1}, g_{2^{t-1}+2}, \ldots, g_{2^t}$ while talking to B_{2^t}. A_1 will acquire this information in the next round since she will be talking to B_{2^t}. This completes the inductive argument. In particular, in the k-th round, B_1 talks to A_{m+1}, who knows $g_{2^{k-2}+m+2}, g_{2^{k-2}+m+3}, \ldots, g_{2^{k-1}+m} = g_n, g_1$.

In the $(k+1)$-st round, the calls in the first round are repeated. B_1, who has just been talking to A_{m+1} in the preceding round, can supply A_1 with all the pieces she is missing. Hence $f(2^k + 2m) = k + 1$.

* * * * * * * * * * * *

Here is a closely related problem. Each of n girls has a different juicy piece of gossip, and is eager to share it with all the other girls. In each phone conversation between two girls, they tell each other all the pieces they have heard. What is the minimum number of phone calls required in order for every girl to have heard all the pieces of gossip?

This is a well-known problem, often known as the Telephone Disease Problem. Let $g(n)$ be the minimum number of phone calls required. Trivially, $g(1) = 0$.

Question 3.
Prove that $g(n) \le 2n - 3$ for $n \ge 2$.

Solution:
We appoint one of the girls as the Head Gossip. She calls everyone to learn their pieces, and return the calls to all the girls except the last one to whom she has been talking. It follows that $g(n) \leq 2n - 3$.

For $n \geq 4$, we can save one call by appointing three other girls as Associate Head Gossips. In the first $n - 4$ calls, the Head Gossip gathers the information from all the other girls. In the next four calls, she exchanges information with the three Associate Head Gossips by calling one another in pairs in two rounds. In the last $n - 4$ calls, the Head Gossip will spread the information to all the other girls. It follows that $g(n) \leq 2n - 4$.

It would be reasonable to expect that further saving should not be hard to find. As it turns out, this is a Mission Impossible. For $n \geq 4$, we actually have $g(n) = 2n - 4$, though none of the known proofs can be described as simple. Here, we will only consider some cases involving small values of n for which special arguments may be improvised.

Question 4.
Prove that $g(6) = 8$.

Solution:
Construct a graph with six vertices, each representing one of the girls. Two vertices are connected by an edge if they telephone each other. Then we have a sequence of graphs G_i with exactly i edges, such that each graph contains the preceding one as a subgraph. The graph G_4 is not connected since it has 6 vertices and only 4 edges. Thus after the fourth telephone call, nobody knows everything. Suppose $g(6) \leq 7$, so that the task can be accomplished by three more calls. Then each girl must be involved in at least one further call. Hence they must be engaged in phone conversations in three separate pairs. Moreover, G_4 must consist of two components each with three vertices and two edges, and each of the last three calls is between two girls represented by vertices in different components. However, among the three girls represented by vertices in same component, consider the one not involved in the second call within that group. Then this girl does not know everything within that group, and the missing information cannot be supplied in the last call by a girl in the other group. It follows that $g(6) = 8$.

Problem 3.

(a) Prove that $g(5) = 6$.

(b) Prove that $f(7) = 10$.

The following is posed as an open problem.

Problem 4.
Find a simple proof that $g(n) = 2n - 4$ for all $n \geq 4$.

Section 3. A Problem on Divisors

This is based on Problem 1 in the Junior A-Level paper of Spring 1982.

Determine all positive multiples of n each of which has exactly n positive divisors, for

(a) $n = 30$;

(b) $n = 24$.

The number of positive divisors of a positive integer is a basic concept in Number Theory, underlying the fundamental classification of positive integers.

- A positive integer is called a **unit** if it has exactly one positive divisor.

- A positive integer is called a **prime number** if it has exactly two positive integers.

- A positive integer is called a **composite number** if it has more than two positive integers.

Clearly, every positive integer is divisible by 1 and by itself. When they coincide, meaning that the positive integer is 1, then there is exactly one positive divisor. It follows that 1 is the only unit. If an integer greater than 1 has no positive divisors other than 1 and itself, then it is a prime number. The first few prime numbers are 2, 3, 5 and 7. The following is a useful result about prime numbers.

Question 1.
The second smallest positive divisor of a positive integer is a prime number.

Solution:
Let d be the second smallest positive divisor of a positive integer n. If d is not a prime number, then it has a positive divisor c where $1 < c < d$. It follows that c is also a positive divisor of n, so that d cannot be the second smallest one. This contradiction shows that d must be a prime number.

Let us refine the classification of composite numbers.

Question 2.
Describe in terms of the prime numbers all positive integers each of which has exactly three positive divisors.

A Problem on Divisors

Solution:
The first few examples are 4, 9 and 25, indicating that we have the squares of prime numbers. In general, let n be a positive integer with exactly three positive divisors, namely, $d_1 < d_2 < d_3$. Then $d_1 = 1$, $d_3 = n$ and d_2 is a prime number by Question 1. Note that $\frac{n}{d_1} > \frac{n}{d_2} > \frac{n}{d_3}$ are also distinct positive divisors of n. Hence $\frac{d}{d_1} = n$, $\frac{n}{d_3} = 1$ while $\frac{n}{d_2} = d_2$. It follows that $n = d_2^2$. Since d_2 is a prime number, n is indeed the square of a prime number.

It has been proved by Euclid that there are infinitely many prime numbers. The argument is rather ingenious. On the other hand, it is obvious that there are infinitely many composite numbers. For instance, by Question 2, the squares of the prime numbers are composite numbers, and there is one for each of the infinitely many prime number

Question 3.
Describe in terms of the prime numbers all positive integers each of which has exactly four positive divisors.

Solution:
The first few examples are 6, 8, 10, 14, 15, 21, 22, 26 and 27. The pattern is not obvious since it is really two patterns merged into one. Let n be a positive integer with exactly four positive divisors $d_1 < d_2 < d_3 < d_4$. Then $d_1 = 1$, $d_4 = n$ and d_2 is a prime number by Question 1. Note that $\frac{n}{d_1} > \frac{n}{d_2} > \frac{n}{d_3} > \frac{n}{d_4}$ are also distinct positive divisors of n. Hence $\frac{d}{d_1} = n$, $\frac{n}{d_4} = 1$ while $\frac{n}{d_2} = d_3$ and $\frac{n}{d_3} = d_2$. It follows that $n = d_2 d_3$. We consider two cases.

Case 1. d_3 is a prime number.
Then $n = d_2 d_3$ is the product of two distinct prime numbers. The first few examples are 6, 10, 14, 15, 21, 22 and 26.

Case 2. d_3 is not a prime number.
Then d_3 has a positive divisor other than 1 and itself. Since any positive divisor of d_3 is a positive divisor of n, d_3 has exactly three positive divisors $d_1 < d_2 < d_3$. By Question 2, $d_3 = d_2^2$ so that $n = d_2^3$ is the cube of a prime number. The first few examples are 8 and 27.

Problems 1.
Describe in terms of the prime numbers all positive integers each of which has exactly five positive divisors.

Problems 2.
Describe in terms of the prime numbers all positive integers each of which has exactly six positive divisors.

Obviously, we cannot continue in this manner forever.

We now appeal to the Fundamental Theorem of Arithmetic, which states that every positive integer is expressible as a product of prime numbers. Moreover, this prime factorization is unique up to the order of the prime numbers. Note that 1 is expressed as an empty product of prime numbers.

Question 4.
Determine the number of positive divisors of
(a) 30;
(b) 24.

Solution:

(a) By the Fundamental Theorem Arithmetic, $30 = 2 \times 3 \times 5$. The prime factors of any positive divisor of 24 must be 2, 3 or 5. We may have 0 or 1 copy of each of 2, 3 and 5, yielding 3+1 choices in each case. The total number of positive divisors of 30 is therefore $2 \times 2 \times 2 = 8$. They are 1, 2, 3, 5, 6, 10, 15 and 30.

(b) By the Fundamental Theorem Arithmetic, $24 = 2^3 \times 3$. The prime factors of any positive divisor of 24 must be 2 or 3. We may have 0, 1, 2 or 3 copies of 2 and 0 or 1 copy of 3. Thus we have 3+1 choices regarding the prime factor 2 and 1+1 choices regarding the prime factor 3. The total number of positive divisors of 24 is therefore $4 \times 2 = 8$. They are 1, 2, 3, 4, 6, 8, 12 and 24.

In general, let n be a positive integer and let its prime factorization be $n = p_1^{k_1} p_2^{k_2} \cdots p_m^{k_m}$, where $p_1, p_2, \ldots, p_m$ are distinct prime numbers and $k_1, k_2, \ldots, k_m$ are positive integers. Then the number of positive divisors of n is exactly $(k_1 + 1)(k_2 + 1) \cdots (k_m + 1)$.

Problem 3.
Find all positive integers under 1000 with exactly eight positive divisors.

We are now ready to tackle part (a) of our problem. Consider a positive multiple of 30: $n = 2^{k_1} 3^{k_2} 5^{k_3} p_4^{k_4} \cdots p_m^{k_m}$, where $p_4, \ldots, p_m$ are distinct prime numbers greater than 5 and $k_1, k_2, \ldots, k_m$ are positive integers with $k_1, k_2, k_3 \geq 1$. Then $(k_1 + 1)(k_2 + 1) \cdots (k_m + 1) = 30$. There are at least three factors on the left side, each at least 2. There is only one such prime factorizations of 30, namely, $30 = 2 \times 3 \times 5$. Thus we may have $n = 2 \times 3^2 \times 5^4 = 11250$, $n = 2^2 \times 3 \times 5^4 = 7500$, $n = 2 \times 3^4 \times 5^2 = 4050$, $n = 2^2 \times 3^4 \times 5 = 1620$, $n = 2^4 \times 3 \times 5^2 = 1200$ and $n = 2^4 \times 3^2 \times 5 = 720$.

For (b), consider a positive multiple of 24: $n = 2^{k_1} 3^{k_2} p_3^{k_3} \cdots p_m^{k_m}$, where $p_3, \ldots, p_m$ are distinct prime numbers greater than 3 and $k_1, k_2, \ldots, k_m$ are positive integers with $k_1 \geq 3$ and $k_2 \geq 1$. Then

$$(k_1 + 1)(k_2 + 1) \cdots (k_m + 1) = 24.$$

There are at least two factors on the left side, one at least 4 and the other at least 2. There are five such prime factorizations of 24. We consider them separately.

Case 1. $24 = 12 \times 2$.
This is not a factorization of the required form.

Case 2. $24 = 8 \times 3$.
We must have $n = 2^7 \times 3^2 = 1152$.

Case 3. $24 = 6 \times 4$.
We may have $n = 2^5 \times 3^3 = 864$ or $n = 2^3 \times 3^5 = 1944$.

Case 4. $24 = 6 \times 2 \times 2$.
We may have $n = 2^5 \times 3p = 96p$, where p is some prime number such that $5 \leq p \leq \frac{10000}{96} < 104.2$.

Case 5. $24 = 4 \times 3 \times 2$.
We may have $n = 2^3 \times 3^2 p = 72p$, where p is some prime number such that $5 \leq p \leq \frac{10000}{72} < 138.9$. We may also have $n = 2^3 \times 3p^2 = 24p^2$ for some prime number p where $5 \leq p \leq \sqrt{\frac{10000}{24}} < 20.5$.

Problem 4.
Determine all positive multiples of n each of which has exactly n positive divisors, for

(a) $n = 12$;

(b) $n = 18$.

Section 4. A Problem on Complex Numbers in Geometry

This is based on Problem 2 in the Senior A-Level paper of Fall 1991.

G is the centroid of triangle ABC. P is the point on the opposite side of the line BG to A such that $GB = GP$ and $\angle BGP = 120°$. Q is the point on the opposite side of CG to A such that $GC = GQ$ and $\angle CGQ = 120°$. Prove that APQ is an equilateral triangle.

We shall give two solutions to this problem. The first one is via geometric transformations.

Let R be the point on the extension of BG such that $BG = GR$. Let S be the point on the same side of AG as B such that GAS is an equilateral triangle. Under a rotation about G counter clockwise through a 60° angle, the image of A is S. Since $\angle BGP = 120°$ and $BG = BP$, GPR is an equilateral triangle, so that the image of P under the rotation is R, as shown in Figure 4.8.

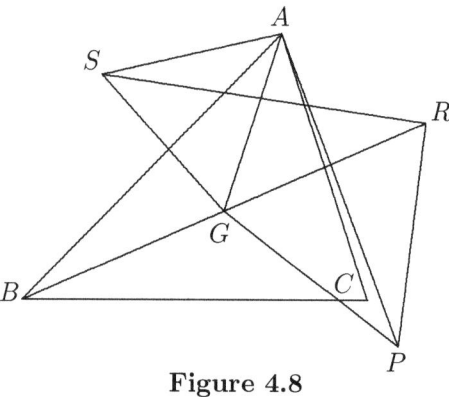

Figure 4.8

Under a rotation about G counter clockwise through a 120° angle, the image of Q is C, and let the image of A be T, as shown in Figure 4.9.

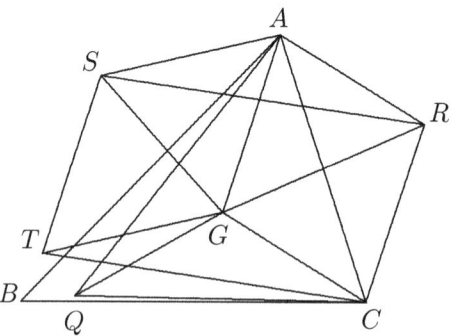

Figure 4.9

Then $GAST$ is a rhombus since GST is an equilateral triangle. $CRAG$ is a parallelogram since its diagonal bisect each other. Hence $RC = AG = ST$. Moreover, these three segments are parallel. It follows that $CRST$ is a parallelogram, so that $AP = RS = CT = AQ$. Moreover, since RS and CT are parallel, the angle between AP and AQ is 60°. Hence APQ is an equilateral triangle.

The second solution makes use of complex numbers. A basic result in algebra is the Quadratic Formula which states that the two solutions to the quadratic equation $ax^2 + bx + c = 0$, where a, b and c are real numbers, are given by $x = \frac{-b \pm \sqrt{b^2 - 4ac}}{2a}$. A problem arises when $b^2 - 4ac < 0$. Since the square of 0 is 0 and the square of a real number other than 0 is positive, this expression is not equal to any real number. In particular, the quadratic equation $x^2 + 1 = 0$ has no solutions in real numbers.

In order to provide solutions to such equations, we need to expand our system of the real numbers into the system of complex numbers. A complex number is a number of the form $a + bi$, where a and b are real numbers and i is a number such that $i^2 = -1$.

Problem 1.
Find all four solutions of $x^4 - 1 = 0$.

Note that every real number r is also a complex number since it has the form $r + 0i$. The arithmetic operations for the complex numbers are carried over from those for the real numbers.

Problem 2.
Let $a + bi$ and $c + di$ be complex numbers. Prove that

(a) $(a + bi) \pm (c + di) = (a \pm c) + (b \pm d)i$;

(b) $(a + bi)(c + di) = (ac - bd) + (ad + bc)i$;

(c) $(a + bi) \div (c + di) = \frac{ac+bd}{c^2+d^2} + \frac{bc-ad}{c^2+d^2}i$, provided that $c^2 + d^2 \neq 0$.

Question 1.
Find all six solutions of $x^6 - 1 = 0$.

Solution:
Since $x^6 - 1 = (x^3 + 1)(x^3 - 1) = (x + 1)(x^2 - x + 1)(x - 1)(x^2 + x + 1)$, two of the solutions are ± 1. The other four are complex solutions, coming from $x^2 \pm x + 1 = 0$. We have $x = \frac{\pm 1 \pm \sqrt{(-1)^2 - 4 \times 1 \times 1}}{2} = \pm \frac{1}{2} \pm \frac{\sqrt{3}}{2}i$ from the Quadratic Formula.

A standard notation is to denote $\frac{1}{2} + \frac{\sqrt{3}}{2}i$ by ω. Then the other five solutions are $\omega^2 = -\frac{1}{2} + \frac{\sqrt{3}}{2}i$, $\omega^3 = -1$, $\omega^4 = -\frac{1}{2} - \frac{\sqrt{3}}{2}i$, $\omega_5 = \frac{1}{2} - \frac{\sqrt{3}}{2}i$ and $\omega^6 = 1$. Note that since ω is a solution of $x^2 - x + 1 = 0$, $\omega = \omega^2 + 1$.

Our primary interest in complex numbers is their applications to geometry. We may regard the complex number $x + yi$ as the point (x, y) in the coordinate plane. We may also regard it as the vector from the origin $(0,0)$ to (x, y).

When we multiply a vector by a positive scalar multiple λ, the product is another vector in the same direction whose length is λ times the length of the original vector.

Question 2.
Let C be a point on the segment AB such that $AC : BC = \alpha : \beta$. If the points A and B are represented by the complex numbers a and b respectively, then C is represented by $\frac{\beta a + \alpha b}{\alpha + \beta}$.

Solution:
Let O be the origin. By parallel-shift if necessary, we may assume that the segment AB so that A lies on the segment OB. Let the vector **OA** be represented by $a - 0 = a$ and the vector **OB** by b. Then $AC = \frac{\alpha}{\alpha+\beta} AB$. Hence $c - a = \frac{\alpha}{\alpha+\beta}(b - a)$, which is equivalent to the desired result.

Question 3.
Let O be the origin and let $OACB$ be a parallelogram. If the points A and B are represented by the complex numbers a and b respectively, then C is represented by $a + b$.

Solution:
The origin is represented by 0. Let C be represented by c. Then the vector **OA** is represented by $a - 0 = a$ and the vector **BC** is represented by $c - b$. Since $OACB$ is a parallelogram, these two vectors are parallel and equal in length. Hence $a = c - b$, which is equivalent to $c = a + b$.

Multiplication of complex numbers is more revealing when performed in polar-coordinates. We have $a + bi = r(\cos\theta + i\sin\theta)$ where $r = \sqrt{a^2 + b^2}$ and $\theta = \arctan\frac{b}{a}$. The former is called the **modulus** and the latter is called the **argument** of the complex number $a + bi$.

Question 4.
Prove that $r(\cos\theta + i\sin\theta)s(\cos\phi + i\sin\phi) = rs(\cos(\theta + \phi) + i\sin(\theta + \phi))$.

Solution:
By the Compound Angle Formulae, we have

$$r(\cos\theta + i\sin\theta)s(\cos\phi + i\sin\phi)$$
$$= rs((\cos\theta\cos\phi - \sin\theta\sin\phi) + i(\sin\theta\cos\phi + \cos\theta\sin\phi))$$
$$= rs(\cos(\theta+\phi) + i\sin(\theta+\phi)).$$

In other words, when multiplying two complex numbers in polar coordinates, we multiply their moduli and add their arguments.

Problem 3.
Prove de Moivre's Formula: $(r(\cos\theta \pm i\sin\theta))^n = r^n(\cos n\theta) \pm i\sin n\theta)$.

When a complex number is multiplied by another whose modulus is 1, the product is the point obtained by rotating the point representing the original number about the origin through an angle equal to the argument of the multiplier. In particular, if the multiplier is w, we are performing $60°$ rotations.

We are now in position to present the second solution to our problem. Let G, A, B, C, P and Q be represented by the complex numbers 0, a, b, c, p and q respectively. By Question 2, the midpoint of BC is represented by $\frac{b+c}{1+1}$ and G is represented by $\frac{2\frac{b+c}{2}+a}{1+2} = \frac{a+b+c}{3} = 0$. Now $p = bw^2$ since P is obtained from B by two $60°$ rotation about G. Similarly, $q = cw^4$. We claim that P can be obtained from Q by a $60°$ rotation about A. This follows from

$$w(q-a) = cw^5 - aw$$
$$= -cw^2 - aw$$
$$= (a+b)w^2 - a(w^2+1)$$
$$= bw^2 - a$$
$$= p - a.$$

Hence APQ is an equilateral triangle.

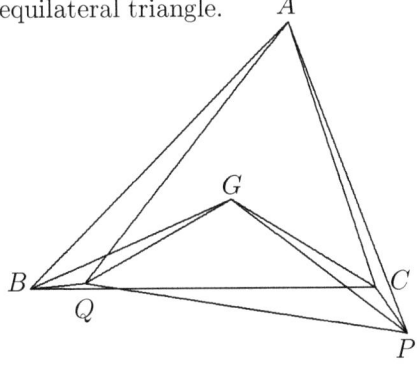

Figure 4.10

There are many applications of complex numbers in geometry. Try the following.

Problem 4.
The hexagon $ABCDEF$ is inscribed in a circle. The sides AB, CD and EF are all equal in length to the radius. Prove that the midpoint of the other three sides determine an equilateral triangle.

Section 5. A Problem on Polyhedra

This is based on Problem 6 in Senior A-Level paper of Fall 1992.

Consider a polyhedron having 100 edges. Determine the maximum number of its edges which can be intersected by a plane not passing through any of its vertices, if the polyhedron is

(a) **convex;**

(b) **non-convex.**

A polyhedron is a solid with polygons as faces in which two faces share at most one edge in common. We use V, E and F to denote respectively its numbers of vertices, edges and faces.

Question 1.
Prove that in a polyhedron, $2E \geq 3V$ and $2E \geq 3F$.

Solution:
Consider a skeleton of the polyhedron consisting only of the vertices and the edges. Cut each edge in halves at its midpoint. Then the total number of half edges is exactly $2E$. On the other hand, there must be at least three half edges attached to each vertex. Hence the total number of half edges is at least $3V$. It follows that $2E \geq 3V$. If instead we cut each edge into halves along its length, then the total number of half edges is again exactly $2E$. Now each face must be surrounded by at least three half edges. Hence the total number of half edges is at least $3F$. It follows that $2E \geq 3F$.

Question 2.
Find a convex polyhedron with 10 edges such that there exists a plane which does not pass through any of its vertices and intersects 6 of its edges.

Solution:
We first choose an arbitrary plane Π. Let $ABCDE$ be a pentagon with A on one side of Π and the other four vertices on the other side of Π. Using $ABCDE$ as the base, construct a pentagonal pyramid with vertex V on the same side of Π as A. Of the ten edges, Π intersects VB, VC, VD, VE, AB and AE.

We now present a solution to part (a) of our problem. We start with the polyhedron in the solution to Question 1. Take any of its faces apart from the pentagon $ABCDE$. Then this face is some triangle XYZ with one vertex, say X, on one side of Π and the other two, namely Y and Z, on the other side of Π. Choose a point W outside the polyhedron and on the same side of Π as X, such that attaching the pyramid $W-XYZ$ to the original polyhedron results in another convex polyhedron. This can be done by choosing W very close to X.

All the edges of the original polyhedron are still edges of the new polyhedron. In addition, we have three new edges, namely, WX, WY and WZ. Note that Π intersects WY and WZ, so that it intersects 8 of the 13 edges of the new polyhedron. Repeating this construction, we can add 3 edges to the polyhedron at a time, with 2 of them intersecting Π. After another steps, we arrive at a convex polyhedron with $13 + 3 \times 29 = 100$ edges, with Π intersecting $8 + 2 \times 29 = 66$ of them.

We now prove that this is indeed maximum. By Question 1, $2E \geq 3F$. Hence $3F \leq 2 \times 100$ so that $F \leq 66$. Each face is surrounded by half edges obtained by cutting the edges in halves along their lengths. Since it is a convex polygon, Π can intersect at most two of these half edges. Hence it can intersect at most $2 \times 66 = 132$ half edges, or equivalently, $132 \div 2 = 66$ edges of the convex polyhedron.

Note that for a convex polyhedron with $E = 10$, we have $F \leq 6$ so that a plane can intersect at most 6 of its edges. Thus the result in Question 2 is optimal.

Problem 1.
Prove that in a polyhedron, $V - E + F = 2$.

This result is the famous **Euler's Formula** for polyhedra.

Problem 2.
Use the result in Question 1 and Euler's Formula to prove that there does not exist a polyhedron with exactly 7 edges.

Problem 3.
In a polyhedron, every vertex lies on h edges and every face has k edges. Use the result in Question 1 and Euler's Formula to prove that there are only five such pairs (h, k) of positive integers.

Problem 4.
Use the result in Question 1 and Euler's Formula to prove that there does not exist a polyhedron

(a) with five vertices every two of which are joined by an edge;

(b) with six vertices such that each of three of them is joined to each of the other three.

Question 3.
Find a non-convex polyhedron with 10 edges such that there exists a plane which does not pass through any of its vertices and intersects 8 of its edges.

A Problem on Polyhedra

Solution:
We start off with a tetrahedron $AB_0C_0D_0$, as shown in Figure 4.11. Our plane Π is positioned so that A and B_0 are on its right side while C_0 and D_0 are on its left side. The only two edges not cut by Π are C_0D_0 and AB_0.

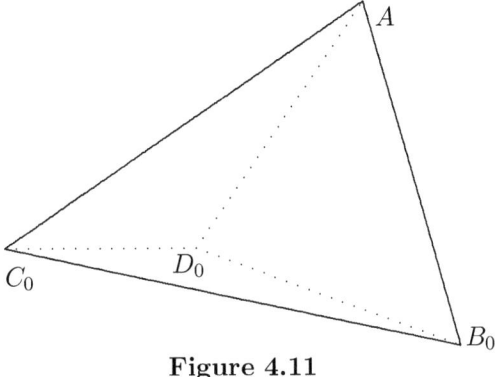

Figure 4.11

We choose a point C_1 on AB_0C_0 and a point D_1 on AB_0D_0 such that they lie on the left side of Π. Then we cut the tetrehedron $AB_0C_1D_1$ from $AB_0C_0D_0$, leaving behind a non-convex polyhedron with 10 edges, as shown in Figure 4.12. The only two edges not cut by Π are C_0D_0 and C_1D_1.

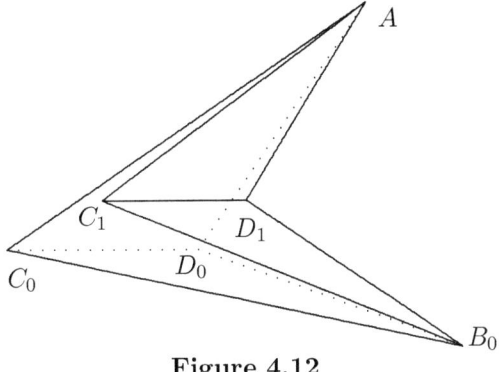

Figure 4.12

We claim that the answer to part (b) of our problem is 98. We first prove, more generally, that there exists an arbitrarily large integer n and a non-convex polyhedron with n edges such that there exists a plane which does not pass through any of its vertices and intersects $n-2$ of its edges.

Starting with the polyhedron constructed in Question 3, we shall modify it to increase the number of edges while retaining the property that the plane Π misses only two of them. We do this via two constructions which are repeated alternately. The first one is exactly what we have used in the solution to Question 3.

Construction 1.
Suppose the tetrahedron $AB_{n-1}C_{n-1}D_{n-1}$ has been defined. We choose a point C_n on $AB_{n-1}C_{n-1}$ and a point D_n on $AB_{n-1}D_{n-1}$ such that they lie on the left side of Π. Then we cut the tetrahedron $AB_{n-1}C_nD_n$ from $AB_{n-1}C_{n-1}D_{n-1}$, leaving behind a polyhedron with 4 more edges than before. The only two edges not cut by Π are C_0D_0 and C_nD_n.

Construction 2.
Suppose the tetrahedron $AB_{n-1}C_nD_n$ has been removed. We choose a point B_n on $B_{n-1}C_nD_n$ which lies on the right side of Π. Then we add the tetrahedron $AB_1C_1D_1$ to the preceding polyhedron, obliterating the edge C_nD_n. This generates a polyhedron with 2 more edges than before. The only two edges not cut by Π are C_0D_0 and AB_n.

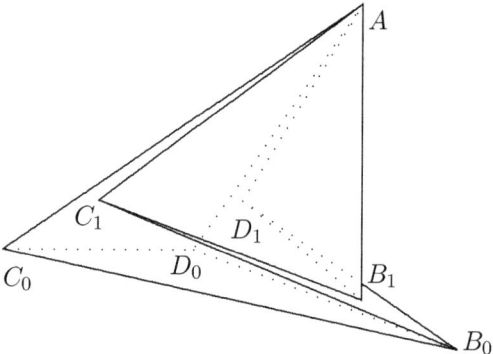

Figure 4.13

The case $n = 1$ is illustrated in Figure 4.13, resulting in the non-convex polyhedron in Figure 4.14.

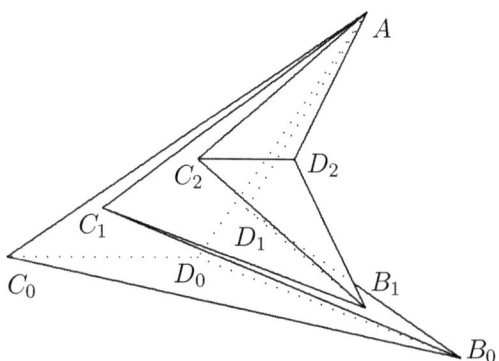

Figure 4.14

A Problem on Polyhedra

We now introduce the concept of the *signed angle-sum* of a polygon. An interior angle of a polygon which exceeds 180° is called a *reflex angle*. The signed angle-sum of a polygon is equal to its angle-sum minus 360° times the number of its reflex angles.

Lemma 1.
If a line cuts every edge of a polygon, then the signed angle-sum of the polygon is 0°.

Proof:
Since a line cuts every edge of the polygon, adjacent vertices must lie on opposite sides of this line. It follows that it must have an even number of vertices. We use induction on m where $2m$ is the number of vertices. For $m = 2$, the quadrilateral must be non-convex in order for a line to cut all four edges. Its angle-sum is 360° and it has exactly one reflex angle. Hence its signed angle-sum is indeed 0°. Suppose the result holds for some $m \geq 2$. Consider a $2(m+1)$-gon $P_{m+1} = A_0 B_0 A_1 B_1 \ldots A_m B_m$, with A_i on one side of a line ℓ and B_i on the other for $0 \leq i \leq m$. Join $A_0 B_2$. Then ℓ still cuts every edge of the $2m$-gon $P_m = A_0 B_0 A_2 \ldots A_m B_m$. By the induction hypothesis, the signed angle-sum of P_m is 0°. Now P_{m+1} has two more edges so that its angle-sum is 360° more than that of P_m. On the other hand, it has one more reflex angle than P_m, namely, $\angle B_0 A_1 B_1$ in Figure 4.15 on the left, and $\angle A_1 B_1 A_2$ in Figure 4.15 on the right. Hence its signed angle-sum is also 0°. This completes the inductive argument.

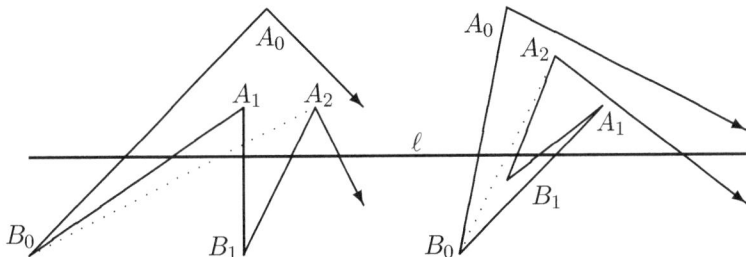

Figure 4.15

A vertex V of a polyhedron is said to be a *saddle point* if $\angle AVB$ and $\angle CVD$ are reflex angles on two faces of the polyhedron. Figure 4.16 on the left illustrates a saddle point.

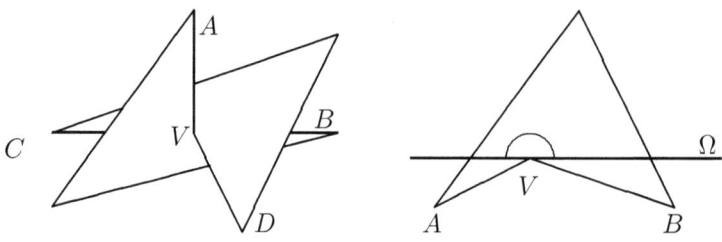

Figure 4.16

Lemma 2.
No plane can cut all of VA, VB, VC and VD where $\angle AVB$ and $\angle CVD$ are reflex angles at a saddle point of a polyhedron. In particular, VA, VB, VC and VD are four distinct edges.

Proof:
Suppose there is a plane Π which cuts all of VA, VB, VC and VD. Translate it so that it passes through V. Then all of A, B, C and D are on the same side of Π. Construct a small hemisphere with center V on the other side of Π. Then it cuts each of the two faces in a semicircular disc, as shown in Figure 4.16 on the right. These two discs must cut each other along a radius of the hemisphere, or else coincide with each other. In either case, the two faces will have common interior points, which is a contradiction. No two of VA, VB, VC and VD can coincide as otherwise the plane determined by one interior point from each of the three edges will cut all of them.

Question 4.
Prove that a plane cannot cut every edge of a polyhedron.

Solution:
Suppose we have a polyhedron Q such that a plane Π cuts each of its edges. Then all edges of each face P is cut by the line of intersection of Π and P. By Lemma 1, the signed angle-sum of P is $0°$. Suppose P' is the image of P under an orthogonal projection onto a plane not perpendicular to P. Then P' is a polygon, and its signed angle-sum is $0°$ since an orthogonal projection preserves planar convexity. Let Q' be the polygon which is the image of Q under an orthogonal projection onto a plane which is not perpendicular to any face of Q, such that the vertices of Q have distinct images. Then the sum S of the signed angle-sums of the images of all faces of Q is $0°$. Let us calculate S in a different way. Consider each vertex V of Q. By Lemma 2, it is not a saddle point, so that at most one face can have a reflex angle at V. The same holds for the image V' of V.

Suppose ∠$A'V'B'$ is a reflex angle. Now the other angles form a chain from $V'A'$ to $V'B'$. If these angles fill in the void in the reflex angle, then the contribution of the angles at V to S is 0°. If they duplicate the reflex angle instead, then the contribution of the angles at V to S is positive. Any overlap would only increase their contribution. In any case, this contribution is non-negative. If V is on the convex hull of Q', then there are no reflex angles at V, and the contribution of the angles at V to S is positive. Since there are at least three vertices on the convex hull of Q', we have $S > 0°$, which contradicts our earlier observation that $S = 0°$. It follows that a plane cannot cut all edges of any polyhedron.

Lemma 3.
Suppose all but one edge UV of a polygon is cut by a line. If neither of the interior angles at U and V is a reflex angle, then the signed angle-sum of the polygon is 180°. If at least one of them is a reflex angle, then the signed angle-sum is −180°.

Proof:
Let ℓ be a line which cuts all edges of the polygon apart from UV. Let the edges adjacent to UV be TU and VW. Suppose both ∠TUV and ∠UVW are less than 180°. Extend TU to X and choose a point Y on the opposite side of ℓ to X such that neither XY nor YV cuts any edge of the polygon, as shown in Figure 4.17. Consider the polygon obtained from the original one by replacing the edges TU and UV with TX, XY and YV. Then ℓ cuts every edge of the new polygon, so that by Lemma 1, its signed angle-sum is 0°. The original polygon has one less side but one less reflex angle, so that its signed angle-sum is 180°.

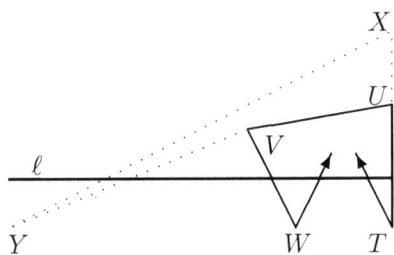

Figure 4.17

Suppose ∠UVW is a reflex angle. Join UW, as shown in Figure 4.18. Consider the polygon obtained from the original one by replacing UV and VW with UW. Then ℓ cuts every edge of the new polygon, so that by Lemma 1, its signed angle-sum is 0. The original polygon has one more side but one more reflex angle, so that its signed angle-sum is −180°.

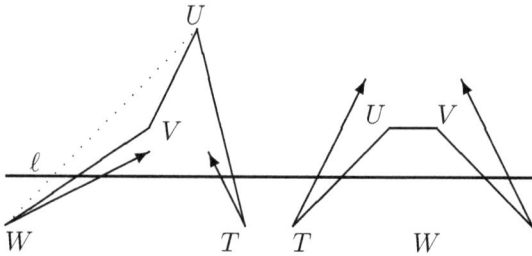

Figure 4.18

Lemma 4.
Suppose a plane cuts every edge but one of a polyhedron. Then the polyhedron has at most one saddle point, and the angles of at most two faces at this point are reflex angles.

Proof:
Suppose UV is the only edge of the polyhedron not cut by a plane Ω. By Lemma 2, no vertices other than U and V can be saddle points. By symmetry, we may assume that Ω is closer to V than to U. If these two distances happen to be equal, we can rotate Ω slightly to make them unequal, without destroying the property that Ω cuts all edges at U and V other than UV itself. Now translate Ω towards U. When it passes beyond V, it will cut every edge at U. By Lemma 2, U cannot be a saddle point. The only possible saddle point is therefore V. Suppose the angles of three faces at V are reflex angles. By Lemma 2 again, UV is an edge of at most one of them. It follows that Ω will cut all four edges on the other two faces forming the reflex angles. This contradicts Lemma 2.

We can now complete the solution to part (b) of our problem. We prove more generally that a plane cannot cut all but one of the edges of a polyhedron.

Suppose we have a polyhedron Q such that a plane cuts all but one of its edges. Let this edge UV be the boundary between the faces P_1 and P_2. Let the polygon Q' be the image of Q under an orthogonal projection onto a plane which is not perpendicular to any face of Q, such that the vertices of Q have distinct images. By Lemma 1, the signed angle-sum of the image P' of any face P other than P_1 or P_2 is $0°$. It follows that the sum S of the signed angle-sums of the images of all faces of Q is equal to the sum of the signed angle-sums of images P_1' and P_2' of P_1 and P_2.

We now consider two cases.

Case 1. Q has a saddle point.

By Lemma 4, Q has at most one saddle point, which is either U or V. We may assume that V is the saddle point. By Lemma 4 again, the angles of exactly two faces at V are reflex angles. One of these faces must be P_1 or P_2, and we may assume that it is P_1. By Lemma 3, its signed angle-sum is $-180°$, while the signed angle-sum of P_2 is $\pm 180°$. It follows that $S \leq 0°$. We now calculate S vertex by vertex. Since there are two reflex angles at the image V' of V, the contribution to S of these angles may be as low as $-360°$ but no lower. Let W be a vertex on the convex hull of Q. We can choose the plane of orthogonal projection so that the image W' of W is in the interior of Q'. Then the contribution of the angles at W' is at least $360°$. Taking into consideration of vertices that are on the convex hull of Q', we have $S > 0°$ as in the solution to Question 4. This contradicts our earlier observation that $S \leq 0°$.

Case 2. Q has no saddle points.

Here $S \leq 360°$ since the signed angle-sum of each of P'_1 and P'_2 may be as high as $180°$, but no higher. Since Q has no saddle points, the contribution of the angles at each vertex of Q' is non-negative. Let W' be chosen as in Case 1. Then the contribution of the angles at W' is at least $360°$. It follows that $S > 360°$, which contradicts our earlier observation that $S \leq 360°$.

Section 6. A Problem on Sequences

This is based on Problem 6 in the Senior A-Level paper of Fall 1994.

Let d_n be the first digit of 2^n in base 10 representation. Prove that the number of different subsequences of $\{d_n\}$ consisting of 13 consecutive terms is equal to 57.

We call such a sequence a *power* sequence. We first explore the situation for shorter length. For power sequences of length 1, any non-zero digit may be the first digit of a power of 2. Hence there are 9 such sequences. For power sequences of length 2, observe that if a power of 2 starts with 1, then the next power of 2 may start with 2 or 3. For instance, 1 is followed by 2 and 16 is followed by 32. Similar analysis yields Figure 4.19, where the arrows show the 13 sequences of length 2.

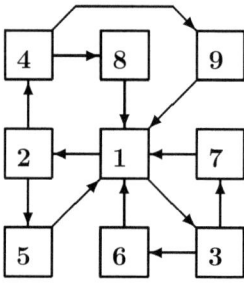

Figure 4.19

Question 1.

(a) Find all 21 sequences of length 3 generated by Figure 4.19.

(b) Find the 4 which are not power sequences.

Solution:

(a) From the Figure 4.19, we can read off 21 sequences of length 3.

$$
\begin{array}{ccccccc}
(1,2,4) & (1,2,5) & (1,3,6) & (1,3,7) & (2,4,8) & (2,4,9) & (2,5,1) \\
(3,6,1) & (3,7,1) & (4,8,1) & (4,9,1) & (5,1,2) & \mathbf{(5,1,3)} & (6,1,2) \\
\mathbf{(6,1,3)} & (7,1,2) & (7,1,3) & \mathbf{(8,1,2)} & (8,1,2) & \mathbf{(9,1,2)} & (9,1,3)
\end{array}
$$

(b) In the list in (a), the 4 entries in boldface are not power sequences. We first consider (5,1,3) and (6,1,3). Let x be a power of 2 such that $5 \times 10^k < x < 7 \times 10^k$ for some k, so that its first digit is 5 or 6. Then $2 \times 10^{k+1} < 2^2 x < \frac{14}{5} \times 10^{k+1}$. Hence the first digit of $2^2 x$ cannot be 3. Consider now (8,1,2) and (9,1,2). Let x be a power of 2 such that $8 \times 10^k < x < 10^{k+1}$ for some k, so that its first digit is 8 or 9. Then $\frac{16}{5} \times 10^{k+1} < 2^2 x < 4 \times 10^{k+1}$. Hence the first digit of $2^2 x$ cannot be 2.

A Problem on Sequences

Problem 1.

(a) Find all 37 sequences of length 4 generated by Figure 4.19.

(b) Find the 16 which are not power sequences.

In Question 1, how do we know that the remaining 17 entries in (a) are indeed power sequences? Taking the block from 2^0 to 2^{11}, namely, 1, 2, 4, 8, 16, 32, 64, 128, 256, 512, 1024 and 2048, we see that (1,2,4), (2,4,8), (4,8,1), (8,1,3), (1,3,6), (3,6,1), (6,1,2), (1,2,5), (2,5,1) and (5,1,2) are power sequences. However, we have to use the block from 2^{44} to 2^{55} to get the power sequences (1,3,7), (3,7,1), (7,1,2), (1,2,5), (2,5,1), (5,1,2), (1,2,4), (2,4,9), (4,9,1) and (9,1,3), four of which having occurred already. It is not until we reach the block from 2^{76} to 2^{78} before we get the final power sequence (7,1,3). We need a better approach.

Question 2.
Let x be a power of 2 whose first digit is 1. Then $10^k \leq x < 2 \times 10^k$ for some k, and $y = \frac{x}{10^k}$ is in the interval $[1,2)$. Subdivide this into intervals so that if y lies within a specific interval, the first digit of $2^i x$ has a specific value for $1 \leq i \leq 5$.

Solution:
Since $2 \times 10^k \leq 2x < 4 \times 10^k$, the first digit of $2x$ is 2 or 3. It is 2 if y is in $[1, \frac{3}{2})$ and 3 if y is in $[\frac{3}{2}, 2)$. Since $4 \times 10^k \leq 2^2 x < 8 \times 10^k$, the first digit of $2^2 x$ is 4, 5, 6 or 7. It is 4 if y is in $[1, \frac{5}{4})$, 5 if y is in $[\frac{5}{4}, \frac{3}{2})$, 6 if y is in $[\frac{3}{2}, \frac{7}{4})$ and 7 if y is in $[\frac{7}{4}, 2)$. Continuing to $2^5 x$, the result is summarized in the following table.

Range of y	\multicolumn{6}{c}{First digit of $2^i x$}					
	0	1	2	3	4	5
[1,9/8)	1	2	4	8	1	3
[9/8,5/4)	1	2	4	9	1	3
[5/4,3/2)	1	2	5	1	2	
[3/2,7/4)	1	3	6	1	2	
[7/4,15/8)	1	3	7	1	2	
[15/8,2)	1	3	7	1	3	

We are not interested in the 1s under the column $2^3 x$ as the power sequences they start must have occurred already. Hence the corresponding entries in the column under $2^5 x$ are left blank.

Problem 2.
Let x be a power of 2 whose first digit is 1. Then $10^k \le x < 2 \times 10^k$ for some k, and $y = \frac{x}{10^k}$ is in the interval $[1,2)$. Subdivide this into intervals so that if y lies within a specific interval, the first digit of $2^i x$ has a specific value for $1 \le i \le 6$.

The 17 entries in bold face in the table in Question 2 are starters of different power sequences of length 3. To show that all can be realized, we just have to exhibit an x with $\frac{x}{10^k}$ in the correct range. That such an x always exists is guaranteed by the following general result.

Question 3.
Prove that given any finite sequence of digits, there exists a power of 2 whose base 10 expression begins with that sequence.

Solution:
We shall prove that for any positive integer m, there exist positive integers n and k such that $10^k m \le 2^n < 10^k(m+1)$. This may be rewritten as $k + \log m \le n \log 2 < k + \log(m+1)$. On the positive x-axis, mark the points $\log 2$, $2 \log 2$, $3 \log 2$, ... We claim that $n \log 2$ falls within the interval $[k + \log m, k + \log(m+1))$ for some n and k. We roll up the positive x-axis into a circle with circumference 1. The marked points are distinct since $\log 2$ is irrational. With infinitely many of them, the distance along the circle between two of them is less than $\log(m+1) - \log m = \log(1 - \frac{1}{m})$. It follows that at least one of these two points must lie between the two numbers $\log m - \lfloor \log m \rfloor$ and $\log(m+1) - \lfloor \log(m+1) \rfloor$. This justifies our claim.

Question 4.
Let x be a power of 2 whose first digit is 1. Then $10^k \le x < 2 \times 10^k$ for some k, and $y = \frac{x}{10^k}$ is in the interval $[1,2)$. Subdivide this into intervals so that if y lies within a specific interval, the first digit of $2^i x$ has a specific value for $1 \le i \le 15$.

Solution:
We extend the table in Question 2 to the one on the following page. The entries in bold face are starters of different power sequences of length 13. Question 3 guarantees that all can be realized. To answer our problem, all we have to do now is to verify that there are indeed 57 entries in boldface.

Problem 3.
Suppose we use powers of 3 instead of powers of 2. Construct a diagram giving all power sequences of length 2.

A Problem on Sequences   145

Range of y	First digit of $2^i x$															
	0	1	2	3	4	5	6	7	8	9	10	11	12	13	14	15
$[1, \frac{35}{32})$	1	2	4	8	1	3	6	1	2	5	1	2	4	8	1	3
$[\frac{35}{32}, \frac{281}{256})$	1	2	4	8	1	3	7	1	2	5	1	2	4	8	1	3
$[\frac{281}{256}, \frac{9}{8})$	1	2	4	8	1	3	7	1	2	5	1	2	4	9	1	3
$[\frac{9}{8}, \frac{75}{64})$	1	2	4	9	1	3	7	1	2	5	1	2	4	9	1	3
$[\frac{75}{64}, \frac{625}{512})$	1	2	4	9	1	3	7	1	3	6	1	2	4	9	1	3
$[\frac{625}{512}, \frac{5}{4})$	1	2	4	9	1	3	7	1	3	6	1	2	5	1	2	4
$[\frac{5}{4}, \frac{175}{128})$	1	2	5	1	2	4	8	1	3	6	1	2	5	1	2	
$[\frac{175}{128}, \frac{45}{32})$	1	2	5	1	2	4	8	1	3	7	1	2	5	1	2	
$[\frac{45}{32}, \frac{375}{256})$	1	2	5	1	2	4	9	1	3	7	1	2	5	1	2	
$[\frac{375}{256}, \frac{3}{2})$	1	2	5	1	2	4	9	1	3	7	1	3	6	1	2	
$[\frac{3}{2}, \frac{25}{16})$	1	3	6	1	2	4	9	1	3	7	1	3	6	1	2	
$[\frac{25}{16}, \frac{875}{512})$	1	3	6	1	2	5	1	2	4	8	1	3	6	1	2	
$[\frac{875}{512}, \frac{7}{4})$	1	3	6	1	2	5	1	2	4	8	1	3	7	1	2	
$[\frac{7}{4}, \frac{225}{128})$	1	3	7	1	2	5	1	2	4	8	1	3	7	1	2	
$[\frac{225}{128}, \frac{1875}{1024})$	1	3	7	1	2	5	1	2	4	9	1	3	7	1	2	
$[\frac{1875}{1024}, \frac{15}{8})$	1	3	7	1	2	5	1	2	4	9	1	3	7	1	3	
$[\frac{15}{8}, \frac{125}{64})$	1	3	7	1	3	6	1	2	4	9	1	3	7	1	3	
$[\frac{125}{64}, 2)$	1	3	7	1	3	6	1	2	5	1	2	4	8	1	3	

Problem 4.
Suppose we use powers of 3 instead of powers of 2.

(a) Find all sequences of length 3 generated by the diagram in Problem 3.

(b) Find all power sequences of length 3.

Section 7. A Problem on Lottery

This is based on Problem 6 in the Junior A-Level paper of Fall 1996 and Problem 6 in the Senior A-Level paper of Fall 1996.

A buyer of a Math-Lotto ticket puts on it n of the first n^2 positive integers for some integer $n > 1$. On the draw date, n of these n^2 numbers are announced. A ticket wins if none of the numbers on it is announced. What is the minimum number of tickets required to guarantee that at least one of them wins, if

(a) $n = 6$;
(b) $n = 10$?

We first explore the situation for small values of n.

Question 1.
Solve the Math-Lotto problem for $n = 2$.

Solution:
There are 6 possible tickets, namely, (1,2), (1,3), (1,4), (2,3), (2,4) and (3,4). Whichever ticket we do not buy, the two numbers not on it may be announced. Then we will not have a winning ticket. Hence the minimum is 6. Of course, buying all 6 tickets guarantees that one of them will win.

Question 2.
Solve the Math-Lotto problem for $n = 3$.

Solution:
We first show that 7 tickets are sufficient. They may be (1,2,3), (1,2,4), (3,4,5), (3,4,6), (5,6,1), (5,6,2) and (7,8,9). Suppose we do not have a winning ticket. Then one of the three numbers announced must be 7, 8 or 9. By symmetry, we may assume that another number announced is 1. However, no third number can eliminate all of (3,4,5), (3,4,6) and (5,6,2). This is a contradiction. We now prove that 7 tickets are necessary. Suppose we buy only 6 tickets. If one number, say x, appears on 3 of them, the 9 numbers on the remaining 3 tickets cannot be distinct. Hence one of them, say y, appears on 2 tickets. Between x and y, they eliminate 5 tickets. If a third number which appears on the remaining ticket is announced along with x and y, we will not have a winning ticket.

Question 3.
Solve the Math-Lotto problem for $n = 4$.

Solution:
We first show that 7 tickets are sufficient. They may be (1,2,3,4), (1,2,5,6), (3,4,5,6), (7,8,9,10), (7,8,11,12), (9,10,11,12) and (13,14,15,16). Then 2 numbers are needed to eliminate the first 3 tickets, 2 more numbers to eliminate the next 3 tickets, and 1 more number to eliminate the last ticket. Since only 4 numbers are announced, we have a winning ticket.

We now prove that 7 tickets are necessary. Suppose we buy only 6 tickets. If one number appears on 3 of them, announcing it along with one number on each of the remaining 3 tickets will prevent us from winning. Hence each number appears on at most 2 tickets. Let x appear on 2 tickets. Then the 16 numbers on the remaining 4 tickets cannot be distinct. Hence one of them, say y, appears on 2 tickets. Between x and y, they eliminate 4 tickets. If one number on each of the remaining 2 tickets is announced along with x and y, we will not have a winning ticket.

Problem 1.
Solve the Math-Lotto problem for $n = 5$.

The argument for the lower bound in Question 3 can be generalized.

Question 4.
Prove that $n + 2$ tickets are not sufficient to guarantee a win in Math-Lotto.

Solution:
If one number appears on 3 of them, announcing it along with one number on each of the remaining $n - 1$ tickets will prevent us from winning. Hence each number appears on at most 2 tickets. Let x appear on 2 tickets. Then the n^2 numbers on the remaining n tickets cannot be distinct. Hence one of them, say y, appears on 2 tickets. Between x and y, they eliminate 4 tickets. If one number on each of the remaining $n - 2$ tickets is announced along with x and y, we will not have a winning ticket.

We can now answer part (a) of our problem. By Question 4, we need 9 tickets. They may be

$$(1, 2, 3, 4, 5, 6), (1, 2, 3, 7, 8, 9), (4, 5, 6, 7, 8, 9),$$

$$(10, 11, 12, 13, 14, 15), (10, 11, 12, 16, 17, 18), (13, 14, 15, 16, 17, 18),$$

$$(19, 20, 21, 22, 23, 24), (25, 26, 27, 28, 29, 30), (31, 32, 33, 34, 35, 36).$$

Then 2 numbers are needed to eliminate the first 3 tickets, 2 more numbers to eliminate the next 3 tickets, and 1 more number to eliminate each of the last 3 tickets. Since only 6 numbers are announced, we have a winning ticket.

Problem 2.
Solve the Math-Lotto problem for $n = 7$.

Problem 3.
Solve the Math-Lotto problem for $n = 8$.

Problem 4.
Solve the Math-Lotto problem for $n = 9$.

Finally, we answer part (b) of our problem. By Question 4, we need 13 tickets. They may be (1–10), (1–5,11–15), (6–15), (16–25), (16–20,26–30), (21–30), (31–40), (41–50), (51–60), (61–70), (71–80), (81–90) and (91–100). Then 2 numbers are needed to eliminate the first 3 tickets, 2 more numbers to eliminate the next 3 tickets, and 1 more number to eliminate each of the last 7 tickets. Since only 10 numbers are announced, we have a winning ticket.

Section 8. A Problem on Angles

This is based on Problem 6 in the Junior A-Level paper of Spring 1996 and Problem 3 in the Senior A-Level paper of Spring 1996.

ABC is an isosceles triangle with a right angle at A. D is the point on AB such that $AD = \frac{1}{n}AB$. $P_1, P_2, \ldots, P_{n-1}$ are points on BC which divide it into n equal segments.

(a) **Prove that $\angle AP_1D + \angle AP_2D = 45°$.**

(a) **For all $n > 2$, determine $\angle AP_1D + \angle AP_2D + \cdots + \angle AP_{n-1}D$.**

The smallest value of n for which the current problem makes sense is $n = 2$. There is only one term in the summation, and it is clear from Figure 4.20 on the left that $\angle AP_1D = 45°$.

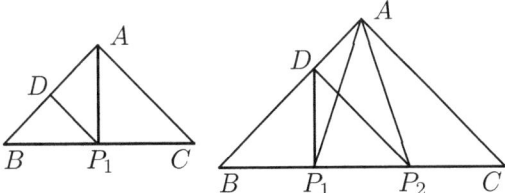

Figure 4.20

We now solve part (a) of our problem, which is illustrated in Figure 4.20 on the right. Note that DP_1 is perpendicular to BC, DP_2 is parallel to AC and $\angle P_2AC = \angle P_1AB$ by symmetry. By the Alternate Angle Theorem, $\angle AP_2D = \angle P_2AC$. Also, $\angle AP_1D + \angle P_1AB = \angle BDP_1$ by the Exterior Angle Theorem. It follows that $\angle AP_1D + \angle AP_2D = 45°$.

Problem 1.
Solve our problem for the case $n = 4$.

Problem 2.
Solve our problem for the case $n = 5$.

It is perhaps too early to conjecture that the answer to part (b) is also $45°$. We should have some more confirming evidence. We examine how the case $n = 3$ is related to the case $n = 2$. Figure 4.21 is the same as Figure 4.20 on the right except that there is an extra dotted line and there are some minor changes in labeling. It is not difficult to see that Figure 4.20 on the left is also incorporated in it. We have $\angle D_2P_1D_1 = 45°$. To deduce the case $n = 3$, we need to prove that $\angle AP_1D_1 + \angle AP_2D_2 = 90°$. Anticipating later application, we shall restate this result in a more general form.

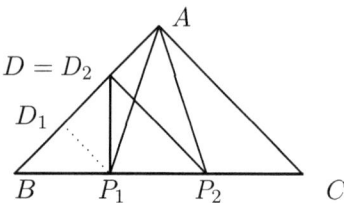

Figure 4.21

Question 1.
Let ABC be a right isosceles triangle. Let P_i and P_j be points on BC such that $BP_i = P_jC$. Lines through P_i and P_j parallel to AC intersect AB at D_i and D_j respectively. Prove that $\angle AP_iD_i + \angle AP_jD_j = 90°$.

Solution:
By symmetry, $\angle P_jAC = \angle P_iAB$. By the Alternate Angle Theorem, we have $\angle AP_jD_j = \angle P_jAC$. By the Exterior Angle Theorem,

$$\angle AP_iD_i + \angle AP_jD_j = \angle AP_iD_i + \angle P_iAD_i = \angle AD_iP_i = 90°$$

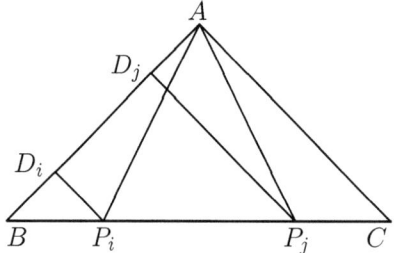

Figure 4.22

We now solve part (b) of our problem by induction on $n \geq 2$. For $1 \leq i \leq n-1$, let the line through P_i parallel to AC intersect AB at D_i. In particular, D_{n-1} coincides with D. We also designate A as D_n. The case $n = 2$ is trivial. The cases $n = 4$ and $n = 5$ illustrated in the Figures 4.23 and 4.24.

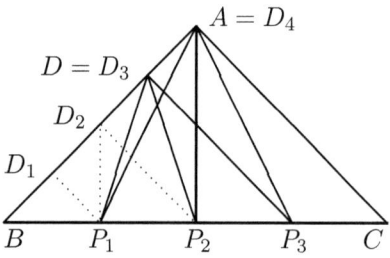

Figure 4.23

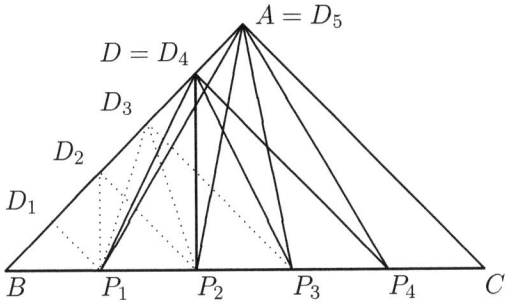

Figure 4.24

Suppose the result holds for 2, 3, ..., $n-1$. Consider the next case n. Note that $\sum_{i=1}^{n-1} \angle AP_iD_i = (n-1)45°$. This follows immediately from Question 1 if n is odd, and if $n = 2k$, the result still holds since $\angle AP_kD_k = 45°$ as in the case $n = 2$. Now

$$\sum_{i=1}^{n-1} \angle AP_iD_i = \sum_{i=1}^{n-1}\sum_{j=i}^{n-1} \angle D_j P_i D_{j+1} = \sum_{j=1}^{n-1}\sum_{i=1}^{j} \angle D_j P_i D_{j+1}.$$

For $1 \leq j \leq n-2$, $\sum_{i=1}^{j} \angle D_j P_i D_{j+1} = 45°$ by the induction hypothesis. It follows that $\sum_{i=1}^{n-1} \angle AP_iD = 45°$.

We now explore non-inductive solutions to part (b).

Question 2.
Solve the case $n = 4$ of our problem by cutting out triangles AP_iD for $1 \leq i \leq 3$, as shown in Figure 4.25 on the left and in the middle.

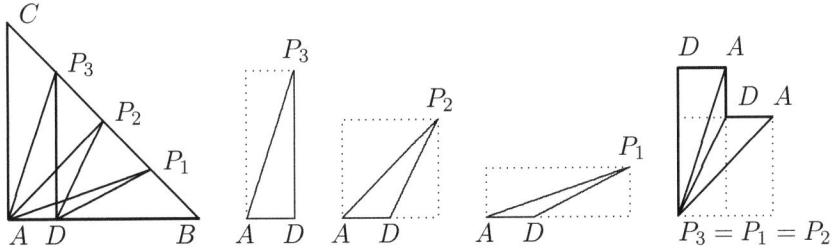

Figure 4.25

Solution:
Arrange the triangles in the order AP_3D, AP_1D and AP_2D as shown in Figure 4.25 on the right, with AP_3D and AP_1D reflected. The triangles are then translated and rotated so that the vertices of the desired angles come together at a common point, using the fact that $AP_3 = AP_1$ and $DP_2 = DP_1$. Clearly, the sum of these angles is $45°$.

Problem 3.
Solve the case $n = 5$ of our problem by cutting out triangles AP_iD for $1 \le i \le 4$.

Question 2 is reminiscent of the following well-known problem.

Question 3.
$ABCD$ is a rectangle with $AB = 3AD$. E and F are points on AB such that $AE = EF = FB$. Determine $\angle CAB + \angle CEB + \angle CFB$.

Solution:
Expand the given 3×1 rectangle into a 3×2 rectangle as shown in Figure 4.26. Triangles CEB, GCK, GAF and AGH are all congruent to one another. Hence $\angle AGC = 90°$ and $GA = GC$. It follows that we have $\angle CAB + \angle CEB + \angle CFB = \angle CAB + \angle GAF + 45° = \angle GAC + 45° = 90°$.

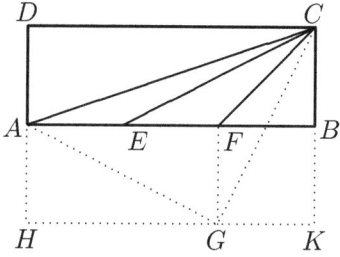

Figure 4.26

Question 4.
Solve the case $n = 4$ of our problem by drawing a line through A parallel to BC, intersecting the extension of P_iD at Q_i, and then cutting out the triangles AP_iQ_i for $1 \le i \le 3$, as shown in Figure 4.27 on the left.

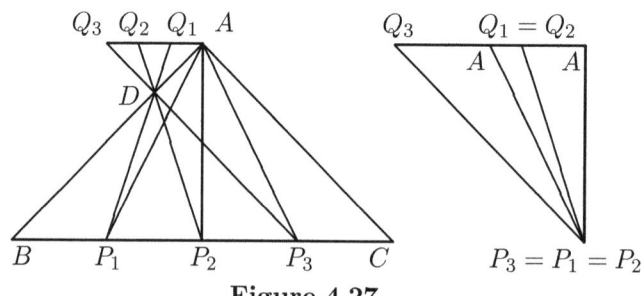

Figure 4.27

Solution:

Arrange the triangles in the order $Q_3 P_3 A$, $Q_1 P_1 A$ and $Q_2 P_2 A$ as shown in Figure 4.27 on the right, with $Q_1 P_1 A$ reflected. The triangles are then translated so that the vertices of the desired angles come together at a common point, using the fact that $AP_3 = AP_1$ and $Q_2 P_2 = Q_1 P_1$. Clearly, the sum of these angles is $45°$.

Problem 4.

Solve the case $n = 5$ of our problem by drawing a line through A parallel to BC, intersecting the extension of $P_i D$ at Q_i, and then cutting out the triangles $AP_i Q_i$ for $1 \leq i \leq 4$.

Finally, we give a solution to part (b) of our problem which does not distinguish between the odd and even cases. We present the general argument. Figure 4.28 illustrates the case $n = 5$.

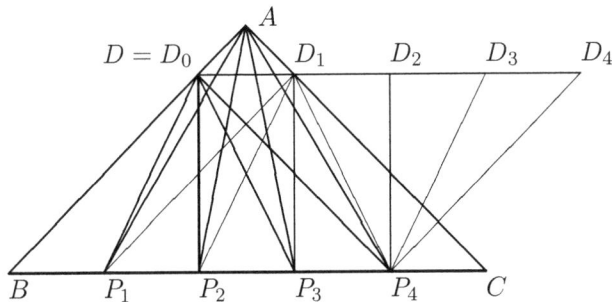

Figure 4.28

Let $D_0 = D$ and draw $D_0 D_1$ parallel to BC, with D_1 on AC. Denote the desired sum by S. By symmetry, $S = \sum_{i=1}^{n-1} \angle A P_i D_1$. Hence

$$2S = \sum_{i=1}^{n-1} (\angle A P_i D_0 + \angle A P_i D_1) = \sum_{i=1}^{n-1} \angle D_0 P_i D_1.$$

Translate $D_0 D_1$ to $D_1 D_2$, $D_2 D_3$, $\ldots$, $D_{n-2} D_{n-1}$. Then

$$\begin{aligned}
S &= \frac{1}{2} \sum_{i=1}^{n-1} \angle D_0 P_{n-i} D_1 \\
&= \frac{1}{2} \sum_{i=1}^{n-1} \angle D_{i-1} P_{n-1} D_i \\
&= \frac{1}{2} \angle D_0 P_{n-1} D_{n-1} \\
&= \frac{1}{2} \angle CAB \\
&= 45°.
\end{aligned}$$

Section 9. A Problem on Balance

This is based on Problem 7 in the Junior A-Level paper of Spring 1997 and Problem 6 in the Junior A-Level paper of Fall 1993.

An object of integral weight is to be balanced using a set of tokens of integral weights.

(a) If the tokens have weights 1, 2, 4, 8, 16, 32, 64, 128, 256 and 512, determine the maximum number of different ways in which an object may be balanced. Each token may be placed on either pan, or left off the balance.

(b) Construct a set of 10 tokens such that an object of weight up to 88 may be balanced, even if an arbitrary token is withheld just prior to balancing. No tokens may be placed on the same pan as the object.

We first focus on part (a) of our problem. Let $b_k(n)$ be the number of different ways of balancing an object of weight n using tokens of weights 1, 2, ..., 2^k. Clearly, $b_k(n) = 0$ if $n \geq 2^{k+1}$. Henceforth, we assume that $n \leq 2^{k+1} - 1$. Let $m_k = \max\{b_k(n) : 1 \leq n \leq 2^{k+1} - 1\}$.

We first explore the situation with small values of k.

Problem 1.
Determine $b_k(n)$ for $1 \leq n \leq 2^{k+1}$, where

(a) $k = 0$;

(b) $k = 1$;

(c) $k = 2$.

Question 1.
Determine $b_k(n)$ for $1 \leq n \leq 2^{k+1}$, where

(a) $k = 3$;

(b) $k = 4$.

Solution:

(a) For $k = 3$, the tokens have weights 1, 2, 4 and 8.

n	Ways of Balancing	$b_3(n)$
1	$1,\ 2-1,\ 4-2-1,\ 8-4-2-1$	4
2	$2,\ 4-2,\ 8-4-2$	3
3	$2+1,\ 4-1,\ 4-2+1,\ 8-4-1,\ 8-4-2+1$	5
4	$4,\ 8-4$	2
5	$4+1,\ 4+2-1,\ 8-2-1,\ 8-4+1,\ 8-4+2-1$	5
6	$4+2,\ 8-2,\ 8-4+2$	3
7	$4+2+1,\ 8-1,\ 8-2+1,\ 8-4+2+1$	4
8	8	1
9	$8+1,\ 8+2-1,\ 8+4-2-1$	3
10	$8+2,\ 8+4-2$	2
11	$8+2+1,\ 8+4-1,\ 8+4-2+1$	3
12	$8+4$	1
13	$8+4+1,\ 8+4+2-1$	2
14	$8+4+2$	1
15	$8+4+2+1$	1

(b) For $k = 4$, the tokens have weights 1, 2, 4, 8 and 16. Clearly, $b_4(16)=1$. For $17 \le n \le 31$, $b_4(n) = b_3(n-16)$. This is because the token of weight 16 must be placed on the opposite pan. For $1 \le n \le 15$, if the token of weight 16 is not used, the number of ways is clearly $b_3(n)$. If it is used, it must be placed on the opposite pan. The number of ways is then $b_3(16-n)$. Hence $b_4(n) = b_3(n) + b_3(16-n)$. In summary,

n	1	2	3	4	5	6	7	8	9	10	11
$b_4(n)$	5	4	7	3	8	5	7	2	7	5	8
n	12	13	14	15	16	17	18	19	20	21	22
$b_4(n)$	3	7	4	5	1	4	3	5	2	5	3
n	23	24	25	26	27	28	29	30	31		
$b_4(n)$	4	1	3	2	3	1	2	1	1		

Problem 2.
Determine $b_5(n)$ for $1 \le n \le 63$.

The Fibonacci numbers $\{f_n\}$ are defined by $f_0 = 1$, $f_1 = 2$ and for $k \ge 2$, $f_k = f_{k-1} + f_{k-2}$. They now make an unexpected appearance.

Question 2.
Prove that $m_k \le f_k$ for all $k \ge 0$.

Solution:
We use induction on k. We have $m_0 = 1 = f_0$ and $m_1 = 2 = f_1$. Suppose the result holds for some $k \geq 1$. Note that $b_{k+1}(2^{k+1}) = 1$ and $b_{k+1}(2^k) = 2$, both of which are obviously less than f_{k+1}. There are three other cases.
Case 1. $2^{k+1} + 1 \leq n \leq 2^{k+2} - 1$.
We have $b_{k+1}(n) = b_k(n - 2^{k+1}) \leq m_k \leq f_k < f_{k+1}$.
Case 2. $2^k + 1 \leq n \leq 2^{k+1} - 1$.
We have $b_{k+1}(n) = b_k(n) + b_k(2^{k+1} - n) = b_k(n) + b_{k-1}(2^{k+1} - n)$, which is less than or equal to $m_k + m_{k-1} \leq f_k + f_{k-1} = f_{k+1}$.
Case 3. $1 \leq n \leq 2^k - 1$.
We have $b_{k+1}(n) = b_k(n) + b_k(2^{k+1} - n) = b_{k-1}(n) + b_k(2^{k+1} - n)$, which is less than or equal to $m_{k-1} + m_k \leq f_{k-1} + f_k = f_{k+1}$.
This completes the inductive argument.

We can now answer part (a) of our problem. We have $m_9 \leq f_9 = 89$ by Question 2. We need find n such that $b_9(n) = 89$. Then we will have $m_9 = 89$.

$$
\begin{array}{rclclclcl}
b_2(1) &=& b_2(3) &=& b_1(3) &+& b_1(1) &=& 3, \\
b_3(3) &=& b_3(5) &=& b_2(3) &+& b_2(1) &=& 5, \\
b_4(5) &=& b_4(11) &=& b_3(5) &+& b_3(3) &=& 8, \\
b_5(11) &=& b_5(21) &=& b_4(11) &+& b_4(5) &=& 13, \\
b_6(21) &=& b_6(43) &=& b_5(21) &+& b_5(11) &=& 21, \\
b_7(43) &=& b_7(85) &=& b_6(43) &+& b_6(21) &=& 34, \\
b_8(85) &=& b_8(171) &=& b_7(85) &+& b_7(43) &=& 55, \\
b_9(171) &=& b_9(341) &=& b_8(171) &+& b_8(85) &=& 89.
\end{array}
$$

We now turn our attention to part (b) of our problem. The effective range of a set of token is the maximum value n such that an object of weight up to n may be balanced, even if an arbitrary token is withheld just prior to balancing. Thus we seek a set of 10 tokens with effective range 88.

We first explore the situation for small number k of tokens. If $k = 1$, it does not matter what the token is. The effective range is 0 since the lone token will be withheld. If $k = 2$, both tokens must have weight 1 in order to have an effective range of 1. Any subsequent set of tokens must contain the preceding set.

For $k = 3$, if one of the tokens of weight 1 is withheld, we can only balance an object of weight 2 if the third token has weight 1 or 2. Clearly, 2 is the better choice. If this token is withheld, we cannot balance an object of weight 3. Hence the effective range is 2. For $k = 4$, the fourth token must have weight 3 in order to stretch the effective range to 4. For $k = 5$, the fifth token may have weight 5, which will stretch the effective range to 7. The set $\{1,1,2,3,5\}$ suggests that Fibonacci strikes again. We reset the indices so that now $f_1 = f_2 = 1$ instead.

Question 3.
Find a set of 6 tokens with effective range 12.

Solution:
We take tokens of weights 1, 1, 2, 3, 5 and 8. The table below shows that the effective range is indeed 12.

Weight of Object	Token Withheld				
	1	2	3	5	8
1	1	1	1	1	1
2	2	1+1	2	2	2
3	3	3	2+1	3	3
4	3+1	3+1	2+1+1	3+1	3+1
5	5	5	5	3+2	5
6	5+1	5+1	5+1	3+2+1	5+1
7	5+2	5+1+1	5+2	3+2+1+1	5+2
8	8	8	8	8	5+3
9	8+1	8+1	8+1	8+1	5+3+1
10	8+2	8+1+1	8+2	8+2	5+3+2
11	8+3	8+3	8+2+1	8+3	5+3+2+1
12	8+3+1	8+3+1	8+2+1+1	8+3+1	5+3+2+1+1

Problem 3.
Find a set of 7 tokens with effective range 20.

Question 4.
Let there be k tokens with weights f_1, f_2, ..., f_k. If no tokens are withheld, the maximum effective range is $f_{k+2} - 1$.

Solution:
Let S be the total weight of the tokens. Then

$$\begin{aligned} 2S &= f_1 + (f_1 + f_2) + (f_2 + f_3) + \cdots + (f_{k-2} + f_{k-1}) + (f_{k-1} + f_k) + f_k \\ &= (f_2 + f_3 + f_4 + \cdots + f_k) + (f_{k+1} + f_k) \\ &= (S - f_1) + f_{k+2}. \end{aligned}$$

It follows that $S = f_{k+2} - 1$. Hence the effective range is at most $f_{k+2} - 1$. We now prove by induction on k that the effective range is indeed $f_{k+2} - 1$. For $k = 1$, the lone token must have weight 1 and the effective range is 1, which is equal to $f_3 - 1$. Suppose the result holds for some $k \geq 1$. Then the effective range without the token of weight f_{k+1} is $f_{k+2} - 1$. Since $f_{k+1} \leq f_{k+2} - 1$, we only need to extend the effective range, This can be done by adding the new token, taking the effective range to $f_{k+2} - 1 + f_{k+1} = f_{k+3} - 1$. This completes the inductive argument.

We are now ready to answer part (b) of our problem. We shall prove by induction on k that that by taking tokens with weights $f_1, f_2, \ldots, f_n$, we have an effective range of $f_{n+1} - 1$. For $k = 1$, $f_2 - 1 = 0$ and the result holds trivially. Suppose it holds for some $k \geq 1$. We add the token with weight f_{k+1}. If it is withheld, then none of the others is. By Question 4, we have an effective range of $f_{k+2} - 1$. Suppose some other token is withheld instead. By the induction hypothesis, the effective range is $f_{k+1} - 1$ without the new token. Adding it stretches the effective range to $f_{k+1} - 1 + f_{k+1} \geq f_{k+2} - 1$ too. This completes the inductive argument. In particular, with tokens of weights 1, 1, 2, 3, 5, 8, 13, 21, 34 and 55, we have an effective range of $89 - 1 = 88$.

Problem 4.
Suppose two arbitrarily tokens may be withheld. Define $\{g_k\}$ by setting $g_1 = g_2 = g_3 = 1$. For $k \geq 4$, set $g_k = g_{k-1} + g_{k-3}$. Prove that by taking tokens with weights $g_1, g_2, \ldots, g_k$, the effective range is $g_{k+1} - 1$.

Section 10. A Problem on Magic Tricks

This problem is based on Problem 3 of the Senior A-Level Problem 3 of Spring 1998.

The audience chooses five cards from a standard deck of fifty-two cards. The assistant arranges them in a row in an order of his choice.

(a) He turns one of the cards face down.

(b) Instead, he puts one of the cards in his pocket.

The magician, who has been out of the room, is brought in. She looks at the row of cards and named the fifth card which she does not see. What strategy can she and her assistant use?

Problem 1.
The magic trick in part (a) is played with a deck of six cards, with three cards chosen by the audience. What strategy can the magician and her assistant use?

Question 1.
The magic trick in part (a) is played with a deck of fifteen cards, with four cards chosen by the audience. What strategy can the magician and her assistant use?

Solution:
Number the cards from 1 to 15. Suppose the audience chooses cards 3, 7, 10 and 13. The assistant may decide to have any of them as the face-down card. Suppose it is card 3. Then it is the third card in the front half of the sequence of twelve invisible cards, namely, (1,2,3,4,5,6,8,9,11,12,14,15). The assistant places card 3 face down in the first position to signify the front half. Had the hidden card been in the back half, it will be placed in the last position. The three visible cards can be placed in six different ways. In "alphabetical" order, these permutations are (7,10,13), (1,13,10), (10,7,13), (10,13,7), (13,7,10) and (13,10,7). The assistant chooses (10,7,13) to signify the third position, so that the row of four cards are (?,10,7,13). If card 7 is face down, the row will be (?,13,10,3). If it is card 10, the row will be (3,13,7,?). If it is card 13, the row will be (7,10,3,?).

This strategy can be generalized for the full trick with fifty-two cards. Since $52 - 4 = 48 = 2 \times 4!$, the placement of the face-down card signifies whether it is in the front half or the back half of the sequence of forty-eight invisible cards, and the permutation of the four visible cards signifies its actual position in that half.

We now turn our attention to part (b) of our problem.

Question 2.
The magic trick in part (b) is played with a deck of six cards, with three cards chosen by the audience. What strategy can the magician and her assistant use?

Solution:
There are $\binom{6}{3} = 20$ possible choices by the audience, and $3 \times 2! = 6$ possible placements for each choice. The following chart is prepared in advance.

Choices	Displays	Choices	Displays
1,2,3	**12**,13,21,23,31,32	2,3,4	23,24,**32**,34,42,43
1,2,4	12,**14**,21,24,41,42	2,3,5	23,25,32,35,**52**,53
1,2,5	12,**15**,21,25,51,52	2,3,6	23,26,32,36,**62**,63
1,2,6	12,**16**,21,26,61,62	2,4,5	24,25,**42**,45,52,54
1,3,4	**13**,14,31,34,41,43	2,4,6	24,26,42,46,62,**64**
1,3,5	13,15,31,**35**,51,53	2,5,6	25,26,52,56,62,**65**
1,3,6	13,16,31,**36**,61,63	3,4,5	34,35,43,45,**53**,54
1,4,5	14,15,41,**45**,51,54	3,4,6	34,36,**43**,46,63,64
1,4,6	14,16,41,**46**,61,64	3,5,6	35,36,53,56,**63**,65
1,5,6	15,16,51,**56**,61,65	4,5,6	45,46,**54**,56,64,65

All twenty boldface entries are distinct. Thus if the audience chooses cards 2, 3 and 5, the assistant will pocket card 3 and display (5,2). In fact, the displayed not in boldface may be omitted.

Problem 2.
The magic trick in part (b) is played with a deck of fifteen cards, with four cards chosen by the audience. What strategy can the magician and her assistant use?

If we wish to generalize the strategy in Question 2 for the full trick with fifty-two cards, we need to that a distinct display of four cards exists for each choice of five cards. We need some tools from Graph Theory. A *bipartite graph* consists of two sets X and Y of vertices such that every edge joins a vertex in X to a vertex in Y.

Problem 3.
A famous bipartite graph is the Utility Graph. Three vertices in a set represent three utilities, and three vertices in the other set represent three feuding families. Each family is to be connected to each utility, but the edges may not cross. Prove that this is impossible in the plane.

A set of edges which join every vertex in X to a distinct vertex in Y is called a *matching*. Note that it is not necessary for every vertex in Y be to in the matching. Let A be a subset of X. The subset of Y consisting of vertices joined to some vertex in A is denoted by $N(A)$.

Question 3.
Prove that a bipartite graph with vertex sets X and Y has a matching if $|A| \leq |N(A)|$ for every subset A of X.

Solution:
We use induction on $|X|$. The case where $|X| = 1$ is trivial. Suppose the result holds whenever $|X| < n$. Now let $|X| = n$, and we consider two cases.

Case 1.
Suppose $|A| < |N(A)|$ for every subset A of X such that $A \neq \emptyset$ and $A \neq X$. We can pick an arbitrary x and match it with any y joined to it. Remove these two vertices. In the resulting bipartite graph, it is still true that $|A| \leq |N'(A)|$ for any subset A of $X - \{x\}$, where $N'(A) = N(A)$ if y does not belong to $N(A)$, and $N'(A) = N(A) - y$ if y belongs to $N(A)$. By the induction hypothesis, we have a matching from $(X - \{x\})$ to Y. Together with the edge (x, y), we have a matching in the original graph.

Case 2.
Suppose $|B| = |N(B)|$ for some subset B of X such that $B \neq \emptyset$ and $B \neq X$. Then we split the graph into two bipartite subgraphs, one with vertices in B and $N(B)$, and the other with vertices in $X - B$ and $Y - N(B)$. Note that some edges may be lost. Since $N(A)$ is a subset of $N(B)$ whenever A is a subset of B, the first subgraph has a matching from B to Y by the induction hypothesis. Suppose in the second subgraph, there exists a subset C of $X-B$ such that $|N'(C)| < |C|$, where $N'(C)$ is the subset of $Y-N(B)$ consisting of all y joined to some x in C. We need this new notation since some vertices of $N(C)$ may not be in $Y - Y(B)$. In the original graph, let $D = B \cup C$. Then $N(D) = N(B) \cup N(C)$ so that $|N(D)| = |N(B)| + |N'(C)| < |B| + |C| = |D|$. This is a contradiction. Hence $|A| < |N'(A)|$ for every subset A of $Y - B$, and the second subgraph has an matching from $X - B$ to Y by the induction hypothesis. The matchings in the two subgraphs can be combined into a matching in the original graph.

This result is known as **Hall's Theorem**.

Question 4.
Without referring to the table in Question 2, show that a distinct display of two cards exists for each choice of three cards.

Solution:
Construct a bipartite graph with $\binom{6}{3} = 20$ vertices in X representing the choices, and $2\binom{6}{2} = 30$ vertices in Y representing the displays. Now each vertex in X has degree $3 \times 2! = 6$ while each vertex in Y has degree 4. Let A be any subset of X. Then there are $6|A|$ edges going from A and $4|N(a)|$ edges going to $N(A)$. Since all edges from A goes to $N(A)$ but not necessarily all edges to $N(A)$ come from A, we have $6|A| \leq 4|N(A)|$. It follows that $|A| \leq |N(A)|$. By Hall's Theorem, the desired matching exists.

Problem 4.
Construct a table, analogous to the one in Question 2, for the full trick with fifty-two cards in part (b) of our problem.

We can now complete the solution to part (b) of our problem. The argument is essentially the same as the one in Question 4. The only change is that $6|A| \leq 4|N(A)|$ is replaced by $(5 \times 4!)|A| \leq (52-4)|N(A)|$.

Section 11. A Problem on Polyomino Dissections

This problem is based on Problem 6 in the Senior A-Level paper of Fall 1999.

Prove that every polyomino consisting of $2n-1$ or $2n$ squares may be dissected into n rectangles.

A polyomino, which means many squares, is a figure consisting of unit squares joined edge to edge. The name is derived from " domino" where the prefix do- or di- means two. There is one monomino and one domino, both of which are rectangles. There are two trominoes. The I-tromino is a rectangle. The V-tromino may be dissected into 2 rectangles, as shown in Figure 4.29.

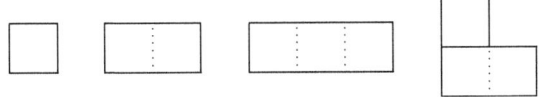

Figure 4.29

Problem 1.
Show that each of the five tetrominoes may be dissected into 2 rectangles.

Problem 2.
Show that each of the twelve pentominoes may be dissected into 3 rectangles.

As the size of the polyomino increases, the number of varieties increases significantly. For instance, there are thirty-five hexominoes, one hundred and eight heptominoes and three hundred and sixty-nine octominoes. Thus it is not feasible to deal with them one at a time.

A systematic approach is mathematical induction. We have already verified the basis $n = 1$. Assume now that the result holds for some $n \geq 1$.

Question 1.
Using the induction hypothesis, prove that every polyomino consisting of $2n+1$ squares may be dissected into $n+1$ squares.

Solution:
Dissect the polyomino in any way into two smaller polyominoes. One of them will consist of $2k+1$ squares for some $k < n$ and the other $2(n-k)$ squares. By the induction hypothesis, the first can be divided into $k+1$ rectangles and the second into $n-k$ rectangles, so that $(k+1)+(n-k) = n+1$ rectangles are sufficient for the original polyomino.

Consider now a polyomino consisting of $2n + 2$ squares. If it can be dissected into two smaller polyominoes each consisting of an even number of squares, we can appeal to the induction hypothesis.

Question 2.
Prove that if a polyomino contains a square with exactly two neighbors, then the polyomino may be dissected into two smaller polyominoes each consisting of an even number of squares.

Solution:
Removing the square with exactly two neighbors breaks the remaining part of the polyomino into two pieces. One of them will consists of an odd number of squares while the other consists of an even number of squares. By restoring the removed square and join it to the former, we have dissected the original polyomino into two smaller polyominoes each consisting of an even number of squares.

Question 3.
Suppose a polyomino does not contain any square s with exactly two neighbors but a square with all four neighbors. Prove that if the polyomino does not contain any of the four squares diagonally adjacent to s, then the polyomino cannot consist of an even number of squares.

Solution:
Since none of the squares diagonally adjacent to s belongs to the polyomino, each of the four neighbors of s must be a dead-end since it does not have exactly two neighbors. Hence the polyomino is the X-pentomino, as shown in Figure 4.30 on the left. It cannot have $2n + 2$ squares for any positive integer n.

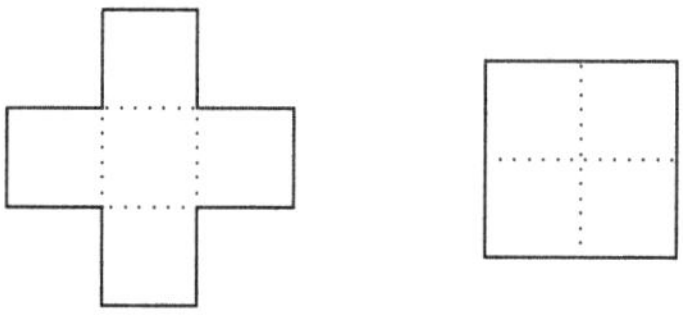

Figure 4.30

If a polyomino consisting of $2n+2$ squares does not contain any square with exactly two neighbors but a square with all four neighbors, then it must contain an O-tetromino, as shown in Figure 4.30 on the right.

Question 4.
Suppose a polyomino consisting of $2n + 2$ squares contains an O-tetromino but no squares with exactly two neighbors. Prove that it can be dissected into $n + 1$ rectangles.

Solution:
The removal of the O-tetromino breaks the remaining part of the polyomino into several pieces. If one of the pieces has an even number of squares, dissect the original polyomino into two, one of which coincides with this piece. Hence we may assume that all pieces are odd, and there are either two or four of them. Since there are no squares with exactly two neighbors, each square in the O-tetromino is adjacent to at least one of the pieces.

We consider two cases.

Case 1. There are only two odd pieces A and B.
Then there exist two squares a and b in the O-tetromino such that a is adjacent to A and b to B. We can dissect the original polyomino into two pieces, one of which consists of a and A.

Case 2. There are four odd pieces A, B, C and D.
Then each piece is adjacent to at most two neighboring squares in the O-tetromino. If there exists a square a in the block which is adjacent to only one piece, say A, we can dissect the original polyomino into two pieces, one consisting of a and A. Otherwise, we can dissect the original polyomino into two pieces, each consisting of a domino in the O-tetromino plus the two pieces adjacent to this domino.

Problem 3.
Find a polyomino which contains an O-tetromino but no squares with exactly two neighbors, such that the removal of the O-tetromino breaks it into four odd pieces of different sizes.

We can now complete the solution to our problem. The only remaining case covers polyominoes not containing an O-tetromino and each square has either exactly one or exactly three neighbors. It is easy to see that they form a sequence, the beginning of which is shown in Figure 4.31. If we take the middle row as one rectangle, we have exactly $n+1$ rectangles as desired.

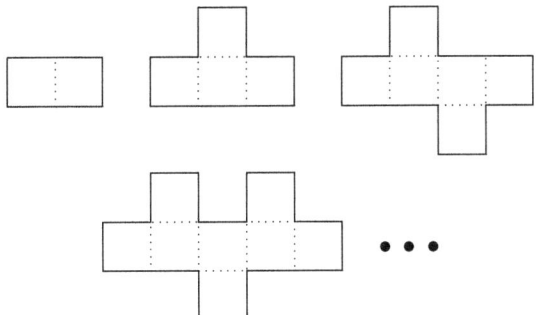

Figure 4.31

Problem 4.
A polyiamond is a figure consisting of unit equilateral triangles joined edge to edge. The name is derived from "diamond" where the prefix do- or di- means two. We have one moniamond, one diamond, one triamond and three tetriamonds, as shown in Figure 4.32.

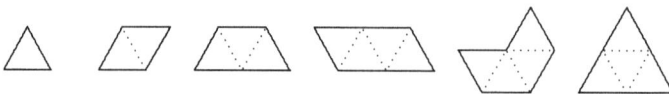

Figure 4.32

Investigate the dissection of polyiamonds into quadrilaterals with at least one pair of parallel sides.

Section 12. A Problem on Chess Tournaments

This problem is based on Problem 6 in the Junior A-Level paper of Spring 2000, Problem 6 in the Senior A-Level paper of Spring 2000 and Problem 5 in the Senior A-Level paper of Spring 2001.

In a chess tournament, every two participants play each other exactly once. A win is worth one point, a draw is worth half a point and a loss is worth zero points.

(a) For each participant X, the sum of the points earned by the participants who were beaten X is computed, as is the sum of the points earned by the participants who beat X. Is it possible that the same sum is greater than the other sum for all participants?

(b) A game is called an upset if the total number of points obtained by the winner of that game is less than the total number of points obtained by the loser of that game. What is the maximum value of the fractions of games which are upsets?

We first deal with part (a). Let there be n participants. For participant k, denote the first sum by w_k, the second sum by ℓ_k and the participant's own score by s_k. Let $T = \sum_{k=1}^{n} s_k(w_k - \ell_k)$. We claim that $T = 0$. Consider the game between participants i and j. If it is a tie, it contributes nothing to T. Suppose i beats j. Then it contributes $s_i s_j$ to the term $s_i(w_i - \ell_i)$ while it contributes $-s_j s_i$ to the term $s_j(w_j - \ell_j)$. The net contribution to T is still nothing, and the same holds by symmetry if j beats i. This justifies the claim. If the same sum is greater than the other for every participant, then $T \neq 0$, which is a contradiction.

Problem 1.
Is it possible for the same sum to be greater than the other sum for all but one participant?

We now turn our attention to part (b) of our problem.

Problem 2.
Construct a chess tournament with at least one upset.

Problem 3.
Construct a chess tournament with at least two upsets.

Question 1.
Prove that if there are $2n$ participants, the fraction of games which are upsets is less than $\frac{3}{4}$.

Solution:
The total number of games played is $n(2n-1)$. The average score of each participant is $\frac{2n-1}{2}$, so that the top n participants have among them more than $\frac{n(2n-1)}{2}$ points. Since they score exactly $\frac{n(n-1)}{2}$ against one another, they score against the bottom n participants more than $\frac{n(2n-1)}{2} - \frac{n(n-1)}{2} = \frac{n^2}{2}$ points. These points are scored in games which cannot be upsets. It follows that the fraction of non-upsets is greater than $\frac{n^2}{2n(2n-1)} > \frac{1}{4}$, so that the fraction of upsets is less than $\frac{3}{4}$.

Question 2.
Prove that if there are $2n+1$ participants, the fraction of games which are upsets is less than $\frac{3}{4}$.

Solution:
The total number of games played is $n(2n+1)$. The average score of each participant is n points, so that the top $n+1$ participants have among them more than $n(n+1)$ points. Since they score exactly $\frac{n(n+1)}{2}$ points against one another, they score against the bottom n participants more than

$$n(n+1) - \frac{n(n+1)}{2} = \frac{n(n+1)}{2}$$

points. These points are scored in games which cannot be upsets. It follows that the fraction of non-upsets is greater than $\frac{n(n+1)}{2n(2n+1)} > \frac{1}{4}$, so that the fraction of upsets is less than $\frac{3}{4}$.

Question 3.
Construct a chess tournament in which the fraction of games which are upsets is almost $\frac{1}{3}$.

| Parti- | Opponents ||||||||||| # of |
cipants	A	B	C	D	E	F	G	H	I	J	K	points
A	–	1	0	0	0	0	1	1	1	1	1	6
B	0	–	1	0	0	0	1	1	1	1	1	6
C	1	0	–	0	0	0	1	1	1	1	1	6
D	1	1	1	–	1	0	0	0	0	1	0	5
E	1	1	1	0	–	1	0	0	0	1	0	5
F	1	1	1	1	0	–	0	0	0	0	1	5
G	0	0	0	1	1	1	–	1	0	0	0	4
H	0	0	0	1	1	1	0	–	1	0	0	4
I	0	0	0	1	1	1	1	0	–	0	0	4
J	0	0	0	0	0	1	1	1	1	–	1	5
K	0	0	0	1	1	0	1	1	1	0	–	5

A Problem on Chess Tournaments

Solution:
Let the participants be A, B, C, D, E, F, G, H, I, J and K. The preceding table shows the results of all 55 games, with the 18 upsets in boldface. The fraction $\frac{18}{55}$ of upsets is almost $\frac{1}{3}$.

Question 4.
Construct a chess tournament in which the fraction of games which are upsets is almost $\frac{1}{2}$.

Solution:
Let there be 55 participants, divided into eight groups $A_1, A_2, \ldots, A_7$ and B, with 7 participants in each A_i and 6 participants in B. Within each A_i, everyone scores exactly 3 points. The internal scores of B is irrelevant and can be arbitrarily assigned. Every participant in A_i beats every participant in A_j for the pairs $(i,j) = (1,7), (7,6), (6,5), (5,4), (4,3), (3,2), (2,1), (1,6), (6,4), (4,2), (2,7), (7,5), (5,3), (3,1), (1,5), (5,2), (2,6), (6,3), (3,7), (7,4)$ and $(4,1)$. This means that every participant other than those in B scores the same number of points against one another. For $1 \le i \le 7$, each participant in A_i beats exactly $7-i$ participants in B, so that a participant in A_i has a higher score than a participant in A_j if and only if $i < j$. It follows that all games between participants in A_i and A_j are upsets for the pairs $(i,j) = (7,6), (6,5), (5,4), (4,3), (3,2), (2,1), (6,4), (4,2), (7,5), (5,3), (3,1), (5,2), (6,3), (7,4)$ and $(4,1)$. Thus the total number of upsets is at least $15 \times 49 = 735$. Since the total number of games is $\binom{55}{2} = 1485$, fraction $\frac{735}{1485}$ of upsets is almost $\frac{1}{2}$.

Problem 4.
Construct a chess tournament in which the fraction of games which are upsets is almost $\frac{2}{3}$.

We now present the solution to part (b) of our problem. Let there be $4n^2 + 6n + 1$ participants, with a total of $n(2n+3)(4n^2+6n+1)$ games played. Divide the participants into groups labeled $A_1, A_2, \ldots, A_{2n+1}$ and B, with $2n+1$ participants in each A_i and $2n$ participants in B. Within each A_i, everyone scores exactly n points. The internal scores of B is irrelevant and can be arbitrarily assigned. For $1 \le i \le 2n+1$, each participant in A_i beats exactly $2n+1-i$ participants in B, every participant in A_{i+j} for $n+1 \le j \le 2n$ but loses to every participant in A_{i+j} for $1 \le j \le n$, where the subscript is reduced by $2n+1$ if it exceeds $2n+1$. This means that a participant in A_i has a higher score than a participant in A_j if and only if $i < j$. For $2 \le i \le n+1$, the participants in A_i beat the participants in A_j for $1 \le j \le i-1$, resulting in $(1+2+\cdots+n)(2n+1)^2 = \frac{n(n+1)(2n+1)^2}{2}$ upsets. For $n+2 \le i \le 2n+1$, the participants in A_i beat the participants in A_j for $i-n \le j \le i-1$, resulting in $n^2(2n+1)^2$ upsets.

Thus the total number of upsets is at least

$$\frac{n(n+1)(2n+1)^2}{2} + n^2(2n+1)^2 = \frac{n(3n+1)(2n+1)^2}{2},$$

and the fraction of upsets is at least $\frac{(3n+1)(2n+1)^2}{2(2n+3)(4n^2+6n+1)}$. Since the limiting value of this fraction is $\frac{3}{4}$ as n approaches infinity, we cannot replace $\frac{3}{4}$ by any smaller number.

Section 13. A Problem on Electrical Networks

This problem is based on Problem 7 in the Junior A-Level paper of Fall 2002 and Problem 7 in the Senior A-Level paper of Fall 2002.

An electrical network consists of n^2 nodes in an $n \times n$ array. Each node is connected by wire to all adjacent nodes in the same row or the same column. Some of the wires may be burned out, but as it turns out, any pair of nodes is joined by a chain of intact wires. However, this needs to be verified, by performing tests between pairs of nodes. Find in terms of n the minimum number of tests required.

We first give a lower bound on the number of tests required.

Question 1.
Prove that at least $\lceil \frac{n^2}{2} \rceil$ tests are required.

Solution:
There are n^2 nodes, each of which must be involved in at least one test. Otherwise, we cannot tell if all wires connected to an untested node may be burnt out. Hence the minimum number of tests required is $\lceil \frac{n^2}{2} \rceil$.

We now turn our attention to finding upper bounds.

Question 2.
Prove that if $n = 2$, then 2 tests are sufficient.

Solution:
Let the 4 nodes be the vertices of a square $ABCD$. We will test the pairs (A, C) and (B, D). If some pair of nodes is not joined by a chain of intact wires, we may assume by symmetry that it is (A, B). This means that the wire connecting A and B must be burnt out, along with at least one other wire. However, this will either isolate A from C, or B from D, or both. This is a contradiction.

Question 3.
Prove that if $n = 3$, then 5 tests are sufficient.

Solution:
Let the 9 nodes be labeled as shown in Figure 4.33. By Question 2, if we test the pairs (B, F) and (C, E), we will know that each pair of B, C, E and F is joined by a chain of intact wires. If we also test the pairs (D, H) and (E, G), so is each pair of D, E, G and H. Since E is common to both sets, all seven are pairwise joined. We now test the pair (A, I). Since A and I are separated by the other seven, the chain of intact wires joining them just pass through at least one of those seven nodes. It follows that the whole network is interconnected.

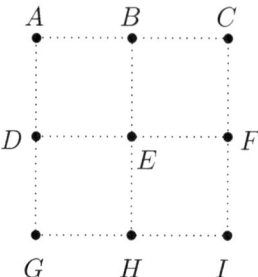

Figure 4.33

Problem 1.
For $n = 3$, let the 9 nodes be labeled as shown in Figure 4.33. Does it work if we test the pairs (A, I), (B, H), (C, G), (D, F) and (B, E)?

Problem 2.
For $n = 4$, let the 16 nodes be labeled as shown in Figure 4.34. Does it work if we test the pairs (A, P), (B, O), (C, N), (D, M), (E, L), (F, K), (G, J) and (H, I)?

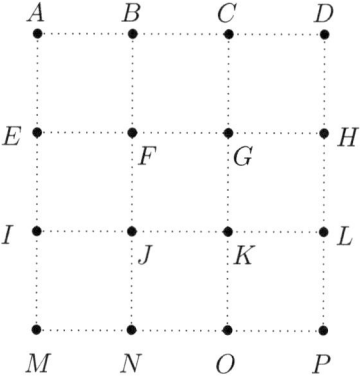

Figure 4.34

Problem 3.
Prove that if $n = 4$, then 8 tests are sufficient.

We now prove that for n^2 nodes, the lower bound $\lceil \frac{n^2}{2} \rceil$ is sharp. Labels the nodes (i, j) where $1 \leq i \leq n$ and $1 \leq j \leq n$. On the main diagonal, we perform the tests between (i, i) and $(k + i, k + i)$ for $1 \leq i \leq k$ if $n = 2k$, and for $1 \leq i \leq k + 1$ if $n = 2k + 1$. Off the main diagonal, we perform the tests between (i, j) and (j, i) for all $i \neq j$. color the node (1,1) red. All nodes connected to (1,1) will also be colored red.

We claim that (2,2) must be red. Let (t,t) be the first red node on the main diagonal after (1,1). This node exists since our test result indicates that $(k+1, k+1)$ is red. If $t = 2$, there is nothing to prove. Suppose $t > 2$. Then there is a path lying on one side of the main diagonal, connecting (1,1) to (t,t). Its nodes form a red chain. Moreover, our test results indicate that the nodes symmetric to them with respect to the main diagonal also form a red chain, even though they may not be lying on a path connecting (1,1) to (t,t). These two chains form a cycle which encloses the node (2,2) but leaves on the outside the node $(k+2, k+2)$. Since (2,2) and $(k+2, k+2)$ are connected, they must be connected to the cycle above, so that (2,2) is red after all. In the same way, we can prove that all nodes on the main diagonal are red. Since each node off the main diagonal is connected to a node on the other side, all nodes are red and the whole network is interconnected.

Problem 4.
Investigate electrical networks which consists of $\frac{n(n+1)}{2}$ nodes in triangular arrays for $n \geq 2$. The first three are shown in Figure 4.35.

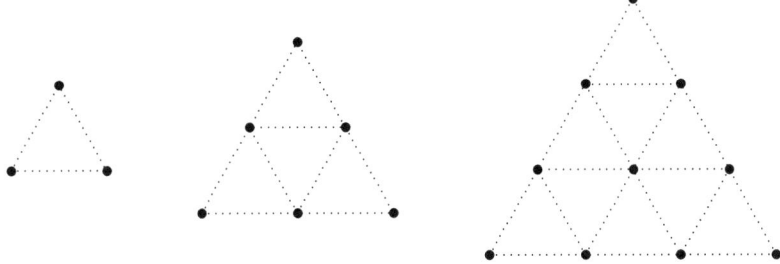

Figure 4.35

Section 14. A Problem on Card Signaling

This problem is based on Problem 6 in the Senior A-Level paper of Spring 2004.

The audience shuffles a deck of 36 cards, containing 9 cards in each of the suits spades, hearts, diamonds and clubs. The design on the back of each card is an arrow. The assistant of a magician examines the deck without changing the order of the cards, and points the arrow on the back of each card either towards or away from himself, according to some system agreed upon in advance with the magician. The magician examines the back of the top card of the deck, and announces the suit of this card. The assistant turns over the card to check against the announcement, while exposing the back of the next card below. Is there such a system which enables the magician to guarantee the correct announcement of the suit of at least

(a) **19 cards;**

(b) **more than 19 cards?**

There is no way the magician can guarantee the correct announcement of the suit of the top card. However, it is possible to guarantee that of the second card.

Denote a card by 0 if the arrow on its back is pointing at the assistant, and by 1 otherwise. Thus the back of the top two cards form one of 00, 01, 10 and 11. By advance agreement, these represent the suits in the natural order, namely, Spades, Hearts, Diamonds and Clubs respectively. The assistant is then able to convey to the magician the suit of the second card.

This is the key idea for the following first attempt at the problem.

Question 1.
Describe a system which enables the magician to guarantee the correct announcement of the suit of at least 18 cards.

Solution:
The magician guesses wildly on the suits of all cards in odd positions. For cards in even positions, correct announcements can be guaranteed using the key idea above.

We can now solve Part (a) of our problem. Using the key idea, the magician can make at least 17 correct announcements among the top 34 cards.

A Problem on Card Signaling

For the remaining two cards, if they are of the same suit, she needs no further assistance to announce correctly their suits. If they are of different suits, she can agree in advance with the assistant that if the suit of the 35th card is higher than that of the 36th card, then the arrow on the back of the 35th card will point towards the assistant. Otherwise, it will point away from him. Thus she can make at least 19 correct announcements.

Instead of dealing with the cards one at a time, the assistant may take into consideration the overall distribution. This leads to the following result which answers Part (b) of our problem.

Question 2.
Describe a system which enables the magician to guarantee the correct announcement of the suit of at least 22 cards.

Solution:
The assistant examines the 17 cards in odd positions except the top card. By the generalized Pigeonhole Principle, at least 5 of them are of the same suit. The assistant uses the first two cards to signal this suit, and the magician always guesses this suit for all odd-numbered cards. This guarantees at least 5 correct announcements among the 17. For each of the 16 cards in even positions except the second and the last cards, a correct announcement can be guaranteed as in Question 1. The suit of the last card can always be announced correctly. Thus the magician can guarantee at least $5+16+1=22$ correct announcements. For each of the 16 cards in even positions except the second and the last cards, a correct announcement can be guaranteed as in Question 1.

A slight improvement can be made.

Question 3.
Describe a system which enables the magician to guarantee the correct announcement of the suit of at least 23 cards.

Solution:
The assistant examines the second card and the 16 cards in odd positions except the top card and the second last card. By the generalized Pigeonhole Principle, at least 5 of these 17 cards are of the same suit. The assistant uses the first two cards to signal this suit, and the magician always guesses this suit for all odd-numbered cards. This guarantees at least 5 correct announcements. For each of the 16 cards in even positions except the second and the last cards, a correct announcement can be guaranteed as in Question 1. The remaining two cards can be handled as in part (a). Thus the magician can make at least $5+16+2=23$ correct announcements.

So far, we have not let the magician utilize what she is allowed to do, namely, examine each card after she has made her announcement. Taking advantage of this leads to a further improvement.

Question 4.
Describe a system which enables the magician to guarantee the correct announcement of the suit of at least 24 cards.

Solution:
Let us start off with a slightly modified solution of Part (a) of our problem. Instead of guessing wildly on cards number $3, 5, \ldots, 21$, the magician makes the same announcements as for cards $2, 4, \ldots, 20$. Thus we can again guarantee 19 correct announcements. Note that if either cards number 2 and 3 or cards number 4 and 5 are of the same suit, we have an additional correct answer. Suppose this is not the case. Now the assistant can arrange for the magician to guess correctly card number 2 or 3, and card number 4 or 5. This flexibility is now used to signal the suit of card number 23. For instance, if card number 23 is Spades, then the correct guesses will occur on cards number 2 and 4. If it is Hearts, they occur on cards number 2 and 5; if Diamonds, 3 and 4; and if Clubs, 3 and 5. Similarly, an extra correct answer can come out of cards number 6 to 9 plus 25, of cards number 10 to 13 plus 27, of cards number 14 to 17 plus 29, and of cards number 18 to 21 plus 31. This guarantees at least 5 extra correct announcements, making the grand total at least 24.

Let us work through a specific example to illustrate the argument in Question 4. Let the suit distribution be

♠♡♡♣◇♠♠◇♡♠♣♠♣◇◇◇♡♣♠♣♡♠♡◇♣♣♠◇◇♡♠♣◇♠♡♡♣.

Here are the arrangements of the arrows.

1. Since cards number 2 and 3 are both ♡, arrows number 1 and 2 should be 01. We arrange arrows number 3 and 4 to convey 11 since card number 4 is ♣. Since cards number 14 and 15 are both ◇, arrows number 13 and 14 should be 10. We arrange arrows number 15 and 16 to convey 01, since card number 16 is ♡.

2. Since cards number 6 and 7 as well as cards number 8 and 9 are of different suits, we note that card number 25 is ♣. This means that we want correct announcements for cards 7 and 9. So we arrange arrows number 5 and 6 to convey 10, since card number 7 is ◇. Then we arrange arrows 7 and 8 to convey 00, since card number 9 is ♠. Similarly, we want correct announcements for cards 11 and 12. So we arrange arrows number 9 and 10 to convey 00, since card number 11 is ♠. Then we arrange arrows 11 and 12 to convey 11, since card number 12 is ♣. In the same way, we arrange arrows number 17 and 18 to convey 11, and arrows 19 and 20 to convey 00.

A Problem on Card Signaling   177

3. We arrange arrows number 21 and 22 to convey 01, since card number 22 is ♡; arrows number 23 and 24 to convey 11, since card number 24 is ♣; arrows number 25 and 26 to convey 00, since card number 26 is ♠; arrows number 27 and 28 to convey 10, since card number 28 is ♢; arrows number 29 and 30 to convey 00, since card number 30 is ♠; arrows number 31 and 32 to convey 10, since card number 32 is ♢; arrows number 33 and 34 to convey 01, since card number 34 is ♡.

4. We arrange arrows number 35 to convey 0, since card number 35 is ♡ and card number 36 is ♣. The arrangement of arrow number 36 is arbitrary.

The overall arrow arrangement is 01111000001110011100011100100010010?, where the final ? is arbitrary. Correct announcements are guaranteed for cards number 2, 3, 4, 7, 9, 11, 12, 14, 15, 16, 19, 21, 22, 24, 25, 26, 27, 28, 30, 31, 32, 34, 35 and 36.

Problem 1.
It is now known that there is a system which enables the magician to guarantee the correct announcement of the suit of at least 25 cards. Try to find one.

Problem 2.
It is now known that there is a system which enables the magician to guarantee the correct announcement of the suit of at least 26 cards. Try to find one.

Problem 3.
It is now known that there is a system which enables the magician to guarantee the correct announcement of the suit of at least 27 cards. Try to find one.

Problem 4.
Investigate the scenario where the distribution of the suits of the cards is not known to the magician, and is not necessarily uniform.

Section 15. A Problem on Tangent Circles

This is based on Problem 5 in the Senior A-Level paper of Spring 2011.

In the convex quadrilateral $ABCD$, BC is parallel to AD. Two circular arcs ω_1 and ω_3 pass through A and B and are on the same side of AB. Two circular arcs ω_2 and ω_4 pass through C and D and are on the same side of CD. The measures of ω_1, ω_2, ω_3 and ω_4 are α, β, β and α respectively. If ω_1 and ω_2 are tangent to each other externally, prove that so are ω_3 and ω_4.

Let X and Y be the respective midpoints of AB and CD. Let O, P, Q' and P' be the respective centers of ω_1, ω_2, ω_3 and ω_4. Let E, F, M, N, M' and N' be points on AD such that YE, XF, OM, PN, $O'M'$ and $P'N'$ are perpendicular to AD. Let K be the point on XF such that OK is perpendicular to XF, L be the points on PN such that LY is perpendicular to PN, K' be the point on XF such that $O'K'$ is perpendicular to XF, and L' be the point on $P'N'$ such that $L'Y$ is perpendicular to $P'N'$.

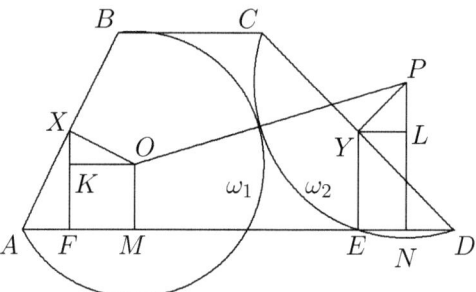

Figure 4.36

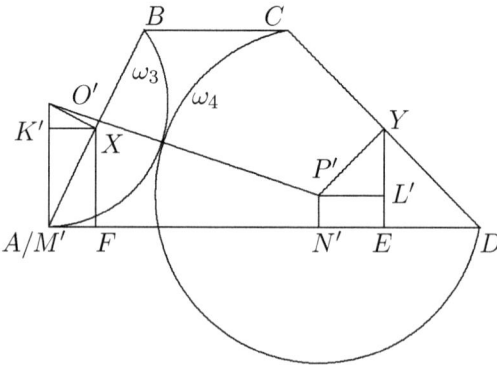

Figure 4.37

A Problem on Tangent Circles

Question 1.
Prove that the product of the radii of w_1 and w_2 is equal to the product of the radii of w_3 and w_4.

Solution:
Since w_1 and w_4 have equal measure, triangles OAB and $P'CD$ are similar. Hence $\frac{OA}{AB} = \frac{P'C}{CD}$. We also have $\frac{O'A}{AB} = \frac{PC}{CD}$, so that $OA \cdot PC = O'A \cdot P'C$.

Question 2.
Prove that $OX \cdot PY = O'X \cdot P'Y$.

Solution:
From halves of the similar triangles in Question 1, we have $\frac{OX}{OA} = \frac{P'Y}{P'C}$ and $\frac{O'X}{O'A} = \frac{PY}{PC}$. Since $OA \cdot PC = O'A \cdot P'C$ by Question 1, we have $OX \cdot PY = O'X \cdot P'Y$.

Question 3. Prove that $KX \cdot LP = K'O' \cdot L'Y$.

Solution:
From similar triangles, $\frac{KX}{OX} = \frac{AF}{AX} = \frac{K'O'}{O'X}$ and $\frac{LP}{PY} = \frac{DE}{DY} = \frac{L'Y}{P'Y}$. It follows from Question 2 that $KX \cdot LP = K'O' \cdot L'Y$.

The horizontal distance between O and P is $MN = EF - MF + EN$. The vertical distance between O and P is $PN - OM = KX + LP$. We are given that w_1 and w_2 are tangent, so that $MN^2 + (KX+LP)^2 = (OB+PC)^2$. In the same way, $M'N' = EF - E'N + F'M$ is the horizontal distance between O' and P' and $O'M' - P'N' = K'O' + L'Y$ is the vertical distance. To have w_3 tangent to w_4, we need $(M'N')^2 + (K'O' + L'Y)^2 = (O'B + P'C)^2$. Note that $FM = \frac{h}{a}OX = \frac{h}{c}P'Y = EN'$ and $FM' = \frac{h}{a}O'X = \frac{h}{c}PY = EN$. Hence $MN = M'N'$. In view of Questions 1 and 3, the desired result now follows from

$$OB^2 - KX^2 + PC^2 - LP^2$$
$$= \frac{a^2}{c^2}P'C^2 - \frac{f^2}{a^2}OX^2 + \frac{c^2}{a^2}O'B^2 - \frac{e^2}{c^2}PY^2$$
$$= \frac{a^2}{c^2}P'C^2 - \frac{f^2}{c^2}P'Y^2 + \frac{c^2}{a^2}O'B^2 - \frac{e^2}{a^2}O'X^2$$
$$= \frac{h^2}{c^2}P'C^2 - f^2 + \frac{h^2}{a^2}O'B^2 - e^2$$
$$= \frac{h^2}{a^2}OB^2 - e^2 + \frac{h^2}{c^2}PC^2 - f^2$$
$$= \frac{a^2}{c^2}PC^2 - \frac{f^2}{c^2}PY^2 + \frac{c^2}{a^2}OB^2 - \frac{e^2}{a^2}OX^2$$
$$= \frac{a^2}{c^2}PC^2 - \frac{f^2}{a^2}O'X^2 + \frac{c^2}{a^2}OB^2 - \frac{e^2}{c^2}P'Y^2$$
$$= O'B^2 - (K'O')^2 + P'C^2 - L'Y^2.$$

Section 16. A Problem on Graph Algorithms

This is based on Problem 7 in the Senior A-Level paper of Spring 2011.

MacroHard and Ogle conduct joint hiring. For every two applicants, either they know each other or they do not. Eleven of the applicants are known to be geniuses. MacroHard hires first, and turns alternate thereafter. There are no restrictions on which applicant a company may hire in the first round. In subsequent rounds, a company may only hire an applicant who knows another applicant already hired by that company. Is it possible for Ogle to hire ten of the geniuses, regardless of the strategy of MacroHard?

A mathematical model of the situation is a graph, where the vertices represent the applicants, and edges represent acquaintance. Vertices representing the geniuses are enlarged. Starting from an arbitrary vertex, each company builds an expanding subgraph which is connected, and tries to include as many of the large vertices as possible.

With only one large vertex, MacroHard wins automatically. With two large vertices, each company gets one. The first interesting case is with three large vertices. It would appear that hiring first, MacroHard should have an advantage. MacroHard cannot be shut out as a large vertex can be taken on the first turn. Yet, counter-intuitively, it is possible for Ogle to get two large vertices.

The graph in Figure 4.38 is a cycle of length 6, represents a talent pool of six applicants. MacroHard has two options in its first turn, either taking a large vertex or a small one. In the latter case, Ogle will take a large vertex adjacent to MacroHard's choice. Each company then moves around the cycle, with Ogle picking up another large vertex. In the former case, Ogle will take the small vertex opposite to MacroHard's choice. Then it will get both adjacent large vertices.

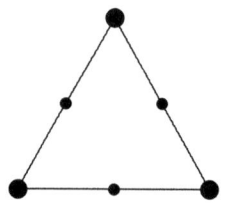

Figure 4.38

Problem 1.
Find a graph with 5 large vertices and 5 small vertices such that Ogle can get 3 of the large vertices.

Question 1.
Find a graph with 11 large vertices such that Ogle can get 6 of them.

Solution:
Following the introductory example, we construct a cycle with 11 large vertices alternating with 11 small vertices. Ogle can use exactly the same strategy as before. If MacroHard takes a large vertex, Ogle takes the opposite small vertex. If MarcoHard takes a small vertex, Ogle takes an adjacent small vertex.

Now that Ogle knows it can get six large vertices, it gets greedy and starts ogling at more large vertices.

Question 2.
Find a graph with 11 large vertices such that Ogle can get 7 of them.

Solution:
Once again, we fall back on the introductory example. We modify the graph to the one in Figure 4.39, with 8 additional large vertices. The same strategy as before, applied to the central cycle, will deliver 7 large vertices to Ogle.

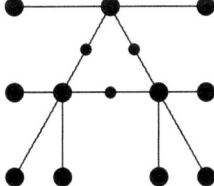

Figure 4.39

Is it possible for Ogle to get 8 large vertices out of 11? Using the idea in Question 2, this can be accomplished if we can get 3 large vertices out of 4. We can then stick 2 additional large vertices to 3 of the large vertices and one additional large vertex to the fourth one.

Question 3.
Find a graph with 4 large vertices such that Ogle can get 3 of them.

Solution:
Once more, our inspiration comes from the introductory example. This time, instead of a triangle, we use a tetrahedron as the base of our graph. We place the 4 large vertices are at the corners of the tetrahedron. We place 3 small vertices along each edge, and 1 small vertex at the center of each face. This central vertex is joined to all three corners of the face, and we place 2 small vertices on each of these three segments. Figure 4.40 shows the net of the tetrahedron $X_1X_2X_3X_4$, cut open along the three edges meeting at X_1. Some of the small vertices are also labeled.

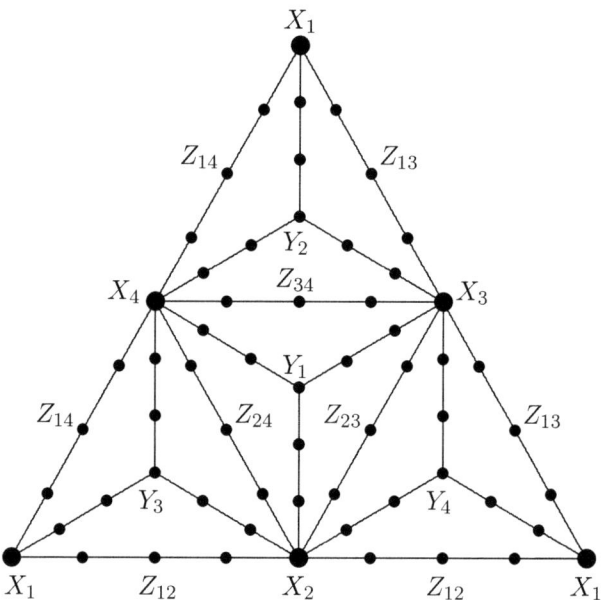

Figure 4.40

Depending on the action of MacroHard in the first round, we have five cases.

Case 1. MacroHard takes an X, say X_1.

Ogle will take Y_1 and win all of the races to X_2, X_3 and X_4.

Case 2. MacroHard takes a Y, say Y_1.

Ogle will take Z_{34} and will win both of the races to X_3 and X_4. Moreover, Ogle will get to one of them before MacroHard can, so that Ogle will also win the race to X_1.

Case 3. MacroHard takes a Z, say Z_{34}.

Ogle can take X_4. By the time MacroHard gets to X_3, Ogle will have taken one step towards each of X_1 and X_2. Hence Ogle will win both of these races.

Case 4. MacroHard takes an unlabeled vertex adjacent to a Y, say the one between Y_1 and X_2.

Ogle can take the unlabeled vertex between X_2 and Z_{24}. Ogle will win both of the races to X_2 and X_4. Moreover, Ogle will get to X_2 before MarcoHard can get to X_3. Hence Ogle will also win the race to X_1.

Case 5. MacroHard takes an unlabeled vertex adjacent to an X.
There are two subcases.
Subcase 5(a). This vertex is between the X and a Y, say between X_2 and Y_1.
Ogle will take X_2. Ogle will win the race to X_3, and get there before MacroHard can get to X_4. Hence Ogle will also win the race to X_1.
Subcase 5(b). This vertex is between the X and a Z, say between X_2 and Z_{24}.
Ogle will take X_2. By the time MacroHard gets to X_4, Ogle will have taken at least one step towards each of X_1 and X_3. Hence Ogle will win both of these races.

Problem 2.
Find a graph with 5 large vertices such that Ogle can get 4 of them.

A solution to Problem 2 will give Ogle 9 large vertices out of 11, but Ogle can do better still, by considering an alternative solution to the introductory example.

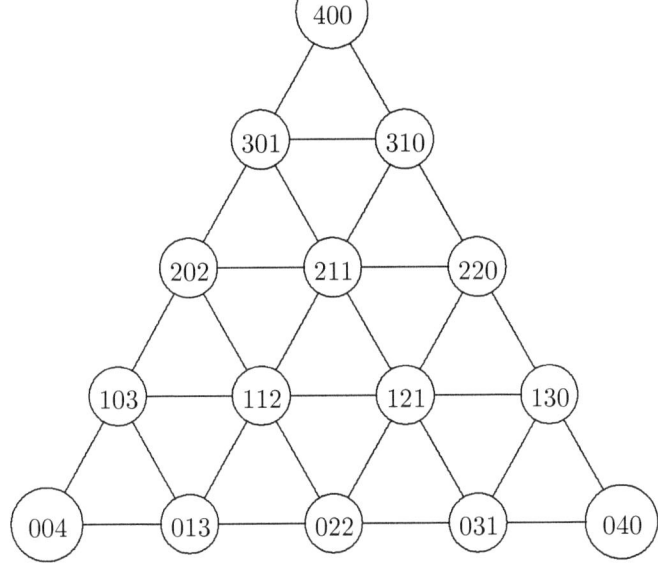

Figure 4.41

Question 4.

The graph in Figure 4.41 consists of 15 vertices each labeled with three numbers from 0 to 4 inclusive. The first number is the level of the vertex from the base of the triangle. The other two numbers are the levels of the vertex with the left side and the right side of the triangle considered as the respective base. The three vertices at the corners are large while the remaining vertices are small. Prove that Ogle can get two of the large vertices.

Solution:

Note that two vertices are connected by an edge if and only if their labels agree in one coordinate and differ by 1 in the other two coordinates. Suppose the first vertex taken by MacroHard is (x, y, z), with $x + y + z = 4$. By the Pigeonhole Principle, at least one of x, y and z, say x, is at least 2. Ogle will take the vertex $(x - 2, y + 1, z + 1)$. MacroHard will get $(4,0,0)$, but Ogle is one step ahead in each of the races to $(0,4,0)$ and $(0,0,4)$. Ogle can maintain this advantage and get both of them.

Using a tetrahedron as the base for a graph analogous to the one in Figure 4.41, we obtain an alternative solution to Question 3. Using a 10-dimensional simplex instead, we will have a graph in which Ogle can get 10 large vertices out of 11.

This is the best possible result, as MacroHard can always get a large vertex in the first turn. The jubilant Ogle team cheered in unison: "Go Ogle!"

Problem 3.

Without using the idea in Question 4, try to find a graph with 11 large vertices such that Ogle can get 9 of them.

Problem 4.

Investigate the situation if YeahWho also gets involved in a three-way joint hiring.

Section 17. A Problem on Fixed Points

This is based on Problem 6 in the Junior A-Level Paper of Fall 2012 and Problem 6 in the Senior A-Level paper of Fall 2012.

(a) A is a fixed point inside a given circle. A regular $2n$-gon with center A is drawn, and the rays from A to its vertices intersect the circle at $2n$ points. Prove that the center of mass of these $2n$ points does not depend on the choice of the regular $2n$-gon.

(b) A is a fixed point inside a given sphere. A regular icosahedron with center A is drawn, and the rays from A to its vertices intersect the sphere at twelve points. Prove that the center of mass of these twelve points does not depend on the choice of the regular icosahedron.

Question 1.
A is a fixed point inside a given circle. Two perpendicular lines are drawn through A, intersect the circle at four points. Prove that the center of mass of these four points does not depend on the choice of the two lines.

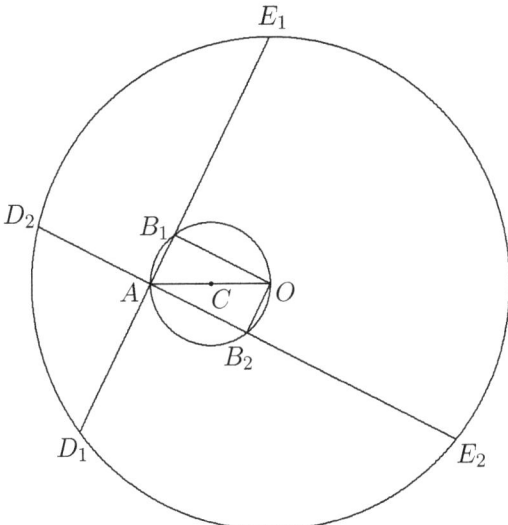

Figure 4.42

Solution:
Let O be the center of the given circle. Let C be the midpoint of OA. Let the two perpendicular chords be D_1E_1 and D_2E_2 as shown in Figure 4.42. Let B_1 and B_2 be the respective midpoints of D_1E_1 and D_2E_2. Then OB_1AB_2 is a rectangle, so that C is the common midpoint of OA and B_1B_2. Assign a weight of 1 to each of D_1, D_2, E_1 and E_2. These may be replaced by a weight of 2 at each of B_1 and B_2, and finally by a single weight of 4 at C, which is thus the center of mass of the four points. Since O and A are fixed points, C is also a fixed point.

Question 2.
Solve part (a) of our problem for a square.

Solution:
The argument is exactly the same as in Question 1.

We now solve part (a) of our problem. Let O be the center of the given circle. Let C be the midpoint of OA. Let the regular $2n$-gon be $D_1D_2\ldots D_nE_1E_2\ldots E_n$. For $1 \leq k \leq n$, let B_k be the midpoint of D_kE_k. Then $\angle OB_kA = 90°$ for $1 \leq k \leq n$. Hence $B_1, B_2, \ldots, B_n$ all lie on the circle with diameter OA. We may assume that A is between B_1 and B_n. Figure 4.43 illustrates the case $n = 5$.

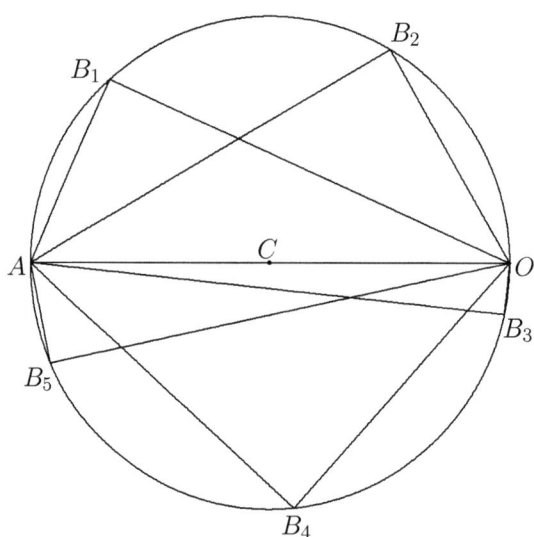

Figure 4.43

Now $\angle B_kAB_{k+1} = \frac{1}{n}180°$ for $1 \leq k \leq n-1$ and $B_nAB_1 = (1 - \frac{1}{n})180°$. Hence $B_1B_2\ldots B_n$ is a regular n-gon whose center is the fixed point C. Assign a weight of 1 to each of D_K and E_k for $1 \leq k \leq n-1$. These may be replaced by a weight of 2 at B_k for $1 \leq k \leq n-1$. Since each B_k has the same weight, the overall center of mass is C, which is a fixed point.

A Problem on Fixed Points

We turn our attention to three-dimensional space.

Question 3.
A is a fixed point inside a given sphere. Three mutually perpendicular lines are drawn through A, intersect the sphere at six points. Prove that the center of mass of these six points does not depend on the choice of the three lines.

Solution:
Let O be the center of the given sphere. Let the three mutually perpendicular chords be D_1E_1, D_2E_2 and D_3E_3. Let their respective midpoints be B_1, B_2 and B_3. Let O_1 be the center of the cross-section of the sphere containing D_2E_2 and D_3E_3, O_2 be the center of the cross-section of the sphere containing D_3E_3 and D_1E_1, and O_3 be the center of the cross-section of the sphere containing D_1E_1 and D_2E_2. Then $AB_1O_3B_2 - B_3O_2OO_1$ forms a rectangular block, as shown in Figure 4.44 on the left.

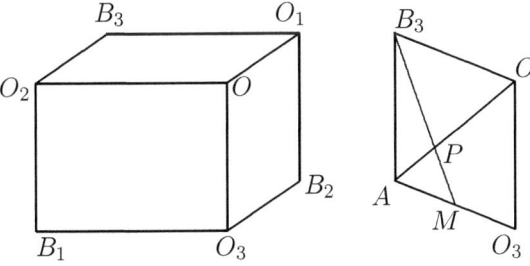

Figure 4.44

Assign a weight of 1 to each of D_K and E_k for $1 \leq k \leq 3$. These may be replaced by a weight of 2 at B_k for $1 \leq k \leq 3$. The overall center of mass C is the centroid of triangle $B_1B_2B_3$. Let M be the midpoint of B_1B_2. Then C lies on MB_3. Since M is also the midpoint of AO_3, C lies on the plane AO_3OB_3, as shown in Figure 4.44 on the right. Let P be the point of intersection of OA and MB_3. Then triangles APM and OPB_3 are similar. Note that $OB_3 = O_3A = 2AM$. Hence $B_3P = 2MP$ so that P coincides with C. Now we have $OC = 2AC$, so that C is a fixed point on OA.

If we have a regular octahedron with center A instead of a regular icosahedron with center A, the argument is exactly as in Question 3. These two regular polyhedra are closely related.

Question 4.

Show that a regular dodecahedron may be obtained from a regular octahedron by splitting each of its vertices into two.

Solution:

Let us call two opposite vertices of the regular octahedron T and B for top and bottom, and the other four N, E, W and S for north, east, west and south. A regular icosahedron may be obtained by splitting each vertex into two, in a direction parallel to a space diagonal of the regular octahedron as follows. T becomes T_n and T_s while B becomes B_n and B_s. N becomes become N_e and N_w while S becomes S_e and S_w. E becomes E_t and E_b while W becomes W_t and W_b. This is shown in Figure 4.45.

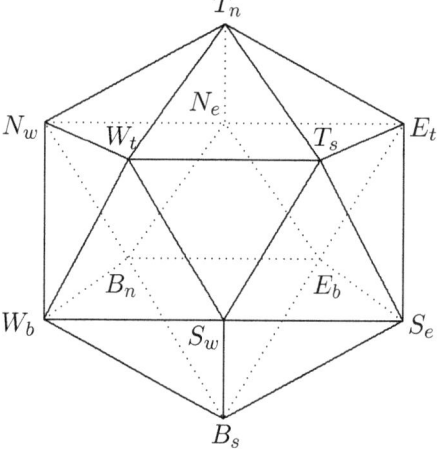

Figure 4.45

Problem 1.

Can a regular dodecahedron be obtained in some way from a cube?

We now tackle part (b) of our problem.

Note that $\angle T_n A T_s = \angle N_e A N_w = \angle E_t A E_b$. Denote the common value by 2θ. Set up a coordinate system with the origin at the center O of the given sphere. Let the x-, y- and z-axes be in the direction of AE, AN and AT. Let the coordinates of A be (a, b, c). Consider the plane $z = c$ which contains the points A, N_e, N_w, S_e and S_w. Let O_c be the center of the circular cross-section, so that its coordinates are $(0, 0, c)$. Let M_e be the midpoint of $N_e S_e$ and M_w be the midpoint of N_w and S_w. Let H be the point with coordinates $(a, 0, c)$. Then AH bisects $\angle M_e A M_w$ and intersects $O_c M_w$ at P, while $O_c H$ bisects $\angle M_e O_c M_w$ and intersects AM_e at Q. Finally, let U, V and K be the feet of perpendicular from M_w, M_e and M_w to AH, OH and OH respectively, as shown in Figure 4.46.

A Problem on Fixed Points

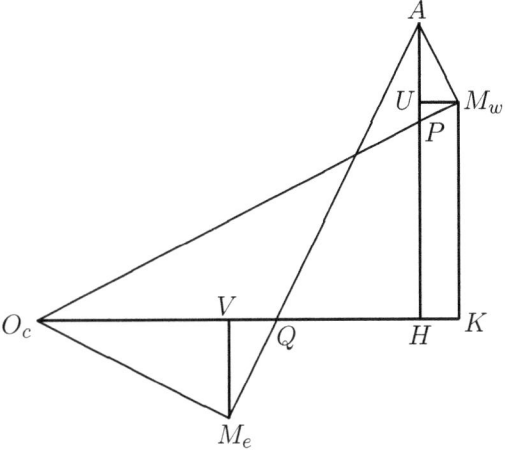

Figure 4.46

Note that $\angle M_e O_c H = \angle M_w O_c H = \angle M_e A H = \angle M_w A H = \theta$. We have

$$\begin{aligned}
PH &= a\tan\theta, \\
AP &= b - a\tan\theta, \\
AM_w &= b\cos\theta - a\sin\theta, \\
AU &= b\cos^2\theta - a\sin\theta\cos\theta. \\
M_w K &= b\sin^2\theta + a\sin\theta\cos\theta, \\
M_w U &= b\sin\theta\cos\theta - a\sin^2\theta, \\
O_c K &= a\cos^2\theta + b\sin\theta\cos\theta.
\end{aligned}$$

Hence the coordinates of M_w are

$$(a\cos^2\theta + b\sin\theta\cos\theta, b\sin^2\theta + a\sin\theta\cos\theta, c).$$

Similarly, the coordinates of M_e are

$$(a\cos^2\theta - b\sin\theta\cos\theta, b\sin^2\theta - a\sin\theta\cos\theta, c).$$

It follows that the center of mass of N_e, N_w, S_e and S_w is at the point whose coordinates are $(a\cos^2\theta, b\sin^2\theta, c)$ with a weight of 4. Similarly, from the plane $y = b$, the center of mass of T_n, T_s, B_n and B_s is at the point whose coordinates are $(a\sin^2\theta, b, c\cos^2\theta)$, and from the plane $x = a$, the center of mass of E_t, E_b, W_t and W_b is at the point whose coordinates are $(a, b\cos^2\theta, c\sin^2\theta)$, also with weight 4. Hence the overall center of mass is at the point whose coordinates are $(\frac{2a}{3}, \frac{2b}{3}, \frac{2c}{3})$. In fact, it is the same fixed point C as in Question 3. If we choose a different regular icosahedron with center A, the coordinate system set up as above will be different, and the coordinates of A will become (a', b', c'). However, the coordinates of C will become $(\frac{2a'}{3}, \frac{2b'}{3}, \frac{2c'}{3})$, which yields the same point.

Problem 2.
Investigate part (b) of our problem for a regular tetrahedron.

Problem 3.
Investigate part (b) of our problem for a cube.

Problem 4.
Investigate part (b) of our problem for a regular dodecahedron.

Section 18. A Problem on Information Extraction

This problem is based on Problem 7 in the Junior A-Level paper of Fall 2002 and Problem 7 in the Senior A-Level paper of Fall 2013.

Anna chooses an interior point of one of the 64 cells of a standard chessboard. Boris draws a subboard consisting of one or more cells such that its boundary is a single closed polygonal line which does not intersect itself. Anna will then tell Boris whether the chosen point is inside or outside this subboard. What is the minimum number of times Boris has to do this in order to determine whether the chosen point is black or white?

Question 1.
Show how Boris can accomplish the task in four questions.

Solution:
Boris uses the fact that all cells along any diagonal are of the same colour. The first subboard is shown in Figure 4.47.

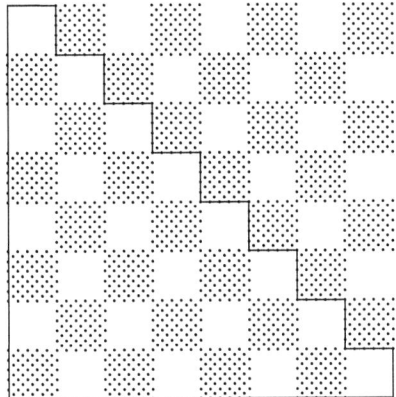

Figure 4.47

We may assume that Anna's announcement is "Inside". This narrows down the chosen cell to eight diagonals. Then the second subboard is shown in Figure 4.48.

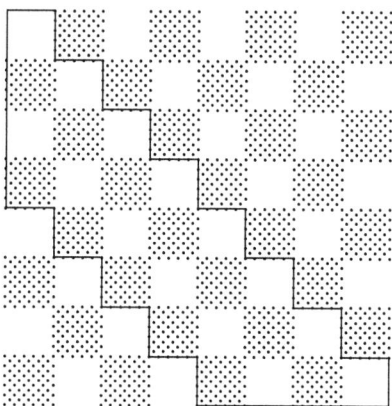

Figure 4.48

Again, we may assume that Anna's announcement is "Inside". This narrows down the chosen cell to four diagonals. Now the third subboard is shown in Figure 4.49.

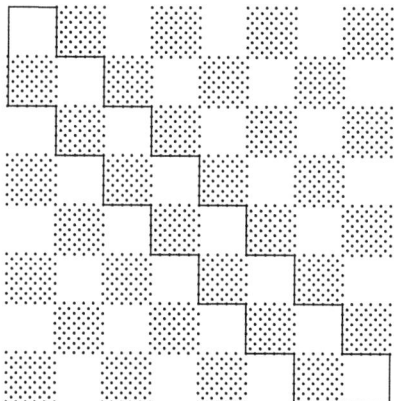

Figure 4.49

We may assume that Anna's announcement is still "Inside". This narrows down the chosen cell to two diagonals. Finally, the fourth subboard is shown in Figure 4.50.

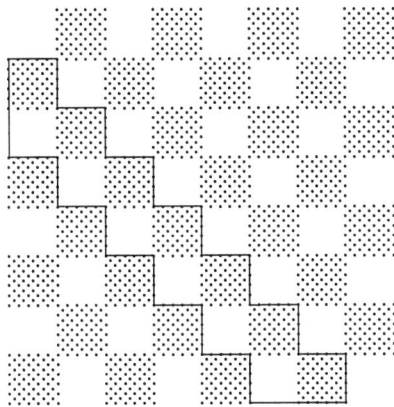

Figure 4.50

If Anna's announcement is "Inside", then her cell is black. Otherwise, it is white.

Problem 1.
Show how Boris should modify his strategy if Anna's announcement been "Outside" in

(a) Figure 4.47;

(b) Figure 4.48;

(c) Figure 4.49.

Rather pleased with what he has accomplished so far, Boris tries to simplify his approach and reduce the number of subboards necessary down to three.

To his chagrin, he is unable to do so. His method only works if two cells in opposite corners are deleted from the chessboard.

Question 2.
Show how Boris can accomplish the task in three questions on such a punctured chessboard.

Solution:
The first subboard is as shown in Figure 4.51.

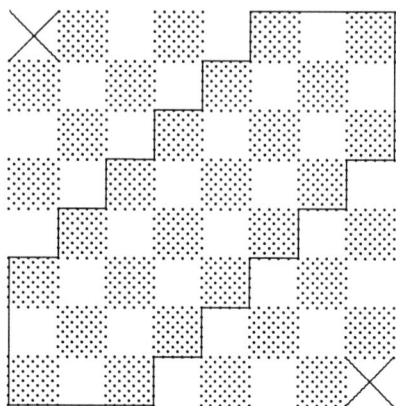

Figure 4.51

If Anna's first announcement is "Outside", then the second subboard is as shown in Figure 4.52.

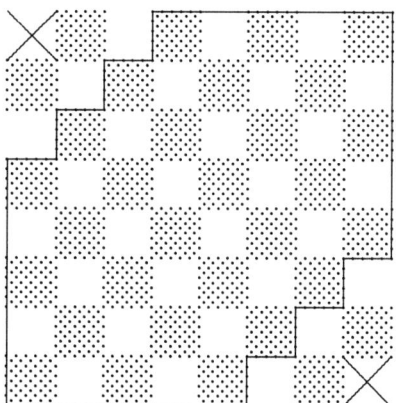

Figure 4.52

If Anna's second announcement is still "Outside", the third subboard is as shown in Figure 4.53. "Outside" means black and "Inside" means white.

A Problem on Information Extraction

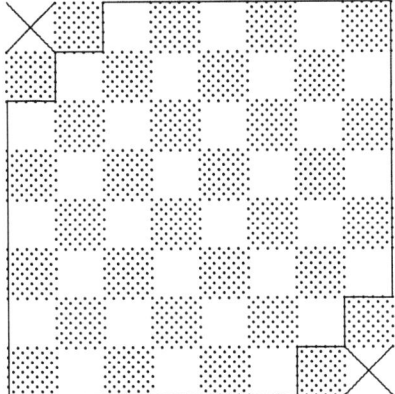

Figure 4.53

If Anna's second announcement is "Inside", the third subboard is as shown in Figure 4.54. "Outside" means black and "Inside" means white.

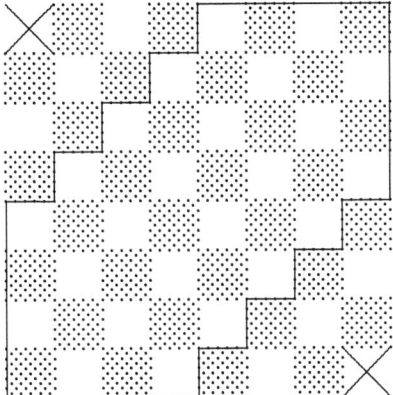

Figure 4.54

Suppose Anna's first announcement is "Inside". Then the second and the third subboards are as shown in Figures 4.55 and 4.56. If Anna's second and third announcements are the same, the cell is black. If they are different, the cell is white.

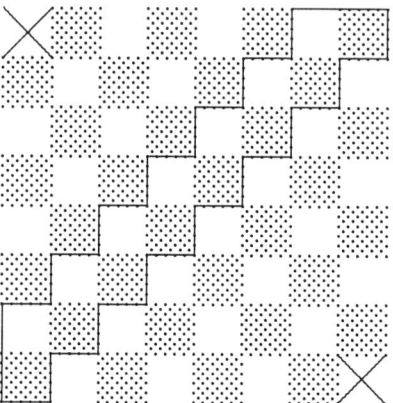

Figure 4.55

Figure 4.56

This is not entirely satisfactory. So Boris abandons the diagonals and turns his attention to the rows and columns.

Question 3.
Use this new approach to accomplish the task in two questions.

Solution:
The first subboard is shown in Figure 4.57.

Figure 4.57

By symmetry, we may assume that Anna's announcement is "Inside". Then the second subboard is shown in Figure 4.58. If Anna's announcement is "Inside", then her cell is black. Otherwise, it is white.

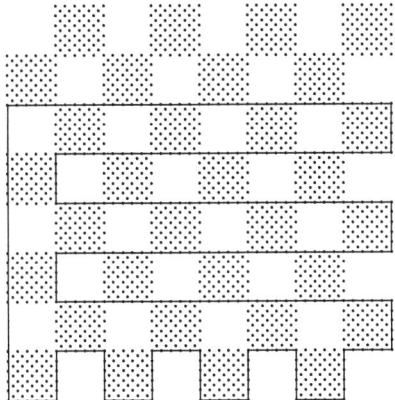

Figure 4.58

Clearly, one question will not be sufficient, since the subboard must separate all the white cells from all the black cells. So the best possible result has been achieved. However, Boris is still not satisfied. In all of his approaches so far, he must wait for Anna's announcement before he can choose his next subboard. This is known as an *adaptive solution*.

What Boris would like is a non-adaptive solution, in which he would present all subboards to Anna at the same time, and make the deduction upon receiving all the announcements simultaneously.

Anna starts with the 2×2 chessboard, and there is indeed a non-adaptive solution using only two questions. The subborads are shown in Figure 4.59. If the two announcements are the same, the cell must be black. Otherwise, it is white.

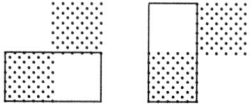

Figure 4.59

Anna now moves onto the 4 × 4 chessboard. There are eight white cells. She constructs the graph in Figure 4.60, where each vertex represents a white cell, and two vertices are joined by an edge if a Bishop can move directly between the two white cells they represent. The longest Bishop path without including four vertices forming a square is indicated by a doubled line with three segments. The five white cells represented by the vertices on this path will be inside both subboards. The reason why a square is forbidden is because the four white cells represented by the vertices forming a square will enclose a black square, which must then appear inside both subboards also.

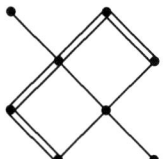

Figure 4.60

These five white cells may be connected by two disjoint sets of black squares, as shown in Figure 4.61.

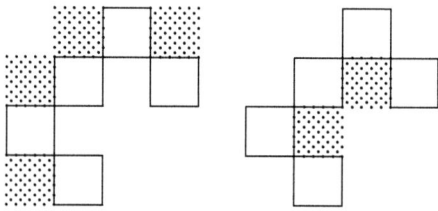

Figure 4.61

A Problem on Information Extraction

These are then expanded into the two subboards in Figure 4.62, each including one of two black squares near the bottom right corner. Now, if the two announcements are the same, then the chosen cell must be white. If the announcements are different, the cell must be black.

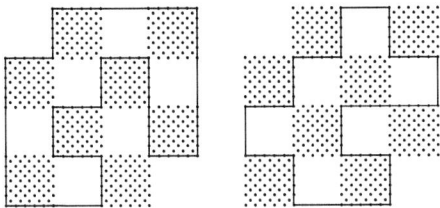

Figure 4.62

Problem 2.
Construct a graph analogous to the one in Figure 4.60 for the white cells of an 8×8 chessboard.

Question 4.
Find a non-adaptive solution to accomplish the task in two questions.

Solution:
The two subboards are shown in Figures 4.63 and 4.64.

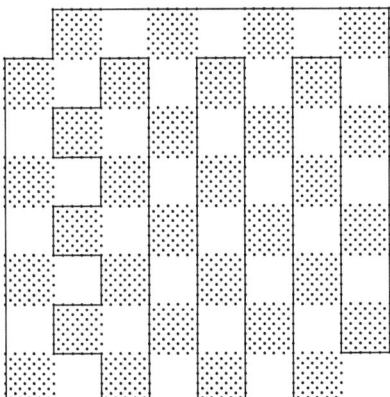

Figure 4.63

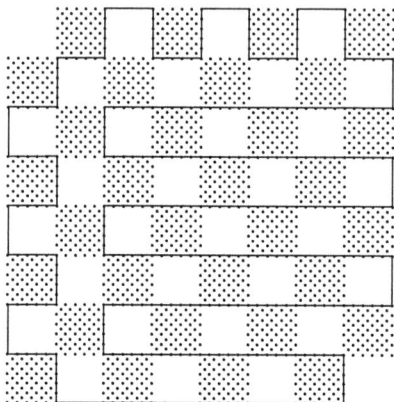

Figure 4.64

Problem 3.
Find a different pair of subboards which also yield a non-adaptive solution.

This scheme may be generalized to work for any $2n \times 2n$ chessboard.

Problem 4.
Find a pair of subboards which yield a non-adaptive solution for a 10×10 chessboard.

PART THREE
MATHEMATICAL CONGREGATIONS

By Mathematics Congregations, we refer to two Mathematics Summer Camps in Edmonton.

The first camp is the Alberta Regional Mathematics Summer Camps, founded by the Canadian Mathematical Society in 1999. There is a National Camp which grooms future members for the Canadian national team which competes in the International Mathematics Olympiad. Regional camps are set up across the nation to serve as feeders into the National Camp. The major sponsor of the Alberta Regional Camps is Alberta Innovates Technology Future, a provincial government think-tank. Thus the camp is known as the CMS/AITF camp. It has been held in Edmonton in late August every second year since 2000. The camp goes to Calgary in odd-numbered years.

Initially, the camp consisted of 24 students and lasted 7 days. It has since grown to 30 students and 13 days. It is by far the largest of the Regional Camps. Eligibility is from Grade 6 to Grade 10. Participation is by invitation only. We look mainly at the results of the Edmonton Junior High Competition and Invitational. Strong recommendations from teachers and good results from other contests are also considered. The majority of the campers are from Edmonton. Others are from elsewhere in Alberta, mainly from Calgary. From time to time, we take in students from neighboring British Columbia and Saskatchewan. To raise the general level of the camp, we also invite some top students from the Chiu Chang Mathematical Circle in Taiwan.

Since 2004, the camp has been held at St. Joseph's College within the University of Alberta campus. The heavy responsibility of looking after the campers overnight falls on the shoulders of the Residence Manager, and we are fortunate to have a sequence of competent and caring people in that capacity: **Gilbert Lee** in 2000, 2002 and 2006, **Tom Holloway** in 2004, **Alan Tsay** in 2008, 2010 and 2012, and **Ryan Morrill** in 2014. Classes have been held at the Decima Robinson Support center of the Department of Mathematics, under the supervision of Director **Sean Graves**.

The three components of the camp are the **Morning Mathematical Maneuvers**, the **Afternoon Academic Activities** and the **Evening Entertaining Endeavors**. The mornings are mostly taken up with three mini-courses. The afternoons are devoted to problem solving sessions, interspersed with an excursion, a team contest and a special event called the Site Visits. The first and the last evenings consist respectively of the Opening Ceremony and the Closing Ceremony. The evening of the Excursion Day is extended to a visit to West Edmonton Mall. In between, board game sessions alternate with nature walks.

The following is a typical schedule.

First Sunday.
14:00 – 17:00	Check in.
17:00 – 18:00	Dinner.
19:00 – 22:00	Opening Ceremony.

First Monday.
07:30 – 08:30	Breakfast.
09:00 – 12:00	Minicourse I.
12:00 – 13:00	Lunch.
13:00 – 16:00	Problem Solving Session.
17:00 – 18:00	Dinner.
19:00 – 22:00	Board Game Session.

First Tuesday.
07:30 – 08:30	Breakfast.
09:00 – 12:00	Minicourse I.
12:00 – 13:00	Lunch.
13:00 – 16:00	Problem Solving Session.
17:00 – 18:00	Dinner.
19:00 – 22:00	Nature Walk.

First Wednesday.
07:30 – 08:30	Breakfast.
09:00 – 12:00	Minicourse I.
12:00 – 13:00	Lunch.
13:00 – 16:00	Problem Solving Session.
17:00 – 18:00	Dinner.
19:00 – 22:00	Board Game Session.

First Thursday.
07:30 – 08:30	Breakfast.
09:00 – 12:00	Minicourse I.
12:00 – 13:00	Lunch.
13:00 – 16:00	From Earth to Moon Contest.
17:00 – 18:00	Dinner.
19:00 – 22:00	Nature Walk.

First Friday.
07:30 – 08:30	Breakfast.
09:00 – 12:00	Minicourse II.
12:00 – 13:00	Lunch.
13:00 – 16:00	Problem Solving Session.
17:00 – 18:00	Dinner.
19:00 – 22:00	Board Game Session.

First Saturday.
07:30 – 08:30	Breakfast.
09:00 – 12:00	Minicourse II.
12:00 – 13:00	Lunch.
13:00 – 16:00	Problem Solving Session.
17:00 – 18:00	Dinner.
19:00 – 22:00	Nature Walk.

Second Sunday.
07:30 – 08:30	Breakfast.
09:00 – 12:00	Minicourse II.
12:00 – 13:00	Lunch.
13:00 – 16:00	Problem Solving Session.
17:00 – 18:00	Dinner.
19:00 – 22:00	Board Game Session.

Second Monday.
07:30 – 08:30	Breakfast.
09:00 – 12:00	Minicourse II.
12:00 – 13:00	Lunch.
13:00 – 16:00	Excursion to Whitemud Drive Amusement Park.
16:00 – 22:00	Excursion to West Edmonton Mall.

Second Tuesday August 24.
07:30 – 08:30	Breakfast.
09:00 – 12:00	Minicourse III.
12:00 – 13:00	Lunch.
13:00 – 16:00	Problem Solving Session.
17:00 – 18:00	Dinner.
19:00 – 22:00	Nature Walk.

Second Wednesday August 25.
07:30 – 08:30	Breakfast.
09:00 – 12:00	Minicourse III.
12:00 – 13:00	Lunch.
13:00 – 16:00	Problem Solving Session.
17:00 – 18:00	Dinner.
19:00 – 22:00	Board Game Session.

Second Thursday August 26.
07:30 – 08:30	Breakfast.
09:00 – 12:00	Minicourse III.
12:00 – 13:00	Lunch.
13:00 – 16:00	Problem Solving Session.
17:00 – 18:00	Dinner.
19:00 – 22:00	Nature Walk.

Second Friday August 27.
07:30 – 08:30   Breakfast.
09:00 – 12:00   Minicourse III.
12:00 – 13:00   Lunch.
13:00 – 16:00   Science and Engineering Site Visits.
17:00 – 22:00   Closing Ceremony.

Second Saturday
07:30 – 08:30   Breakfast.
09:00 – 12:00   Check out.
12:00 – 13:00   Lunch.
13:00 – 14:00   Chuck out!

The Opening Ceremony is primarily an opportunity for the campers to get to know one another. Each is provided with a name list, but the names are anagrams of actual names of themselves. The only clue is that a boy's name becomes a girl's name, and vice versa, in the anagram. The first camper to identify everyone wins a prize.

The minicourses are just like those in Part I. The main difference lies in pace. On the one hand, we have twelve hours instead of eight or nine. On the other hand, these hours are packed into four mornings instead of spreading out over eight or nine weeks. This makes it a slightly difference learning experience.

It has been our tradition that Professor **Charles Leytem**, the leader of the Luxembourg team in the International Mathematics Olympiad, gives the first minicourse, in Euclidean geometry. One of the other two minicourses is on computing, given by Professor *Piotr Rudnicki* of the University of Alberta until his untimely passing. The problem solving sessions reinforce the minicourses, and slack time is taken up by the presentation of investigation topics.

Personal computers and other electronic devices are forbidden in the camp. There is a comfortable common room on the top floor of St. Joseph's College, and campers are encouraged to congregate and engage in multi-player games. The main game played in *Diplomacy*. The campers are divided into six teams, representing the major powers before World War I, and one move is played each day. Other popular games are *Carcassonne, Seven Wonders, Ticket to Ride, Acquire, Rail Baron* and *Wizard's Quest*. In nature walks, our favorite destinations are the Legislature Building and Hawrelak Park.

From Earth to Moon, named after the title of a famous science fiction by the great French author Jules Verne, is a team contest. All members start on earth, and each team is given the first of 25 Earthly problem. A team member is selected to hand in the answer.

If it is correct, the member is sent to the Moon and begins working on the first of 25 Lunar problems. If not, the member remains on Earth. In either case, the team is issued the next Earthly problem if there are still any, and there is at least one team member still on earth.

When the answer to a Lunar problem is handed in, the member is sent back to Earth if it is incorrect. Correct answers are worth points. In either case, the next Lunar problem is issued if there are still any, and there is at least one team member on the Moon.

The first correct answer to a Lunar problem is worth 3 points, and each subsequent correct answer is worth 1 point more than the preceding one. However, if the streak is broken by an incorrect answer, the next correct answer starts over again at 3 points. Earthly problems are not worth any points. They only serve to send members from Earth to the Moon.

The contest ends after three hours. It may end sooner for a team, either because it has run out of Earthly problems and all members are still on Earth, or it has run out of Lunar problems. Two sample papers are given in Chapter 5.

Both the Faculty of Science and the Faculty of Engineering are sponsors to our camp. During the last afternoon, campers are divided into two groups. For one and a half hour, one group attends a special session organized by each faculty. The sessions are then repeated with the two groups trading places. This allows the faculties to showcase their personnel, facilities and programs and gives the campers a more in-depth look.

During the Closing Ceremony, a party game is played, usually *Haggle*. In this game, each of the thirty camper is given an envelope containing some colored cards and some rules. Each card is worth a number of points, and there are special combinations which range from yielding big bonus to disqualifying the camper. These are spelt out in fifteen rules, with two copies of each inside the envelopes. Thus each camper has incomplete information about what the cards are worth. They would trade cards and rules among themselves. When time is up, the score for each camper is determined and a special prize is awarded. Prizes are also awarded for best performances in the minicourses and problem solving sessions. The prize for the team contest is also presented at this time.

The second camp was the Chiu Chang Mathematics Summer Camp in Edmonton, first organized in 1999. It was three weeks in duration. The thirty participants were members of the Chiu Chang Mathematical Circle of Taipei.

There was an English component with classes in the morning, under the direction of **Mimi Hui** of the Faculty of Extension of the University of Alberta. I was responsible for the mathematics component, with classes in the afternoon.

The mathematics classes were also conducted in English. Since the students' language proficiency was relatively low, we made rather slow progress. There was a dire lack of resource material in English at this level. I found a Chinese contest, the Peking University Mathematics Invitational for Youths, organized by Professor **Zonghu Qiu** of **Academia Sinica**. It ran for only six years. The contest papers, which I translated and edited, are given in Chapter 6.

Our class was honored with a visit of **Lois Hole**, who was then the Chancellor of the University of Alberta. Somehow she found out about our camp, and came to welcome the children in person, giving each of them her famed bear hug!

Another highlight of the Camp was a trip to the Jasper and Banff National Parks in the famous Canadian Rocky Mountains. Local programs include visits to West Edmonton Mall, the Telus World of Science, the Muttart Conservatory, the Provincial Museum, Fort Edmonton Park and Sir Winston Churchill Square.

Local Artiste **Jon Charles** also gave a special performance of magic for the Taiwanese kids. He had a rather difficult time because the students' English was not good enough to allow them to be distracted by his conversation. They all stared straight at him and watched every move he made. Nevertheless, they were mesmerized and totally enjoyed the experience.

The camp was repeated in 2000, but enrollment was down to nineteen students. The third time was not until 2004, when the number of participants went back up to thirty again. By then, **Lois Hole** was serving as the Lieutenant Governor of Alberta. Despite already worn down by terminal cancer, she gave an official reception for the children in her office in the Legislature Building.

The Chiu Chang camp has not been offered since 2004 because Chiu Chang Circle members began to come over and attend the CMS/AITF camp. These are top students from all over Taiwan, and their participation raised the level of our camp significantly.

Sven Chou, George Hung, Jason Liao, Neo Lin and Hsin-Po Wang came in 2006. They had played host a year earlier to the Edmonton team which participated in the International Mathematics Competition held in Kaohsiung. David Lin came in 2010, and to Calgary in 2011. Scott Wang and Hans Yu came in 2012. Scott had taken part in the International Mathematics Competition in Indonesia a year earlier, under the Edmonton banner. In 2014, Daniel Chiu and Kelvin Shih were the guest members from Taiwan.

Hsin-Po Wang, David Lin and Hans Yu have all since won Gold Medals for Taiwan in the International Mathematical Olympiad. David Lin also won a Gold Medal in the International Physics Olympiad.

Appendix A: Canadian Geography

To supplement the English Language component of the Chiu Chang Summer Camp and to introduce Canadian geography to the Taiwanese campers, we offered the following word game. *What do the following groups of sentences have to do with Canada?*

1. Can Adam come to the charity dinner?
   Is Scott away from the till?
   What is our total on donations?

2. In the musical, Bert appeared as one of the cats.
   In the same musical, Gary served as the light technician.
   The new band called itself "The Rollicking Stones".

3. We had to cancel the barbeque because of the rain.
   You can expect whirling winds or blinding snowstorms.

4. For better food and shows, ask at "Chew and Views".
   The house specialties are gin and tonic.

5. The demonstrators confront a riot squad.
   Call the marshal if axes are used.

6. No, Vasco, Tia Carmen may not be called a hag!
   She was an inventor on top of being a novelist.

7. Victor, I am afraid I have to bench you.
   Carl is the lead-off man, I to bat second, followed by John, with Alan in the clean-up spot.

Appendix B: Mathematical Jeopardy

As a parallel to Canadian Geography, the word game Mathematical Jeopardy is played in the CMS/AITF camp. There are questions asking for the definitions of mathematical terms. The surprising answers are usually of a non-mathematical nature. From these unorthodox answers, figure out what the questions are. As a first hint, the number of letters in each term is given by the number of asterisks. As a second hint, the terms are in alphabetical order.

Questions	Answers
What is ******** ********?	Food coloring.
What is an ********* ***********?	An improvement on numeracy.
What are *********** ****?	Cases not handled by partners.
What is a ************ ***?	A refund policy.
What is a ******* **********?	A Federal budget.
What is a *********** *****?	A car pool.
What is a ****** *****?	A biker gang.
What is a ************* ********?	An interrupted party.
What is a ********** ********?	A fund raising event.
What is an ******** *****?	A circumstantial evidence.
What is an ******** ******?	A country which has repelled all invasions.
What is an ********** ******?	A measure of insanity.
What is an *********** *******?	The last stronghold.
What is ****** ************?	Separate phone lines.
What are ****-***** ****-****?	Loose blouses and short skirts.
What is a **** *****?	A measure of viciousness.
What is a *************** *******?	A failed pregnancy.
What is a ***-*********** ****?	An odd ball.
What is a ****** *****?	A flight plan.
What are ***** *******?	Health and wealth.
What are ********* ********?	Semi-finalists.
What are ******* *******?	Communist headquarters.
What is a ********** ********?	Hooking up with an old flame.
What is a ******* ***********?	A confirmed bachelor.
What is a ***** *****?	A prince charming.
What is a ****-********* *****?	A brain-scan for dummies.

Appendix C: Answers

Answers to Canadian Geography

1. C̲an Adam come to the charity dinner? Is Sco̲tt away from the till? What is our tota̲l on donations?

2. In the musica̲l, Bert appeared as one of the cats. In the same musica̲l, Gary served as the light technician. The new band called itself "The Rollick̲ing Stones".

3. We had to cancel the barbeq̲ue because of the rain. You can expect whirling w̲inds or blinding snowstorms.

4. For better food and shows̲, ask at "Chew and Views". The house specialties ar̲e gin and tonic.

5. The demonstrators confro̲nt a riot squad. Call the marsh̲al if axes are used.

6. N̲o, Vasco, Tia Carmen may not be called a hag! She was an invento̲r on top of being a novelist.

7. V̲ictor, I am afraid I have to bench you. Carl is the lead-off m̲an, I to bat second, followed by John, with Alan in the clean-up spot.

Answers to Mathematical Jeopardy

additive identity, arithmetic progression, associative laws, cancellation law, central projection, commutative group, cyclic group, discontinuous function, generating function, indirect proof, integral domain, irrational number, irreducible element, linear independence, maxi-flows mini-cuts, mean value, multiplicative inverse, non-homogeneous case, planar graph, prime factors, quadratic residues, radical centers, recurrence relation, regular singularity, right coset, zero-knowledge proof.

Chapter Five
"From Earth to Moon"
Sample Contest I

Earthly Problems

1. Nicolas and his son and Peter and his son were fishing. Nicolas and his son caught the same number of fish while Peter caught three times as many fish as his son. All of them together caught 25 fish. Determine the number of fish caught by Peter's son.

2. Five circles are drawn on a plane, with exactly five points lying on at least two circles. Determine the least possible number of parts into which the plane is divided by these circles.

3. To color the six faces of a $3 \times 3 \times 3$ cube, 6 grams of paint is used. When the paint dried, the cube was cut into 27 small cubes. Determine the amount of paint, in grams, needed to color all the uncolored faces of these cubes.

4. A Knight is placed on a square of an infinite chessboard. This Knight can move either 2 squares North and 1 square East, or 2 squares East and 1 square North. After several moves it lands on a square located 2009 squares North and 2008 squares East from its initial position. Determine the total number of moves.

5. Each of A, B and C writes down 100 words in a list. If a word appears on more than one list, it is erased from all the lists. In the end, A's list has 61 words left and B's list has 80 words left. Determine the least number of words that can be left on C's list.

6. Determine the number of different rectangles which may be added to a 7×11 rectangle and a 4×8 rectangle so that the three can be put together, without overlapping, to form one rectangle.

7. Joshua and his younger brother George attend the same school. It takes 12 minutes for Joshua and 16 minutes for George to reach the school from their home. If George leaves home 1 minute earlier than Joshua, determine the number of minutes Joshua will take to catch up to him.

8. In a box are 6 red balls, 3 white balls, 2 green balls and 1 black ball. Determine the least number of balls one needs to draw at random from the box in order to be sure that balls of at least three different colors are drawn.

9. Every inhabitant on an island is either a Knight who always tells the truth, or a Knave who always lies. Each of 7 inhabitants seated at a round table claims that exactly one of his two neighbors is a Knave. Determine the number of Knights among these 7 inhabitants.

10. To a number 345 add two digits to its right so that 5-digit number is divisible by 36. Determine the average value of all such 5-digit numbers.

11. A rectangle which is not a square has integer side lengths. Its perimeter is n centimeters and its area is n square centimeters. Determine n.

12. In a soccer tournament where each pair of teams played exactly once, a win was worth 3 points, a draw 1 point and a loss 0 points). If one-fifth of the teams finished the tournament with 0 points, determine the number of teams that participated in the tournament.

13. In a target, the center is worth 10 points, the inner rim 8 points and the outer rim 5 points. A trainee hits the center and the inner rim the same number of times, and misses the target altogether one quarter of the time. His total score is 99 points. Determine the total number of shots he has fired.

14. Each of 100 fishermen caught at least 1 fish, but nobody caught more than 7 fish. There were 98 fishermen who caught no more than 6 fish each, 95 fishermen who caught no more than 5 fish each, 87 fishermen who caught no more than 4 fish each, 80 fishermen who caught no more than 3 fish each, 65 fishermen who caught no more than 2 fish each, and 30 fishermen who caught no more than 1 fish each. Determine the total number of fish caught by all fishermen.

15. There is a pile of candies on a table. The first boy takes $\frac{1}{10}$ of the candies. The second boy takes $\frac{1}{10}$ of leftovers plus $\frac{1}{10}$ of what the first has. The third boy takes $\frac{1}{10}$ of leftovers plus $\frac{1}{10}$ of what the first two boys have together. This continues until nothing is left. Determine the number of boys who have taken candies from this pile.

16. Each of A, B and C had candies. A gave some of her candies to B and C so that each had three times as much candy as before. Then B gave some of his candies to C and A so that each of them had three times as much candy as before. Finally, C repeated the same procedure. In the end each child had 27 candies. Determine the number of candies A had in the beginning.

17. Determine the greatest common divisor of all 9-digit numbers in which each of the digits 1, 2, ..., 9 appears exactly once.

18. A farmer was selling milk, sour cream, and cottage cheese. He had one bucket of cottage cheese, and the total number of buckets with milk and sour cream is five. The volumes of the buckets are 15, 16, 18, 19, 20 and 31 liters respectively. The farmer has twice as much milk as he has sour cream. Determine the volume, in liters, of the bucket containing cottage cheese.

19. Determine the smallest positive integer which leaves a remainder of 22 when divided by the sum of its digits.

20. The sum of the first n positive integers is a 3-digit number with all its digits equal. Determine n.

21. In a sequence $\{a_n\}$, $a_1 = 19$, $a_2 = 99$ and for $n \geq 2$, $a_{n+1} = a_n - a_{n-1}$. Determine a_{2009}.

22. The distances between a point inside a rectangle and three of rectangle's vertices are 3, 4 and 5. Determine the distance between this point and the fourth vertex, given that it is the greatest of the four distances.

23. Determine the number of integer solutions of equation $x^2 y^3 = 6^{12}$.

24. There are 30 pikes which can eat one another. A pike is sated if it has eaten 3 other pikes (sated or not). Determine the maximal number of pikes that could be sated, including those which have been eaten.

25. The positive integers x, y and z have been increased by 1, 2, and 3 respectively. Determine the maximal possible value by which the sum of their reciprocals can change.

Lunar Problems

1. Determine $x + y$ if $x^3 + y^3 = 9$ and $yx^2 + xy^2 = 6$.

2. Let x and y be positive integers such than $\gcd\{x, y\} = 999$. Suppose $\text{lcm}\{x, y\} = n!$. Determine the least value of n.

3. Determine the average of all 3-digit numbers which read the same forward and backward.

4. Peter has 8 white $1 \times 1 \times 1$ cubes. He wants to construct a $2 \times 2 \times 2$ cube with all its faces being completely white. Determine the minimal number of faces of small cubes that Basil must paint black in order to prevent Peter from fulfilling his task.

5. $ABCD$ is a quadrilateral with AB parallel to DC. If $AD=11$, $DC=7$ and $\angle D = 2\angle B$, determine AB.

6. We write the integers from 1 to 1000 in order along a circle. Starting from 1, we mark every 15-th number: 1,16,31, and so on. The next marked number after 991 is 6 and we go around the circle until no more numbers are marked. Determine number of unmarked numbers.

7. On an $n \times n$ board, there are 21 dominoes. Each domino covers exactly two squares and no two dominoes touch one another, even at a corner. Determine the minimal value of n.

8. The four sides and one diagonal of a quadrilateral have lengths 1, 2, 2.8, 5 and 7.5, not necessarily in that order. Determine which number was the length of the diagonal.

9. Simplify $8(3^2 + 1)(3^4 + 1) \cdots (3^{2^{10}} + 1) + 1$.

10. Each monkey collected the same number of nuts and threw one of them at every other monkey. Nuts thrown were lost. Each monkey still had at least two nuts left, and among them, they have 33 nuts left. Determine the number of nuts collected by each monkey.

11. From an 8×8 chessboard, the central 2×2 block rises up to form a barrier. Queens cannot be placed on the barrier, and may not attack one another across this barrier. Determine the maximal number of Queens which can be placed on the chessboard so that no two attack each other.

12. P is an interior point of triangle ABC. Perpendiculars are dropped from P to F on the side AB and E on the side AC. The triangles PAF and PCE are congruent, but the vertices are not necessarily in corresponding orders. Determine the ratio $\frac{AE}{AC}$.

13. In a kindergarten, 17 children made an even number of postcards. Any group of 5 children made no more than 25 postcards while any group of 3 children made no less than 14 postcards. Determine the total number of postcards made.

14. Determine the number of integers of the form $\overline{abc}+\overline{cba}$, where $\overline{abc}$ and $\overline{cba}$ are three-digit numbers with $ac \neq 0$.

15. M is the midpoint of side AC of triangle ABC. If $\angle ABM = 2\angle BAM$ and $BC = 2BM$, determine the measure, in degrees, of the largest angle of ABC.

16. A judge knows that among 9 coins that are identical in appearance, there are exactly 3 coins each weighing 3 grams, 3 coins each weighing 2 grams and 3 coins each weighing 1 gram. An expert wants to prove to the judge how much each of the coins weighs. He has a balance that shows which of two groups of coins is heavier, or that they have the same total weight. Determine the least number of weighings needed to persuade the judge.

17. On an island, each inhabitant is either a Knight who always tells the truth, or a Knave who always lies. Each inhabitant earns a different amount of money and works a different number of hours. Each inhabitant says, "Less than 10 inhabitants work more hours than I do." Each also said, "At least 100 inhabitants make more money than I do." Determine the number of inhabitants on this island.

18. Find the second smallest positive multiple of 11 with digit-sum 600.

19. In a chess tournament with an odd number of players, each player plays every other exactly once. A win is worth 1 point and a draw is worth $\frac{1}{2}$ point. The player in the fourth position from the bottom gets more points than the player in the third position from the bottom. Each player except the bottom three gets half of his or her points playing against the bottom three. Determine the number of players in the tournament.

20. $ABCD$ is a parallelogram. M and N are points on the sides AB and AD respectively, such that $AB = 4AM$ and $AD = 3AN$. Let K be the point of intersection of MN and AC. Determine the ratio $\frac{AC}{AK}$.

21. Determine the largest real number t such that the two polynomials $x^4 + tx^2 + 1$ and $x^3 + tx + 1$ have a common root.

22. Determine the smallest positive multiple of 777 such that the sum of its digits is the least.

23. Determine the minimum number of points marked on the surface of a cube so that no two faces of the cube contain the same number of marked points. A point at a corner or on a side of a face is considered to be in that face.

24. In a right triangle, the smallest height is one quarter the length of the hypotenuse. Determine the measure, in degrees, of the smallest angle of this triangle.

25. The real numbers $x < y$ satisfy $(x^2 + xy + y^2)\sqrt{x^2 + y^2} = 185$ and $(x^2 - xy + y^2)\sqrt{x^2 + y^2} = 65$. Determine the smallest value of y.

Sample Contest II

Earthly Problems

1. The little tycoon Johnny says to his fellow capitalist Annie, "If I add $7 to $\frac{3}{5}$ of my funds, I'll have as much money as you do." Annie replies, "So you have only $3 more than I." How much money do they have between them?

2. According to a contract, a worker is to be paid $48 for each day he is at work and $12 is deducted from his pay for each day he is not at work. After 30 days, the worker's net pay is $0. How many days is he at work during these 30 days?

3. Using each of the numbers 1, 2, 3 and 4 twice, I succeeded in writing out an eight-digit number in which there is one digit between the ones, two digits between the twos, three digits between the threes, and four digits between the fours. What was the number if the first digit is larger than the last one?

4. John, Jim, and Gerry went to a baseball game. On the way, John bought five bags of potato chips, Jim bought two bags, and Gerry did not buy any. During the game they all ate the chips, each one eating as much as the others. After the game, Gerry figured out how much the bags of chips cost and handed over $1.40. How much money should John get?

5. A large rectangle is divided into 16 smaller rectangles in a 4×4 array. The areas of some of them are shown in Figure 5.1. Find the area of the small rectangle at the top right corner.

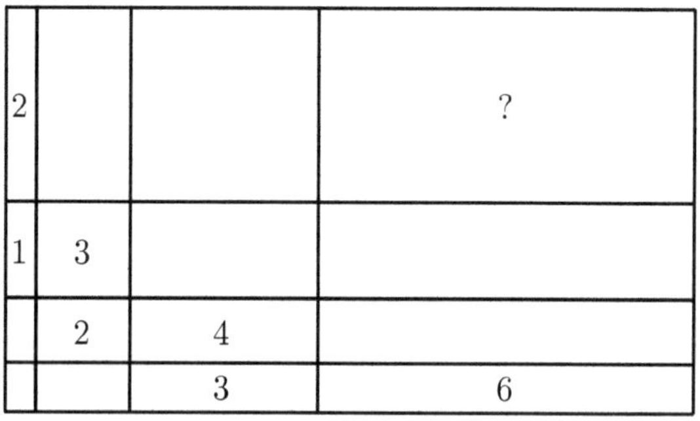

Figure 5.1

6. A checker starts on the lower left square of a 5×5 board. How many different paths are there for it to move from there to the upper right square? The checker may move in only two directions: upward and to the right, and may not visit the central square.

7. In 1992, it was observed that in the year x^2 my nephew will be x years of age. In what year was he born?

8. Find the smallest positive integer such that the sum of its digits and that of the subsequent integer are both divisible by 17.

9. A number of coins are placed on each square of a checkerboard such that the sums on every two squares having a common side differ by one cent. Given that the sum on one of the squares is 3 cents, and on another one 17 cents, find the total amount of money on both diagonals on the checkerboard.

10. Two numbers are called mirror numbers if one is obtained from the other by reversing the order of digits, for example, 123 and 321. Find the smaller of two mirror numbers whose product is 92565.

11. Thirty children ride a carousel swing. Every girl rides behind a boy, half of the boys ride behind a boy, and all the other boys ride behind girls. How many girls are there?

12. The length of a rectangle is twice its width. The circumference of a circle is one half the width of the rectangle. The circle rolls along the perimeter of the rectangle until it comes back to its initial position. How many revolutions will the circle make if it is rolling inside the rectangle?

13. Mademoiselle Dubois has more than two pets. All her pets but two are dogs, all but two are cats, and all but two are hamsters. How many pets does she own which are not dogs, cats or hamsters?

14. A raft and a motorboat left town A simultaneously and traveled downstream to town B. The raft moved at the same speed as the current, which was constant. The motorboat, which traveled at constant speed in still water, arrived at B, immediately turned back, and encountered the raft two hours after they had set out from A. How many hours did it take the motorboat to go from A to B?

15. Nine coins lie on a table. In each move, you turn over any five coins. What is the minimum number of moves needed to turn all coins upside down?

16. Mr. R. A. Scall, president of the Pyramid Bank, lived in a suburb rather far from his office. Every weekday, a car from the bank came to his house, always at the same time and traveling at the same constant speed, so that he would arrive at work precisely when the bank opened. One morning his driver called very early to tell him he would probably be late because of mechanical problems. So Mr. Scall left home one hour early and started walking to his office. The driver managed to fix the car quickly, however, and left the garage on time. He met the banker on the road and brought him to the bank. They arrived 20 minutes earlier than usual. For how many minutes did Mr. Scall walk?

17. There are twelve people in a room. Some of them always tell the truth while the others always lie. One of them said, "None of us is honest." Another said, "There is not more than one honest person here." A third said, "There are not more than two honest persons here." This continued on until the twelfth said, "There are not more than eleven honest persons here." How many honest persons are in the room?

18. Koshchei the Immortal buried his ill-gotten treasure in a hole 1 meter deep. That did not seem safe enough for him, so he dug his treasure up, deepened the hole to 2 meters, and buried his hoard again. He was still worried, so he dug up his hoard, made the hole 3 meters deep, and hid his treasure once more. But he just could not stop — he kept increasing the depth of the hole, to 4 meters, 5 meters, 6 meters, and so on, each time extracting his property and burying it again, until on the 1001-st day he died of exhaustion. Koshchei dug a hole n meters deep in n^2 days, but took no time filling one in. How many meters deep was his treasure buried when he dropped dead?

19. Prince Ivan is fighting the three-headed, three-tailed dragon with a magic sword. If he cuts off one head of the dragon, a new head grows. If he cuts off one tail, two new tails grow. If he cuts off two tails, one new head grows. If he cuts off two heads, nothing grows. These are the only four things Prince Ivan can do. What is the smallest number of strokes Prince Ivan needs to cut off all the dragon's heads and tails?

20. Solve $10 - 9(9 - 8(8 - 7(7 - 6(6 - 5(5 - 4(4 - 3(3 - 2(2 - x)))))))) = x$.

21. I went to the bank to cash a check. As the cashier gave me the money, I put it in my empty wallet without counting it. During the day I spent $6.23. When I checked my wallet in the evening, it contained exactly twice the amount of the check I had cashed. Strange! A little calculation revealed that while making the payment, the cashier had interchanged the figures for dollars and cents. What was the amount of the check?

22. Two cyclists, Kaitlin and Josh, simultaneously started toward each other from two towns 40 kilometers apart. Josh rode at 23 kilometers per hour, and Kaitlin rode at 17 kilometers per hour. Before departure, a fly landed on Josh's nose. At the moment of departure, it started to fly toward Kaitlin at 40 kilometers per hour. When it reached Kaitlin, it immediately flew in the opposite direction at 30 kilometers per hour (the wind blew toward Kaitlin). As soon as the fly reached Josh, it turned back again, and so on. How many kilometers had the fly flown when the cyclists met?

23. Find $\sqrt{12345678987654321}$.

24. Side AD of the quadrilateral $ABCD$ equals its diagonal BD. The three other sides of this quadrilateral are equal to each other. The diagonal BD bisects $\angle ADC$. What is the maximum measure of $\angle BAD$?

25. Find the fraction between $\frac{96}{35}$ and $\frac{97}{36}$ with the smallest denominator.

Lunar Problems

1. I am four times older than my sister was when she was half as young as I was. In 15 years our combined age will be 100. How old am I now?

2. At a certain sporting event, 100 students took part in running, 50 students participated in swimming and 48 in biking. It turned out that the number of students who took part in only one event was twice as high as that who took part in just two events and three times as high as the number of students who participated in all three events. What is the total number of students that participated in the events?

3. In triangle ABC, M is the midpoint of BC and N is the midpoint of AM. The extension of BN intersects AC at K. Compute $\frac{KC}{KA}$.

4. In the equation ONE×9 =NINE, different letters stand for different digits and identical letters stand for identical digits. What is the maximum value of NINE?

5. One day all of Mrs. Brown's grandchildren came to visit her. There was a bowl of apples and pears on the kitchen table. Mrs. Brown gave each child the same number of pieces of fruit without keeping track of which kind. Billy, the youngest grandchild, got $\frac{1}{8}$ of all the apples and $\frac{1}{10}$ of all the pears. How many grandchildren did Mrs. Brown have?

6. Thirty people took part in a shooting match. The first participant scored 80 points, the second scored 60 points, the third scored the average of the number of points scored by the first two, and each subsequent participant scored the average of the number of points scored by the previous ones. How many points did the last participant score?

7. How many numbers end in the four digits 1995 and become an integral number of times smaller when these digits are erased?

8. The working hours of a receptionist at a hotel are either 8 a.m. to 8 p.m., 8 p.m. to 8 a.m. ,or 8 a.m. to 8 a.m. the next day. In the first case, the break before the next shift must be not less than 24 hours; in the second case, not less than 36 hours; and in the third case, not less than 60 hours. What is the smallest number of receptionists that can provide round-the-clock operation of the hotel?

9. Baron Munchhausen says: "Today I shot more ducks than two days ago, but fewer than a week ago." For how many days in a row can the baron say this truthfully?

10. A, B, C and D are four points in that order on a line. Semicircles on AB and AC as diameters are drawn on one side of this line, and semicircles on BD and CD as diameters are drawn on the other. What is the area of the region enclosed by the four semicircles if $AD = 10$ and $BC = 2$?

11. A computer printed out digit by digit two numbers whose values are 2^{1995} and 5^{1995}. How many digits in all were printed?

12. You can see three faces (which meet at a corner of the die) of each of two dice. On each die, the three pairs of opposite faces contain 1 and 6 dots, 2 and 5 dots, and 3 and 4 dots, respectively. The total number of dots on the visible faces of the two dice is 27. The total number of dots on the visible faces of each die is calculated. What is the larger of these two numbers?

13. Several chess players played chess in a park the whole day long. Since they had only one set of pieces, they chose the following rules: The winner of a game skipped the next two games, and the loser skipped the next four. If the game ended in a draw, the player who played white pieces was considered to have lost. How many players took part if they managed to follow these rules?

14. A huge military band was playing and marching in formation on the parade grounds. First the musicians formed a square. Then they regrouped into a rectangle so that the number of rows increased by 5. How many musicians were there in the band?

15. The length of the chord of a circle is 2. Two smaller circles are tangent to the large circle and to each other at the midpoint of the chord. Find the area of the region inside the large circle but outside the small circles.

16. A hundred officials were invited to the annual meeting at their Ministry of Affairs. They were seated in a rectangular hall with ten rows of chairs, ten chairs in each row. The opening was delayed, and the officials could find nothing better to do than compare their salaries. To consider oneself "highly paid", an official had to determine that no more than one person seated to the left, right, front or rear or at a diagonal was paid as much or more. What is the greatest number of officials who could count themselves as "highly paid"?

17. In a chess tournament, each participant plays each of the others once. A win is worth one point, a draw is worth half a point, and a loss is worth no points. Two precocious students from an elementary school took part in a chess tournament at a nearby university. The combined score of the elementary school students was 6.5. The scores of the university students all happened to be the same. How many university students participated in the tournament?

18. The sequence $\{a_n\}$ is defined as follows: $a_1 = 1776$, $a_2 = 1999$ and for $n \geq 0$, $a_{n+2} = \frac{a_{n+1}+1}{a_n}$. Find a_{2002}.

19. Side AE of the pentagon $ABCDE$ equals its diagonal BD. All the other sides of this pentagon are equal to 1. What is the radius of the circle passing through points A, C and E?

20. Find $\sqrt{11111112222222 - 3333333}$.

21. You have two balls of each of several colors. A number of balls of different colors are placed on the left pan of a balance while the other balls of the same colors are put on the right one. The balance tips to the left. If you exchange any pair of balls of the same color, however, the balance either tips to the right or stay even. What are the possible number of pairs of balls on the balance?

22. Sixteen baseball teams, ranked 1 through 16, took part in a knock-out tournament. In the first round, eight teams were eliminated; in the second round, four teams were eliminated; in the third round, two teams were eliminated; and the two remaining teams met in the championship game. It turned out that the team with the higher ranking won every game. What is the minimum possible number of uninteresting games, that is, games in which the ranking of the teams differed by at least five?

23. A student wrote three positive integers on the blackboard that are consecutive terms of an arithmetic progression. Then he erased the commas between them, creating a seven-digit number. What is the maximum possible value of this number?

24. Calculate the following quantity to the fifth decimal place without using your calculator: $(\sqrt[3]{2}+1)\sqrt[3]{\frac{1}{3}(\sqrt[3]{2}-1)}$.

25. All sides of triangle ABC are of length 1. D is a point such that $AD = 7$ and both BD and CD are integral. What is the minimum length of BD?

Answers to Earthly Problems of Sample Contest I

1	5	2	7	3	12	4	1339	5	41
6	4	7	3	8	10	9	0	10	34560
11	18	12	5	13	20	14	245	15	10
16	55	17	9	18	20	19	689	20	36
21	-99	22	$\sqrt{32}$	23	18	24	9	25	$\frac{23}{12}$

Answers to Lunar Problems Sample Contest I

1	3	2	37	3	550	4	2	5	18
6	800	7	11	8	28	9	3^{2048}	10	13
11	10	12	$\frac{1}{2}$	13	84	14	170	15	2
16	120	17	110	18	34989 9…9 9…9	19	9	20	7
21	-2	22	10101	23	6	24	15	25	-3

Answers to Earthly Problems of Sample Contest II

1	47	2	6	3	$\frac{4131}{2432}$	4	1.60	5	24
6	34	7	1980	8	8899	9	160	10	165
11	10	12	$12 - \frac{4}{\pi}$	13	0	14	1	15	3
16	50	17	6	18	13	19	27	20	1
21	21.49	22	$\frac{263}{7}$	23	111111111	24	72	25	$\frac{19}{7}$

Answers to Lunar Problems of Sample Contest II

1	40	2	121	3	2	4	5850	5	9
6	70	7	16	8	3	9	6	10	5π
11	1996	12	15	13	8	14	400	15	$\frac{\pi}{2}$
16	50	17	11	18	1999	19	1	20	3333333
21	2	22	1	23	9995049	24	1	25	7

Solutions to Sample Contest I

Earthly Problems

1. Nicolas and his son and Peter and his son were fishing. Nicolas and his son caught the same number of fish while Peter caught three times as many fish as his son. All of them together caught 25 fish. Determine the number of fish caught by Peter's son.

   Solution:
   Nicolas and his son caught an even number of fish, as did Peter and his son. If there were four people, the total number of fish caught could not have been 25. It follows that Nicolas was Peter's son, and it easy to see he caught 5 fish.

2. Five circles are drawn on a plane, with exactly five points lying on at least two circles. Determine the least possible number of parts into which the plane is divided by these circles.

   Solution:
   We have at least 6 regions since the interior of each circle is one, and there is also the infinite region. Since there are five circles and five points lying on at least two circles, some of the circles must form a closed "ring". This forces an additional region. Figure 5.2 shows that 7 regions can be attained in several ways.

   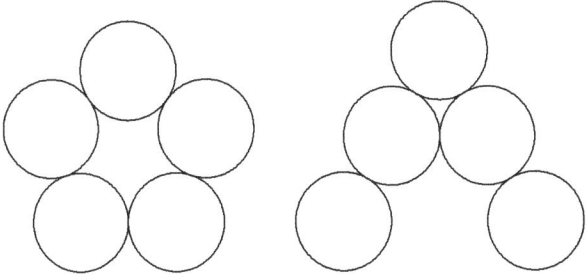

   Figure 5.2

3. To color the six faces of a $3 \times 3 \times 3$ cube, 6 grams of paint is used. When the paint dried, the cube was cut into 27 small cubes. Determine the amount of paint, in grams, needed to color all the uncolored faces of these cubes.

   Solution:
   The faces of the original cube consist of $6 \times 3 \times 3 = 54$ faces of small cubes. The total number of faces of the small cubes is $27 \times 6 = 162$. The number of colored faces is $162 - 54 = 1 - 8 = 2 \times 54$. Hence $2 \times 6 = 12$ grams of paint are needed.

4. A Knight is placed on a square of an infinite chessboard. This Knight can move either 2 squares North and 1 square East, or 2 squares East and 1 square North. After several moves it lands on a square located 2009 squares North and 2008 squares East from its initial position. Determine the total number of moves.

   Solution:
   In each move, the Knight covers three squares towards the North and the East. At the end, 2008+2009=4017 squares are covered. Hence the number of moves is $4017 \div 3 = 1339$.

5. Each of A, B and C writes down 100 words in a list. If a word appears on more than one list, it is erased from all the lists. In the end, A's list has 61 words left and B's list has 80 words left. Determine the least number of words that can be left on C's list.

   Solution:
   A has 39 words crossed off while B has 20. Even if these are all against C, C will still have $100 - 39 - 20 = 41$ words left on the list.

6. Determine the number of different rectangles which may be added to a 7×11 rectangle and a 4×8 rectangle so that the three can be put together, without overlapping, to form one rectangle.

   Solution:
   Since there are four different ways of abutting two rectangles, the third rectangle can be chosen in at most 4 ways. Figure 5.3 shows that they are indeed different, being 3×4, 3×8, 1×11 and 7×8 respectively.

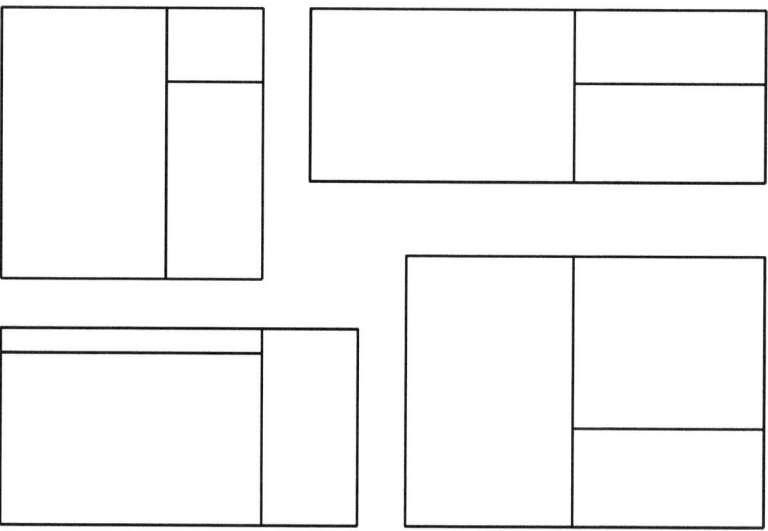

Figure 5.3

Solutions

7. Joshua and his younger brother George attend the same school. It takes 12 minutes for Joshua and 16 minutes for George to reach the school from their home. If George leaves home 1 minute earlier than Joshua, determine the number of minutes Joshua will take to catch up to him.

   Solution:
   The ratio of the brothers' speeds is 4:3. Joshua will catch up with George after 3 minutes because George will have been moving for 4 minutes.

8. In a box are 6 red balls, 3 white balls, 2 green balls and 1 black ball. Determine the least number of balls one needs to draw at random from the box in order to be sure that balls of at least three different colors are drawn.

   Solution:
   The largest number of balls drawn without getting balls of at least three different colors is 9, by drawing all 6 red balls and 3 white balls. Hence if we draw 10 balls, we are guaranteed to have balls of at least three different colors.

9. Every inhabitant on an island is either a Knight who always tells the truth, or a Knave who always lies. Each of 7 inhabitants seated at a round table claims that exactly one of his two neighbors is a Knave. Determine the number of Knights among these 7 inhabitants.

   Solution:
   Each Knight must be seated between a Knight and a Knave and each Knave must be seated either between two Knights or two Knaves. If there is at least one Knight, then the Knights and Knaves must occupy alternate seats, but 7 is an odd number. Hence all 7 are Knaves.

10. To a number 345 add two digits to its right so that 5-digit number is divisible by 36. Determine the average value of all such 5-digit numbers.

    Solution:
    To be divisible by 2, the last digit must be even. In order to be divisible by 9, the sum of the five digits must be a multiple of 9. Hence the sum of the last two digits must be 6 or 15. In the first case, these two digits form the number 06, 24, 42 or 60. This number must be divisible by 4, which eliminates 06 and 42. In the second case, these two digits must be 78 or 96, with 78 eliminated. The average of 34524, 34560 and 34596 is 34560.

11. A rectangle which is not a square has integer side lengths. Its perimeter is n centimeters and its area is n square centimeters. Determine n.

    Solution:
    Let the dimensions of the rectangle be $a \times b$ with $a < b$. The perimeter is $2(a+b)$ and the area is ab. Hence $ab = 2a + 2b$, so that
    $$(a-2)(b-2) = ab - 2a - 2b + 4 = 4.$$
    Hence $a - 2 = 1$ and $b - 2 = 4$, yielding $a = 3$, $b = 6$ and $n = 18$.

12. In a soccer tournament where each pair of teams played exactly once, a win was worth 3 points, a draw 1 point and a loss 0 points). If one-fifth of the teams finished the tournament with 0 points, determine the number of teams that participated in the tournament.

    Solution:
    In any tournament where each pair of teams played exactly once, at most one team can have no wins and no draws. Since one-fifth of the teams finished with 0 points, the number of teams is 5.

13. In a target, the center is worth 10 points, the inner rim 8 points and the outer rim 5 points. A trainee hits the center and the inner rim the same number of times, and misses the target altogether one quarter of the time. His total score is 99 points. Determine the total number of shots he has fired.

    Solution:
    The total score from hitting the center and the inner rim is 18, 36, 54, 72 or 90. The total score from hitting the outer rim will then be 81, 63, 45, 27 or 9 respectively. Of these, only 45 is a multiple of 5. Hence the outer rim is hit 9 times and each of the center and the inner rim is hit 3 times. Since 15 shots hit the target, 5 shots miss. Hence the total number of shots fired is 20.

14. Each of 100 fishermen caught at least 1 fish, but nobody caught more than 7 fish. There were 98 fishermen who caught no more than 6 fish each, 95 fishermen who caught no more than 5 fish each, 87 fishermen who caught no more than 4 fish each, 80 fishermen who caught no more than 3 fish each, 65 fishermen who caught no more than 2 fish each, and 30 fishermen who caught no more than 1 fish each. Determine the total number of fish caught by all fishermen.

    Solution:
    The numbers of fishermen who caught 7, 6, 5, 4, 3, 2 and 1 fish are respectively $100 - 98 = 2$, $98 - 95 = 3$, $95 - 87 = 8$, $87 - 80 = 7$, $80 - 65 = 15$, $65 - 30 = 35$ and $30 - 0 = 30$. Hence the total number of fish caught was $2 \times 7 + 3 \times 6 + 8 \times 5 + 7 \times 4 + 15 \times 3 + 35 \times 2 + 30 \times 1 = 245$.

15. There is a pile of candies on a table. The first boy takes $\frac{1}{10}$ of the candies. The second boy takes $\frac{1}{10}$ of leftovers plus $\frac{1}{10}$ of what the first has. The third boy takes $\frac{1}{10}$ of leftovers plus $\frac{1}{10}$ of what the first two boys have together. This continues until nothing is left. Determine the number of boys who have taken candies from this pile.

   Solution:
   Each boy, including the first, takes $\frac{1}{10}$ of what has already been taken plus $\frac{1}{10}$ of what is left behind. In other words, each boy takes $\frac{1}{10}$ of everything. The number of boys needed to take everything is 10.

16. Each of A, B and C had candies. A gave some of her candies to B and C so that each had three times as much candy as before. Then B gave some of his candies to C and A so that each of them had three times as much candy as before. Finally, C repeated the same procedure. In the end each child had 27 candies. Determine the number of candies A had in the beginning.

   Solution:
   Note that the total number of candies, namely 81, remained constant throughout. Suppose A started with n candies. Then B and C started with $81-n$ together. After A's gift, she had $81-3(81-n) = 3n-162$ candies left. After B's and C's gifts, she had $9(3n-162) = 27$. Hence $n = 55$.

17. Determine the greatest common divisor of all 9-digit numbers in which each of the digits 1, 2, ..., 9 appears exactly once.

   Solution:
   Each of these numbers is divisible by 9 since so is its digit-sum. Their greatest common divisor must divide their pairwise differences, one of which is $123456798 - 123456789 = 9$. Hence their greatest common divisor is 9.

18. A farmer was selling milk, sour cream, and cottage cheese. He had one bucket of cottage cheese, and the total number of buckets with milk and sour cream is five. The volumes of the buckets are 15, 16, 18, 19, 20 and 31 liters respectively. The farmer has twice as much milk as he has sour cream. Determine the volume, in liters, of the bucket containing cottage cheese.

   Solution:
   The total volume, in liters, of milk and sour cream is a multiple of 3. The volumes of the buckets, in liters, are congruent modulo 3 to 0, 1, 0, 1, 2 and 1 respectively. The total volume is congruent modulo 3 to 2. Hence the volume of cottage cheese must also be congruent modulo 3 to 2, so that it must be 20 liters. Indeed, the farmer had 15+18=33 liters of sour cream and 16+19+31=66 liters of milk.

19. Determine the smallest positive integer which leaves a remainder of 22 when divided by the sum of its digits.

   Solution:
   The minimal value of the digit-sum is 23, with digits (5,9,9), (6,8,9), (7,7,9) or (7,8,8). As it happens, the remainder is indeed 22 when 689 is divided by 23, but not when 599 is divided by 23. Hence 689 is the smallest such number.

20. The sum of the first n positive integers is a 3-digit number with all its digits equal. Determine n.

   Solution:
   A 3-digit number with all its digits equal is divisible by 37. The sum of the first n positive integer is $\frac{n(n+1)}{2}$. Since 37 is prime, it divides n or $n+1$. To minimize the sum, we take $n = 36$ so that $\frac{n(n+1)}{2} = 666$.

21. In a sequence $\{a_n\}$, $a_1 = 19$, $a_2 = 99$ and for $n \geq 2$, $a_{n+1} = a_n - a_{n-1}$. Determine a_{2009}.

   Solution:
   Iteration yields $a_3 = 80$, $a_4 = -19$, $a_5 = -99$, $a_6 = -80$, $a_7 = 19$ and $a_8 = 99$. Hence the sequence is of period 6. Since the remainder when 2009 is divided by 6 is 5, we have $a_{2009} = a_5 = -99$.

22. The distances between a point inside a rectangle and three of rectangle's vertices are 3, 4 and 5. Determine the distance between this point and the fourth vertex, given that it is the greatest of the four distances.

   Solution:
   The fourth vertex is opposite to the one at the smallest distance from the given point, namely 3. Let the distances from the given point to the four sides of the rectangle be w, x, y and z as shown in Figure 5.4. Then $w^2 + y^2 = 9$, $w^2 + z^2 = 16$ and $x^2 + y^2 = 25$. It follows that $x^2 + z^2 = 16 + 25 - 9 = 32$, and the distance from the given point to the fourth vertex is $\sqrt{32}$.

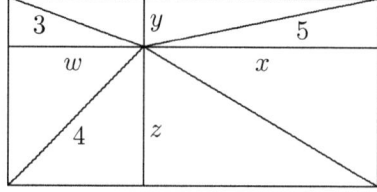

Figure 5.4

23. Determine the number of integer solutions of equation $x^2y^3 = 6^{12}$.

    Solution:
    Since 2 and 3 are relatively prime, each of 2^{12} and 3^{12} is the product of a square and a cube. In the former case, the two factors may be $(2^0)^2(2^4)^3$, $(2^3)^2(2^2)^3$ or $(2^6)^2, (2^0)^3$, with three possibilities for the latter case. Since x can be positive or negative, the total number of integer solutions is $2 \times 3^2 = 18$.

24. There are 30 pikes which can eat one another. A pike is sated if it has eaten 3 other pikes (sated or not). Determine the maximal number of pikes that could be sated, including those which have been eaten.

    Solution:
    Each pike is eaten by at most one other pike, so that the 29 pikes being eaten can make 9 pikes sated. We may have as many as 9 sated pikes by having the 30 pikes line up in a row. Pikes 1, 2, and 3 will be eaten by pike 4, pikes 4, 5 and 6 will be eaten by pike 7, and so on. The sated pikes are therefore pikes 4, 7, 10, 13, 16, 19, 22, 25 and 28, for a total number of 9.

25. The positive integers x, y and z have been increased by 1, 2, and 3 respectively. Determine the maximal possible value by which the sum of their reciprocals can change.

    Solution:
    We have $\frac{1}{x} - \frac{1}{x+1} = \frac{1}{x(x+1)}$, $\frac{1}{y} - \frac{1}{y+2} = \frac{2}{y(y+2)}$ and $\frac{1}{z} - \frac{1}{z+3} = \frac{3}{z(z+3)}$. These are decreasing functions of x, y and z respectively, so that their minima occur at $x = y = z = 1$. Thus the maximum possible value of change is $\frac{1}{2} + \frac{2}{3} + \frac{3}{4} = \frac{23}{12}$.

Lunar Problems

1. Determine $x + y$ if $x^3 + y^3 = 9$ and $yx^2 + xy^2 = 6$.

   Solution:
   We have $(x + y)^3 = x^3 + y^3 + 3(yx^2 + xy^2) = 27$. Hence $x + y = 3$.

2. Let x and y be positive integers such than $\gcd\{x, y\} = 999$. Suppose $\text{lcm}\{x, y\} = n!$, Determine the least value of n.

   Solution:
   Since $999 = 3^3 \times 37$, the least value of n is 37. This can be attained if $x = 999$ and $y = 37!$,

3. Determine the average of all 3-digit numbers which read the same forward and backward.

   Solution:
   Any digit can be the tens-digit so that they appear with equal likelihood. The average of the tens-digit is $(0+10+20+\cdots+90) \div 10 = 45$. Any non-zero digit can be the units-digit. It follows that the average is $(1+2+\cdots+9) \div 9 = 5$. Similarly, the average of the hundreds-digit is $(100 + 200 + \cdots + 900) \div 9 = 500$. The average of the numbers is $500 + 45 + 5 = 550$.

4. Peter has 8 white $1 \times 1 \times 1$ cubes. He wants to construct a $2 \times 2 \times 2$ cube with all its faces being completely white. Determine the minimal number of faces of small cubes that Basil must paint black in order to prevent Peter from fulfilling his task.

   Solution:
   If Basil paints only one face, Peter can hide it easily. However, if Basil paints two opposite faces of a small cube, it is not possible for Peter to hide both. Hence the minimal number of faces Basil must paint is 2.

5. $ABCD$ is a quadrilateral with AB parallel to DC. If $AD=11$, $DC=7$ and $\angle D = 2\angle B$, determine AB.

   Solution:
   Let the bisector of $\angle D$ cut AB at E. Then $BCDE$ is a parallelogram since it has a pair of parallel opposite edges and a pair of equal opposite angles. Hence $BE = CD = 7$. Moreover, BC is parallel to ED, so that $\angle AED = \angle B = \angle ADE$. Hence $AE = AD = 11$ so that $AB = AE + BE = 18$.

6. We write the integers from 1 to 1000 in order along a circle. Starting from 1, we mark every 15-th number: 1,16,31, and so on. The next marked number after 991 is 6 and we go around the circle until no more numbers are marked. Determine number of unmarked numbers.

Solution:
We extend 1000 to 3000 so that the last marked number is 2986. Then $3000 \div 15 = 200$ numbers have been marked. We now replace 1001 and 2001 by 1, 1001 and 2002 by 2, and so on, so that every number from 1 to 1000 appears 3 times. A number cannot be marked more than once since all marked numbers are congruent to 1 modulo 5. Hence the number of unmarked numbers is $1000 - 200 = 800$.

7. On an $n \times n$ board, there are 21 dominoes. Each domino covers exactly two squares and no two dominoes touch one another, even at a corner. Determine the minimal value of n.

Solution:
Expand each domino by half a unit on all four sides so that each becomes a 2×3 tile. Expand the board similarly so that it becomes $(n+1) \times (n+1)$. Any placement of the dominoes on the original board with no touching corresponds to a placement of the tiles on the expanded board with no overlapping, and vice versa. The area of the expanded board is $(n+1)^2$ while the total area of the 21 tiles is 126. It follows that $n \geq 11$. Figure 5.5 shows that an 11×11 board can hold 24 dominoes without touching.

Figure 5.5

8. The four sides and one diagonal of a quadrilateral have lengths 1, 2, 2.8, 5 and 7.5, not necessarily in that order. Determine which number was the length of the diagonal.

   Solution:
   Each of the five segments with given lengths in a triangle with two others. The segment of length 7.5 must be in a triangle with the segments of lengths 2.8 and 5 because only they are the only ones whose total length exceeds 7.5, as required by the Triangle Inequality. Now the segments of length 1 and 2 must form a triangle with the segment of length 2.8 since it is the only one of length less than 1+2, again as required by the Triangle Inequality. It follows that the diagonal is of length 2.8.

9. Simplify $8(3^2+1)(3^4+1)\cdots(3^{2^{10}}+1)+1$.

   Solution:
   We have
   $$\begin{aligned}
   & 8(3^2+1)(3^4+1)\cdots(3^{2^{10}}+1)+1 \\
   =\ & (3^2-1)(3^2+1)(3^4+1)\cdots(3^{2^{10}}+1)+1 \\
   =\ & (3^4-1)(3^4+1)\cdots(3^{2^{10}}+1)+1 \\
   =\ & (3^8-1)(3^8+1)\cdots(3^{2^{10}}+1)+1 \\
   =\ & \cdots \\
   =\ & 3^{2^{11}}-1+1 \\
   =\ & 3^{2048}.
   \end{aligned}$$

10. Each monkey collected the same number of nuts and threw one of them at every other monkey. Nuts thrown were lost. Each monkey still had at least two nuts left, and among them, they have 33 nuts left. Determine the number of nuts collected by each monkey.

    Solution:
    We either have 11 monkeys each with 3 nuts left or 3 monkeys each with 11 nuts left. Since each monkey throws a nut at every monkey, it must have collected $3+11-1=13$ nuts.

11. From an 8×8 chessboard, the central 2×2 block rises up to form a barrier. Queens cannot be placed on the barrier, and may not attack one another across this barrier. Determine the maximal number of Queens which can be placed on the chessboard so that no two attack each other.

    Solution:
    Each of the clear rows can hold at most 1 Queen while each of the divided row can hold at most 2 Queens. It follows that the number of Queens is at most $6\times 1+2=10$. Figure 5.6 shows that 10 non-attacking Queens can be placed.

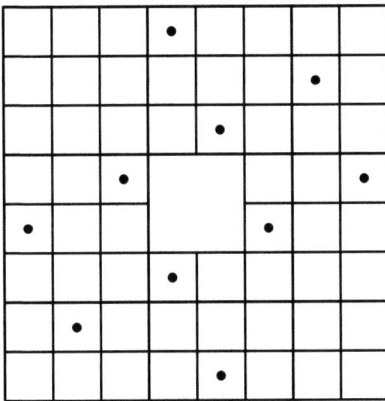

Figure 5.6

12. P is an interior point of triangle ABC. Perpendiculars are dropped from P to F on the side AB and E on the side AC. The triangles PAF and PCE are congruent, but the vertices are not necessarily in corresponding orders. Determine the ratio $\frac{AE}{AC}$.

    Solution:
    Since congruent right triangles must have equal hypotenuses, we have $PA = PC$. Since PE is perpendicular to AC, it must bisect it so that $\frac{AE}{AC} = \frac{1}{2}$.

13. In a kindergarten, 17 children made an even number of postcards. Any group of 5 children made no more than 25 postcards while any group of 3 children made no less than 14 postcards. Determine the total number of postcards made.

    Solution:
    Consider the two children with the smallest total of cards made. Suppose this total is at most 8. Since any group of children made no less than 14, everyone else makes at least 6. This yields a total of at least $8 + 15 \times 6 = 98$, for an average exceeding 5. This contradicts the condition that any group of 5 children made no more than 25 postcards. It follows that the smallest total is at least 9. If it is at least 10, then everyone made 5 postcards for a total of 85, an odd number. Hence the smallest total is 9, with the two children making 4 and 5 postcards respectively. It follows that every other child made 5 postcards, for a total of 84.

14. Determine the number of integers of the form $\overline{abc} + \overline{cba}$, where $\overline{abc}$ and $\overline{cba}$ are three-digit numbers with $ac \neq 0$.

Solution:
The digit b can be any of $0, 1, 2, \ldots, 9$ while the sum of the digits a and c can be any number of $2, 3, 4, \ldots, 18$. Hence the total number of integers of the form $101(a+c)+10b$ is $(9-0+1)(18-2+1) = 170$.

15. M is the midpoint of side AC of triangle ABC. If $\angle ABM = 2\angle BAM$ and $BC = 2BM$, determine the measure, in degrees, of the largest angle of ABC.

Solution:
Let N be the midpoint of BC so that $BN = NC = BM$. Moreover, MN is parallel to AB. Now

$$\begin{aligned} 2\angle BAM &= \angle ABM \\ &= \angle BMN \\ &= \angle BNM \\ &= \angle NMC + \angle NCM \\ &= 2\angle NCM. \end{aligned}$$

It follows that $AB = BC$ so that BM is perpendicular to AC. Hence $90° = \angle BAM + \angle ABM = 3\angle BAM$ so that $\angle ABC$ is the largest angle of triangle ABC, and its measure in degrees is 120.

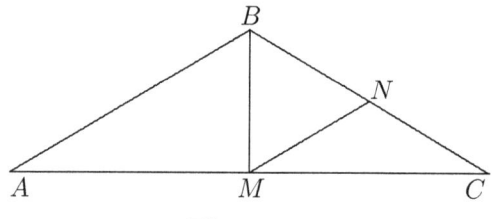

Figure 5.7

16. A judge knows that among 9 coins that are identical in appearance, there are exactly 3 coins each weighing 3 grams, 3 coins each weighing 2 grams and 3 coins each weighing 1 gram. An expert wants to prove to the judge how much each of the coins weighs. He has a balance that shows which of two groups of coins is heavier, or that they have the same total weight. Determine the least number of weighings needed to persuade the judge.

Solution:
Clearly, 1 weighing is not sufficient. However, 2 weighings are. Let A, B and C be the coins each weighing 3 grams, D, E and F be the coins each weighing 2 grams, and G, H and I be the coins each weighing 1 gram. In the first weighing, show that A is equal in weight to G, H and I combined. This is only possible if A indeed weighs 3 grams while each of G, H and I weighs 1 gram.

In the second weighing, show that B and C combined are equal in weight to D, E and F combined. This shows that each of B and C weighs 3 grams while each of D, E and F weighs 2 grams.

17. On an island, each inhabitant is either a Knight who always tells the truth, or a Knave who always lies. Each inhabitant earns a different amount of money and works a different number of hours. Each inhabitant says, "Less than 10 inhabitants work more hours than I do." Each also said, "At least 100 inhabitants make more money than I do." Determine the number of inhabitants on this island.

Solution:
There may be exactly 110 inhabitants on this island. Let inhabitant n work n dollars and make n dollars. For $1 \leq n \leq 10$, both statements of inhabitant n are true, making them Knights. For $11 \leq n \leq 110$, both statements of inhabitant n are false, making them Knaves. Suppose there are at least 111 inhabitants. Then the second statement is true for at least 11 of them, making them Knights. However, for the one working the highest number of hours among them, the first statement will be false, which is a contradiction. Suppose there are at most 100 inhabitants. Then the second statement is false for all of them, making them Knaves. However, for the one working the highest number of hours, the first statement will be true, which is a contradiction. Finally, suppose the number of inhabitants is at least 101 but at most 109. Then the first statement is true for 10 of them, making them Knights. However, for the one making the least amount of money among them, the second statement will be false, which is a contradiction. It follows that the number of inhabitants must be 110.

18. Find the second smallest positive multiple of 11 with digit-sum 600.

Solution:
To minimize the multiple, we should use as many copies of the digit 9 as we can. Since $600 = 9 \times 66 + 6$, we can use 66 copies of 9 plus some digits adding up to 6. Thus the smallest such multiple of 11 is $33999\cdots 9$ and the second smallest is $34989\cdots 9$, where $9\ldots 9$ denotes 64 copies of the digit 9.

19. In a chess tournament with an odd number of players, each player plays every other exactly once. A win is worth 1 point and a draw is worth $\frac{1}{2}$ point. The player in the fourth position from the bottom gets more points than the player in the third position from the bottom. Each player except the bottom three gets half of his or her points playing against the bottom three. Determine the number of players in the tournament.

Solution:
Let there be $n+3$ players. Each of the n players not in the bottom three gets at most 3 points from the bottom three, so that among them, they can get at most $3n$ points this way. Among themselves, $\frac{n(n-1)}{2}$ games are played, so that this many points are obtained. From the given condition, $\frac{n(n-1)}{2} \le 3n$ so that $n \le 7$. Note that n is even. Suppose $n = 6$. Let all the games among themselves be draws, so that each gets $2\frac{1}{2}$ points this way. Each can also get the same number of points from the bottom three with 2 wins and 1 draw. If each of the bottom three has 2 draws against them, and 2 draws among themselves, each will have 2 points, which is less than $2\frac{1}{2}$ points. Suppose $n = 4$. These players get 6 points from one another, so that they must also get 6 points from the bottom three. Since there are 12 games played between the two groups, the bottom three also get 6 points from the top players. Moreover, they get 3 points among themselves. The total score of the top four players is 6+6 for an average of 3. The total score of the bottom three players is 6+3 for an average of 3 also. Thus the player in the fourth position from the bottom cannot get more points than the player in the third position from the bottom. Finally, $n = 2$ also leads to a contradiction. Hence the total number of players is 6+3=9.

20. $ABCD$ is a parallelogram. M and N are points on the sides AB and AD respectively, such that $AB = 4AM$ and $AD = 3AN$. Let K be the point of intersection of MN and AC. Determine the ratio $\frac{AC}{AK}$.

Solution:
Extend NM and CB, intersecting at L. We have $BM = 3AM$. Since triangles BLM and ANM are similar, $BL = 3AN$. Note that we have $BC = AD = 3AN$, so that $LC = 6AN$. Since triangles CLK and ANK are similar, $CK = 6AK$ so that $\frac{AC}{AK} = 7$.

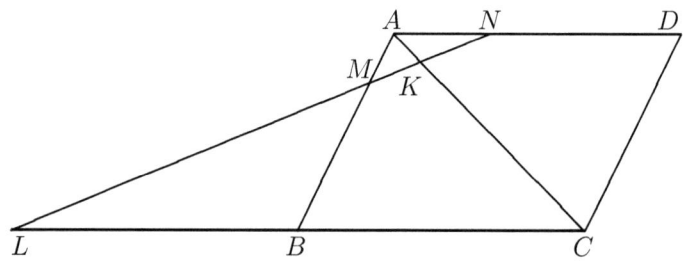

Figure 5.8

21. Determine the largest real number t such that the two polynomials $x^4 + tx^2 + 1$ and $x^3 + tx + 1$ have a common root.

Solution:
Any common root of the two polynomials is also a root of
$$x^4 - x^3 + tx^2 - tx = x(x-1)(x^2 + t).$$
We cannot have $x^2 + t = 0$ since $x^2(x^2 + t) + 1 = 0$. We cannot have $x = 0$ since 0 is not a root of $x^3 + tx + 1 = 0$. Hence $x = 1$. From $1^3 + t + 1 = 0$, we have $t = -2$.

22. Determine the smallest positive multiple of 777 such that the sum of its digits is the least.

Solution:
Since 777 is divisible by 3, the digit-sum of any of its positive multiples is at least 3. It is routine to verify that $777 \times 13 = 10101$ is the smallest such multiple of 777.

23. Determine the minimum number of points marked on the surface of a cube so that no two faces of the cube contain the same number of marked points. A point at a corner or on a side of a face is considered to be in that face.

Solution:
To minimize the total number of points, the numbers of points on the six faces should be 0, 1, 2, 3, 4 and 5. Hence 5 points in total is necessary. This is however not sufficient, because all 5 points must be on the same face. The opposite face will have 0 points, but the other four faces must have 1+2+3+4=10 points in all. Hence each point must belong to two of these four faces, so that they are all corners of the face containing them. This is impossible since a face has only four corners. Figure 5.9 shows a placement of 6 points satisfying the condition of the problem.

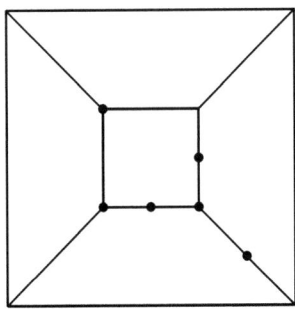

Figure 5.9

24. In a right triangle, the smallest height is one quarter the length of the hypotenuse. Determine the measure, in degrees, of the smallest angle of this triangle.

Solution:
Make four copies of the right triangles OPA, OPB, OQC and OQD as shown in Figure 5.10. Then PQ is twice the length of the smallest height, and $AC = 2PQ$. It follows that $AC = OB$ and OAC is equilateral. Hence the smallest angle of the right triangle, in degrees, is $60 \div 4 = 15$.

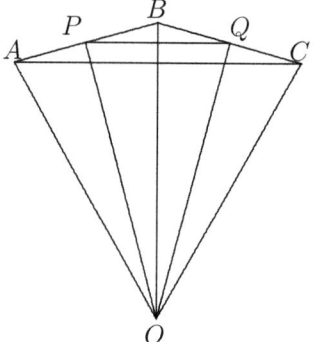

Figure 5.10

25. The real numbers $x < y$ satisfy $(x^2 + xy + y^2)\sqrt{x^2 + y^2} = 185$ and $(x^2 - xy + y^2)\sqrt{x^2 + y^2} = 65$. Determine the smallest value of y.

Solution:
Subtraction yields $2xy\sqrt{x^2 + y^2} = 120$. Adding $xy\sqrt{x^2 + y^2} = 60$ to the first equation, we have $(x + y)^2\sqrt{x^2 + y^2} = 245$. Adding to the second yields $(x - y)^2\sqrt{x^2 + y^2} = 5$. Division yields $\frac{x+y}{x-y} = -7$ so that $4x = 3y$. Substituting into $xy\sqrt{x^2 + y^2} = 60$ yields $(x, y) = (-4, -3)$ or $(3, 4)$. Hence the smallest value of y is -3.

Solution to Sample Contest II

Earthly Problems

1. The little tycoon Johnny says to his fellow capitalist Annie, "If I add $7 to $\frac{3}{5}$ of my funds, I'll have as much money as you do." Annie replies, "So you have only $3 more than I." How much money does each have?

   Solution:
   Since $10 will make up $\frac{2}{5}$ of Johnny's funds, he has $25 while Annie has $22.

2. According to a contract, a worker is to be paid $48 for each day he is at work and $12 is deducted from his pay for each day he is not at work. After 30 days, the worker' net pay is $0. How many days is he at work during these 30 days?

   Solution:
   Since the ratio of the pay and the fine is 4:1, the number of idle days to the number of working days is also 4:1. Hence the worker is at work for $30 \times \frac{1}{1+4} = 6$ days.

3. Using each of the numbers 1, 2, 3 and 4 twice, I succeeded in writing out an eight-digit number in which there is one digit between the ones, two digits between the twos, three digits between the threes, and four digits between the fours. What was the number?

   Solution:
   The unique solution, apart from reversal, is 41312432.

4. John, Jim, and Gerry went to a baseball game. On the way, John bought five bags of potato chips, Jim bought two bags, and Gerry did not buy any. During the game they all ate the chips, each one eating as much as the others. After the game, Gerry figured out how much the bags of chips cost and handed over $1.40. How much money should John get?

   Solution:
   The total cost of the bags was $4.20 so that the cost of each was $0.60. John paid $3 but ate only $1.40 worth. Hence he should get all of Gerry's $1.40 plus another $0.20 from Jim.

5. A large rectangle is divided into 16 smaller rectangles in a 4×4 array. The areas of some of them are shown in Figure 5.11. Find the area of the small rectangle at the top right corner.

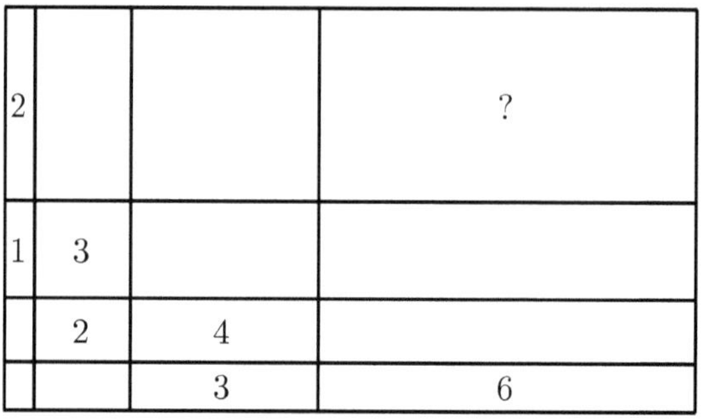

Figure 5.11

Solution:
The area of that rectangle is $6 \times \frac{2}{1} \times \frac{3}{2} \times 43 = 24$.

6. A checker starts on the lower left square of a 5×5 board. How many different paths are there for it to move from there to the upper right square? The checker may move in only two directions: upward and to the right, and may not visit the central square.

Solution:
Figure 5.12 shows the number of different paths the checker can take to reach each square of the 5×5 board except central square.

1	5	9	17	34
1	4	4	8	17
1	3	X	4	9
1	2	3	4	5
1	1	1	1	1

Figure 5.12

7. In 1992, it was observed that in the year x^2 my nephew will be x years of age. In what year was he born?

Solution:
Note that $x^2 - x \leq 1992 \leq x^2$, so that $x = 45$. Since my nephew will be 45 years of age in $45^2 = 2025$, he was born in 1980.

8. Find the smallest positive integer such that the sum of its digits and that of the subsequent integer are both divisible by 17.

Solution:
The larger of two consecutive numbers is obtained from the smaller one by adding 1. If there is no carrying in this addition, their digit-sums will differ by 1 and cannot both be divisible by 17. If there is a carrying from the units digit to the tens digit but not beyond, then the digit-sums differ by 8. Again not both can be divisible by 17. If there is a further carrying from the tens digit to the hundreds digit but not beyond, then the digit-sums differ by 17. It follows that the larger number ends in 00. The smallest such number whose digit-sum is divisible by 17 is 8900. The number preceding it is 8899, and is the number we seek.

9. A number of coins are placed on each square of a checkerboard such that the sums on every two squares having a common side differ by one cent. Given that the sum on one of the squares is 3 cents, and on another one 17 cents, find the total amount of money on both diagonals on the checkerboard.

Solution:
It takes 14 moves to go from one corner square to the opposite one, and since $17 - 3 = 14$, we must place 3 at one corner and 17 at the opposite corner. The numbers on the other squares are then uniquely determined, as shown in Figure 5.13. The sum of the 16 numbers on the diagonals is 160.

10	11	12	13	14	15	16	17
9	10	11	12	13	14	15	16
8	9	10	11	12	13	14	15
7	8	9	10	11	12	13	14
6	7	8	9	10	11	12	13
5	6	7	8	9	10	11	12
4	5	6	7	8	9	10	11
3	4	5	6	7	8	9	10

Figure 5.13

10. Two numbers are called mirror numbers if one is obtained from the other by reversing the order of digits — for example, 123 and 321. Find two mirror numbers whose product is 92565.

Solution:
Since the product of the two numbers ends in a 5, one of them must also end in a 5, and the other starts with a 5. The first number must start with a 1 as otherwise the product has at least six digits. Let the middle digit of either number be x. From the tens-digit 6 in the product, we see that $6x$ ends in a 6, so that either $x = 1$ or $x = 6$. The product of 165 and 561 is indeed 92565.

11. Thirty children ride a carousel swing. Every girl rides behind a boy, half of the boys ride behind a boy, and all the other boys ride behind girls. How many boys and girls are there?

Solution:
Since every girl rides behind a boy, every girl is also followed by a boy. Hence the number of boys riding behind girls is equal to the number of girls. Since this account for half of the boys, there must be 20 boys and 10 girls.

12. The length of a rectangle is twice its width. The circumference of a circle is one half the width of the rectangle. The circle rolls along the perimeter of the rectangle until it comes back to its initial position. How many revolutions will the circle make if it is rolling

    (a) inside
    (b) outside

    the rectangle?

Solution:
The perimeter of the rectangle is twelve times the circumference of the circle.

   (a) If the circle is rolling inside the rectangle, there are eight segments at the corners, each equal in length to the radius of the circle, which the circle cannot touch. Hence it has made $12 - \frac{4}{\pi}$ revolutions when it returns to its original position.

   (b) If the circle is rolling outside the rectangle, it will make 13 revolutions before returning to its initial position, because it has to make an additional quarter turn at each corner of the rectangle.

13. Mademoiselle Dubois loves pets. All her pets but two are dogs, all but two are cats, and all but two are parrots. Those that are not dogs, cats or parrots are cockroaches. She has more than two pets. How many pets of each kind does she own?

Solution:
Suppose Mademoiselle Dubois has k pets. Then she has $k - 2$ dogs, $k - 2$ cats and $k - 2$ parrots. Hence $3(k - 2) \leq k$ or $3 \leq k$.

Since she has more than two pets, $k = 3$. Hence she has 1 dog, 1 cat, 1 parrot and 0 cockroaches.

14. A raft and a motorboat left town A simultaneously and traveled downstream to town B. The raft moved at the same speed as the current, which was constant. The motorboat, which traveled at constant speed in still water, arrived at B, immediately turned back, and encountered the raft two hours after they had set out from A. How much time did it take the motorboat to go from A to B?

    Solution:
    When the motorboat was going downstreams, the rate at which its distance from the raft was increasing was equal to its speed in still water. When the motorboat was going upstreams, the rate at which its distance from the raft was decreasing was also equal to its speed in still water. Hence it spent an equal amount of time going downstreams and upstreams. It follows that it took 1 hour to go from A to B.

15. Nine coins lie on a table.

    (a) Can you turn them all upside down if in each move, you are only allowed to turn over any five coins?

    (b) What is the minimum number of moves if this task is possible?

    Solution:

    (a) We can turn over in succession the following groups: 12345, 12367 and 12389. Each coin has been turned over an odd number of times and is therefore upside down. Thus the task can be accomplished in three moves.

    (b) In one move, some coins will not be turned over. If each coin is turned over at least once in two moves, then one of them will have been turned over exactly twice and will not be upside down. Hence three moves are minimum.

16. Mr. R. A. Scall, president of the Pyramid Bank, lived in a suburb rather far from his office. Every weekday, a car from the bank came to his house, always at the same time and traveling at the same constant speed, so that he would arrive at work precisely when the bank opened. One morning his driver called very early to tell him he would probably be late because of mechanical problems. So Mr. Scall left home one hour early and started walking to his office. The driver managed to fix the car quickly, however, and left the garage on time. He met the banker on the road and brought him to the bank. They arrived 20 minutes earlier than usual. How much time did Mr. Scall walk?

Solution:
Mr. Scall's walking saved the driver a return trip between Mr. Scall's house and where he was picked up. Hence a one-way trip would take 10 minutes. Since Mr. Scall left 1 hour early, he had been walking for 50 minutes.

17. There are twelve people in a room. Some of them always tell the truth while the others always lie. One of them said, "None of us is honest." Another said, "There is not more than one honest person here." A third said, "There are not more than two honest persons here." This continued on until the twelfth said, "There are not more than eleven honest persons here." How many honest persons are in the room?

   Solution:
   Let the number of honest people in the room be k. Then the first k statements are false and the remaining $12 - k$ are true. Since these true statements are made by the k honest people, $12 - k = k$ or $k = 6$.

18. Koshchei the Immortal buried his ill-gotten treasure in a hole 1 meter deep. That did not seem safe enough for him, so he dug his treasure up, deepened the hole to 2 meters, and buried his hoard again. He was still worried, so he dug up his hoard, made the hole 3 meters deep, and hid his treasure once more. But he just could not stop — he kept increasing the depth of the hole, to 4 meters, 5 meters, 6 meters, and so on, each time extracting his property and burying it again, until on the 1001-st day he died of exhaustion. Koshchei dug a hole n meters deep in n^2 days, but took no time filling one in. How deep was his treasure buried when he dropped dead?

   Solution:
   Note that $1^2 + 2^2 + \cdots + n^2 = \frac{n(n+1)(2n+1)}{6}$. For $n = 13$, this is equal to $819 < 1001$, but for $n = 14$, this is equal to $1015 > 1001$. Hence when Koshchei dropped dead, his treasure buried 13 meters deep.

19. Prince Ivan is fighting the three-headed, three-tailed dragon with a magic sword. If he cuts off one head of the dragon, a new head grows. If he cuts off one tail, two new tails grow. If he cuts off two tails, one new head grows. If he cuts off two heads, nothing grows. These are the only four things Prince Ivan can do. What is the smallest number of strokes Prince Ivan needs to cut off all the dragon's heads and tails?

   Solution:
   Clearly, there is no point in cutting off one dragon head. Suppose x, y and z denote respectively the numbers of strokes in which Prince Ivan cuts off one dragon tail, two dragon tails and two dragon heads. If all the heads and tails are cut off, then $3 + y - 2z = 0$ and $3 + x - 2y = 0$. These yield $9 + x = 4z$.

Since z is an integer, $x \geq 3$ and $z \geq 3$. From $2y = 3 + x$, we have $y \geq 3$ as well. Hence $x = y = z = 3$ is an optimal solution. By cutting off one tail, two tails and two heads, Prince Ivan reduces each of the numbers of heads and tails by one. Performing this cycle three times accomplishes the task.

20. Solve $10 - 9(9 - 8(8 - 7(7 - 6(6 - 5(5 - 4(4 - 3(3 - 2(2 - x)))))))) = x$.

    Solution:
    Since $10 - 9(9 - 8(8 - 7(7 - 6(6 - 5(5 - 4(4 - 3(3 - 2(2 - 1)))))))) = 1$, $x = 1$ is a solution. Since the equation is linear, it is the only solution.

21. I went to the bank to cash a check. As the cashier gave me the money, I put it in my empty wallet without counting it. During the day I spent $6.23. When I checked my wallet in the evening, it contained exactly twice the amount of the check I had cashed. Strange! A little calculation revealed that while making the payment, the cashier had interchanged the figures for dollars and cents. What was the amount of the check?

    Solution:
    Suppose the check was for x dollars and y cents. If $x > 23$ and $y < 50$, we have $x - 23 = 2y$ and $y - 6 = 2x$. However, this system does not yield positive integral solutions. If $x \leq 23$ and $y \geq 50$, we have $100 + x - 23 = 2y - 100$ and $y - 7 = 2x + 1$. This system does not yield positive integral solutions either. If $x > 23$ and $y \geq 50$, we have $x - 23 = 2y - 100$ and $y - 6 = 2x + 1$. This system yields $x = 21$ and $y = 29$, but must be rejected they contradict our assumptions. Finally, if $x \leq 23$ and $y < 50$, we have $100 + x - 23 = 2y$ and $y - 7 = 2x$. This system also yields $x = 21$ and $y = 49$, but this time the solution is acceptable. Hence the amount of the check was $21.49.

22. Two cyclists, Kaitlin and Josh, simultaneously started toward each other from two towns 40 kilometers apart. Josh rode at 23 kilometers per hour, and Kaitlin rode at 17 kilometers per hour. Before departure, a fly landed on Josh's nose. At the moment of departure, it started to fly toward Kaitlin at 40 kilometers per hour. When it reached Kaitlin, it immediately flew in the opposite direction at 30 kilometers per hour (the wind blew toward Kaitlin). As soon as the fly reached Josh, it turned back again, and so on. Find the total distance flown by the fly until the cyclists met. The speed of the fly was constant in each direction.

Solution:
When the cyclists met, the fly had been airborne for 1 hour. During this time, Josh had covered 23 kilometers. Independent of the speeds of the fly in either of the directions, as long as they were greater than those of the cyclists, this was the difference in distance between flying with the wind and flying against it. Suppose the fly spent t_1 hours flying with the wind and t_2 hours against it. Then $t_1 + t_2 = 1$ and $40t_1 - 30t_2 = 23$. Hence $t_1 = \frac{53}{70}$ and $t_2 = \frac{17}{70}$, so that the fly covered $40 \times \frac{53}{70} + 30 \times \frac{17}{70} = \frac{263}{7}$ kilometers.

23. Find $\sqrt{12345678987654321}$.

Solution:
We verify directly that $\sqrt{12345678987654321} = 111111111$ in Figure 5.14.

```
                          1 1 1 1 1 1 1 1 1
         ×                1 1 1 1 1 1 1 1 1
                        ─────────────────────
                          1 1 1 1 1 1 1 1 1
                        1 1 1 1 1 1 1 1 1
                      1 1 1 1 1 1 1 1 1
                    1 1 1 1 1 1 1 1 1
                  1 1 1 1 1 1 1 1 1
                1 1 1 1 1 1 1 1 1
              1 1 1 1 1 1 1 1 1
            1 1 1 1 1 1 1 1 1
          1 1 1 1 1 1 1 1 1
        ─────────────────────────────────────
          1 2 3 4 5 6 7 8 9 8 7 6 5 4 3 2 1
```

Figure 5.14

24. Side AD of the quadrilateral $ABCD$ equals its diagonal BD. The three other sides of this quadrilateral are equal to each other. The diagonal BD bisects $\angle ADC$. What are the possible measures of $\angle BAD$?

Solution:
Since $BC = CD$, $\angle DBC = \angle CDB = \angle ADB$. Hence AD is parallel to BC. If AB is also parallel to DC, then

$$\angle ADB = \angle CDB = \angle ABD = \angle BAD$$

since $AD = BD$. Hence $\angle BAD = 60°$. Suppose AB is not parallel to DC. Drop perpendiculars BP and CQ onto AD from B and C respectively. Then $BP = CQ$ since AD is parallel to BC. We also have $AB = CD$. Hence triangles BAP and CDQ are congruent, so that $\angle BAD = \angle CDA$.

Solutions

Now $\angle BAD = \angle BDA$ since $AD = BD$. We have

$$180° = \angle BAD + \angle BAD + \angle ADB = \frac{5}{2}\angle BAD$$

so that $\angle BAD = 72°$. In summary, the possible measures of $\angle BAD$ are $60°$ and $72°$.

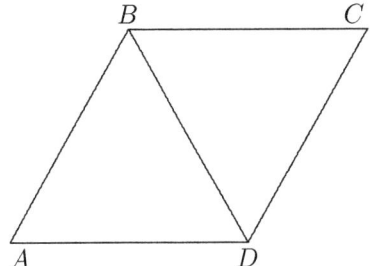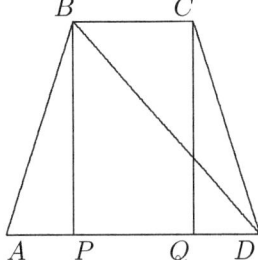

Figure 5.15

25. Find the fraction between $\frac{96}{35}$ and $\frac{97}{36}$ with the smallest denominator.

   Solution:
   We have $\frac{96}{35} > \frac{95}{35} = \frac{19}{7} > \frac{97}{36}$. On the other hand, $6 \times \frac{96}{35} < 17$ while $6 \times \frac{97}{36} > 16$, so that there is no fraction between these two fractions with denominator 6. We can similarly verify that no such fraction with denominator less than 6. Thus $\frac{19}{7}$ is the fraction with the smallest denominator between $\frac{96}{35}$ and $\frac{97}{36}$.

Lunar Problems

1. I am four times older than my sister was when she was half as young as I was. In 15 years our combined age will be 100. How old are we now?

   Solution:
   At some point in the past, my sister was x years old while I was $2x$ years old. Now I am $4x$ years old, so that she is $3x$ years old. In 15 years our combined age will be $4x + 15 + 3x + 15 = 100$, which yields $x = 10$. Hence I am 40 years old and my sister is 30 years old.

2. At a certain sporting event, 100 students took part in running, 50 students participated in swimming and 48 in biking. It turned out that the number of students who took part in only one event was twice as high as that who took part in just two events and three times as high as the number of students who participated in all three events. What is the total number of students that participated in the events?

   Solution:
   Let x, y and z be the number of students who took part in one, two and three events respectively. Then $x + 2y + 3z = 100 + 50 + 48 = 198$. Since $x = 2y = 3z$, $x = 66$, $y = 33$ and $z = 22$, so that $x + y + z = 121$ students participated in the events.

3. In triangle ABC, M is the midpoint of BC and N is the midpoint of AM. The extension of BN intersects AC at K. Compute $\frac{KC}{KA}$.

   Solution:
   Join MK. Since $AN = NM$. triangles BAN and BMN have the same area, as do triangles KAN and KMN. Hence triangles BAK and BMK have the same area. Since $BM = CM$, triangles BMK and CMK also have the same area. It follows that the area of triangle BCK is double that of triangle BAK, so that $\frac{KC}{KA} = 2$.

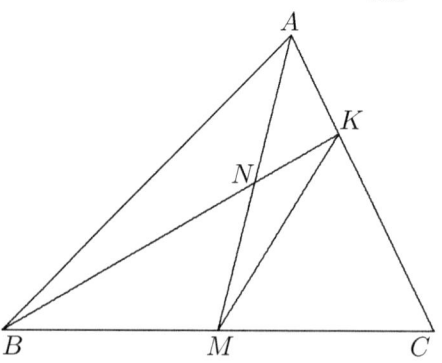

Figure 5.16

4. In the equation ONE×9 =NINE, different letters stand for different digits and identical letters stand for identical digits. What is the maximum value of NINE?

Solution:
From the fourth column, E=0 or 5. If E=0, then N=5 from the third column. We must then have O=6, I=8. Therefore the multiplication is $650 \times 9 = 5850$. If I=5, then N=2 or 7 from the third column. If N=7, then we must have O=8, but then I=8 also, which is forbidden. If N=2, then we must have O=3 and I=9, and the multiplication is $325 \times 9 = 2925$. Hence the maximum value of NINE is 5850.

5. One day all of Mrs. Brown's grandchildren came to visit her. There was a bowl of apples and pears on the kitchen table. Mrs. Brown gave each child the same number of pieces of fruit without keeping track of which kind. Billy got $\frac{1}{8}$ of all the apples and $\frac{1}{10}$ of all the pears. How many grandchildren did Mrs. Brown have?

Solution:
Billy got less than $\frac{1}{8}$ of all the fruit and more than $\frac{1}{10}$ of all the fruit. Hence he got $\frac{1}{9}$ of all the fruit, meaning that Mrs. Brown had 9 grandchildren. This was indeed possible if she had 8 apples and 10 pears, and each grandchild got one each except one who got two pears.

6. Thirty people took part in a shooting match. The first participant scored 80 points, the second scored 60 points, the third scored the average of the number of points scored by the first two, and each subsequent participant scored the average of the number of points scored by the previous ones. How many points did the last participant score?

Solution:
The third participant scored $\frac{80+60}{2} = 70$ points. Adding more scores equal to the average will not change the average. Hence the last participant scored 70 points.

7. How many numbers end in the four digits 1995 and become an integral number of times smaller when these digits are erased?

Solution:
A number ending in 1995 has the form $10000x + 1995$ where x is the number obtained when these four digits are erased. We are given that $10000x + 1995 = kx$ for some integer k, so that $(k - 10000)x = 1995$. It follows that x can be any divisor of $1999 = 3 \times 5 \times 7 \times 19$. Since x can either take or leave each of the four primes 3, 5, 7 and 19, there are $2^4 = 16$ such numbers.

8. The working hours of a receptionist at a hotel are either 8 a.m. to 8 p.m., 8 p.m. to 8 a.m., or 8 a.m. to 8 a.m. the next day. In the first case, the break before the next shift must be not less than 24 hours; in the second case, not less than 36 hours; and in the third case, not less than 60 hours. What is the smallest number of receptionists that can provide round-the-clock operation of the hotel?

Solution:
The hotel can have four receptionists each working 24 hours in rotation. We now prove that three are not enough. Suppose one of them works 24 hours. Then at least three others are needed for the 60 hours during which the round-the-clock receptionist rests. Suppose none of them works 24 hours. Then at least three others are needed for the 48 hours during which a night shift receptionist rests.

9. Baron Munchhausen says: "Today I shot more ducks than two days ago, but fewer than a week ago." For how many days in a row can the baron say this truthfully?

Solution:
Suppose Baron Munchhausen can say that truthfully seven days in a row. Let these be the third to the ninth days, and for $1 \leq i \leq 9$, let a_i be the number of ducks he shots on the i-th day. Then we have $a_1 < a_3 < a_5 < a_7 < a_9 < a_2 < a_4 < a_6 < a_8 < a_1$, which is a contradiction. On the other hand, let Baron Munchhausen shoot 5, 1, 6, 2, 7, 3, 8 and 4 ducks on eight consecutive days, and 10 ducks any other day. Then he will be telling the truth from the third to the eighth day.

10. A, B, C and D are four points in that order on a line. Semicircles on AB and AC as diameters are drawn on one side of this line, and semicircles on BD and CD as diameters are drawn on the other. Can we calculate the area of the region enclosed by the four semicircles, knowing only the lengths of AD and BC?

Solution:
The area of the region in question is given by
$$\frac{\pi}{8}(AC^2 + BD^2 - AB^2 - CD^2)$$
$$= \frac{\pi}{8}(AB^2 + 2AB \cdot BC + BC^2 + BC^2$$
$$+ 2BC \cdot CD + CD^2 - AB^2 - CD^2)$$
$$= \frac{\pi}{4}BC(AB + BC + CD)$$
$$= \frac{\pi}{4}BC \cdot AD,$$

an expression which depends only on the lengths of AD and BC.

Solutions

11. A computer printed out digit by digit two numbers whose values are 2^{1995} and 5^{1995}. How many digits in all were printed?

    Solution:
    Suppose 2^{1995} is an m-digit and 5^{1995} an n-digit number. Then we have $10^{m-1} < 2^{1995} < 10^m$ while $10^{n-1} < 5^{1995} < 10^n$. It follows that $10^{m+n-2} < 2^{1995}5^{1995} < 10^{m+n}$. However, $2^{1995}5^{1995} = 10^{1995}$. Hence $m + n - 1 = 1995$ so that the total number of digits printed is $m + n = 1996$.

12. You can see three faces of each of two dice. On each die, the three pairs of opposite faces contain 1 and 6 dots, 2 and 5 dots, and 3 and 4 dots, respectively. The total number of dots on the visible faces is 27. What is the total number of dots you can see on each die?

    Solution:
    The maximum number of dots on three faces of a die is 4+5+6=15. Thus the total 27 can arise either as 13+14 or 12+15. We claim that 13 is an impossible total. By the Pigeonhole Principle, one of the three faces has more than four dots. If it is a 6, we need 7 more, but only opposite faces add up to 7. Hence it is 5, and we need 8 more. We obviously cannot have 8=4+4, but 8=3+5 must also be rejected since we already have 5. Finally, 8=6+2 has already been ruled out. Hence the two totals are 12 and 15. The former is achievable as 2+4+6 and the latter as 4+5+6.

13. Several chess players played chess in a park the whole day long. Since they had only one set of pieces, they chose the following rules: The winner of a game skipped the next two games, and the loser skipped the next four. If the game ended in a draw, the player who played white pieces was considered to have lost. How many players took part if they managed to follow these rules?

    Solution:
    We need two new players for each of the first three games, and one new player for each of the next two. Thereafter, no new players may be introduced. Hence the total number of players is eight.

14. A huge military band was playing and marching in formation on the parade grounds. First the musicians formed a square. Then they regrouped into a rectangle so that the number of rows increased by 5. How many musicians were there in the band?

    Solution:
    Let the initial square be $n \times n$. Then the length of the rectangle is $n+5$ while its width is $\frac{n^2}{n+5} = n - 5 + \frac{25}{n+5}$. The only divisor of 25 greater than 5 is itself. Hence $n = 20$ and there were $n^2 = 400$ musicians in the band.

15. The length of the chord of a circle is 2. Two smaller circles are tangent to the large circle and to each other at the midpoint of the chord. Find the area of the region inside the large circle but outside the small circles.

Solution:
Let the chord be CD with midpoint T. Let AB be the diameter of the large circle through T. It is perpendicular to CD. Moreover, AT and BT are diameters of the small circles respectively. Let $AT = 2x$ and $BT = 2y$ so that $CT = 2xy$. By Pythagoras' Theorem,

$$AB^2 = AC^2 + BC^2 = AT^2 + BT^2 + 2CT^2.$$

It follows that $4x^2 + 4y^2 + 8xy = 4x^2 + 4y^2 + 2$ so that $xy = \frac{1}{4}$. The area of the region inside the large circle but outside the small circles is given by $\pi(x+y)^2 - \pi x^2 - \pi y^2 = 2\pi xy = \frac{\pi}{2}$.

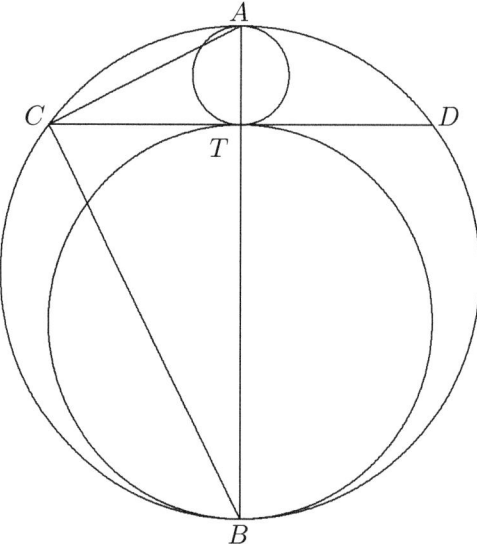

Figure 5.17

16. A hundred officials were invited to the annual meeting at their Ministry of Affairs. They were seated in a rectangular hall with ten rows of chairs, ten chairs in each row. The opening was delayed, and the officials could find nothing better to do than compare their salaries. To consider oneself "highly paid", an official had to determine that no more than one person seated to the left, right, front or rear or at a diagonal was paid as much or more. What is the greatest number of officials who could count themselves as "highly paid"?

Solution:

Partition the one hundred officials into twenty-five 2×2 squares. At most two officials in each square can consider themselves "highly paid". Hence the total number is at most fifty. In the seating arrangement shown in Fgure 5.18, the fifty officials in rows 1, 3, 5, 7 and 9 can consider themselves "highly paid".

2	3	4	5	6	7	8	9	10	11
1	1	1	1	1	1	1	1	1	1
2	3	4	5	6	7	8	9	10	11
1	1	1	1	1	1	1	1	1	1
2	3	4	5	6	7	8	9	10	11
1	1	1	1	1	1	1	1	1	1
2	3	4	5	6	7	8	9	10	11
1	1	1	1	1	1	1	1	1	1
2	3	4	5	6	7	8	9	10	11
1	1	1	1	1	1	1	1	1	1

Figure 5.18

17. In a chess tournament, each participant plays each of the others once. A win is worth one point, a draw is worth half a point, and a loss is worth no points. Two precocious students from an elementary school took part in a chess tournament at a nearby university. The combined score of the elementary school students was 6.5. The scores of the university students all happened to be the same. How many university students participated in the tournament?

Solution:
Let n university students participate in the tournament. Let the score for each be $\frac{m}{2}$, where m is a positive integer. The total number of games is $\binom{n+2}{2} = \frac{(n+2)(n+1)}{2}$ while the total score is $\frac{mn+13}{2}$. Hence $n(n+3-m) = 11$ which implies $n = 1$ or 11. However, $n = 1$ must be rejected as otherwise m would be negative. Hence 11 university students participated in the tournament.

18. The sequence $\{a_n\}$ is defined as follows: $a_1 = 1776, a_2 = 1999$ and for $n \geq 0$, $a_{n+2} = \frac{a_{n+1}+1}{a_n}$. Find a_{2002}.

Solution:
Let $a_1 = x$ and $a_2 = y$. Then we have $a_3 = \frac{y+1}{x}$, $a_4 = \frac{\frac{y+1}{x}+1}{y} = \frac{x+y+1}{xy}$, $a_5 = \frac{\frac{x+y+1}{xy}+1}{\frac{y+1}{x}} = \frac{x+1}{y}$, $a_6 = \frac{\frac{x+1}{y}+1}{\frac{x+y+1}{xy}} = x$ and $a_7 = \frac{\frac{x+1}{y}}{\frac{x+1}{y}} = y$. Hence the sequence is periodic and has period 5, so that $a_{2002} = a_2 = 1999$.

19. Side AE of the pentagon $ABCDE$ equals its diagonal BD. All the other sides of this pentagon are equal to 1. What is the radius of the circle passing through points A, C and E?

   Solution:
   Note that $ABDE$ is a parallelogram. Complete the parallelogram $ABCO$. Then $CDEO$ is also a parallelogram. It follows that we have $OA = OC = OE = 1$, so that the circle passing through A, C and E has center O and radius 1.

   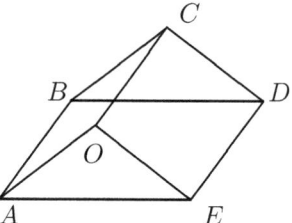

   Figure 5.19

20. Find $\sqrt{11111112222222 - 3333333}$.

    Solution:
    We have

    $$11111112222222 - 3333333 = \frac{10^{14} - 1}{9} + \frac{10^7 - 1}{9} - 3\frac{10^7 - 1}{9}$$
    $$= \frac{10^7 - 1}{9}(10^7 + 1 + 1 - 3)$$
    $$= \left(\frac{10^7 - 1}{3}\right)^2.$$

    Hence $\sqrt{11111112222222 - 3333333} = \frac{10^7 - 1}{3} = 3333333$.

21. You have two balls of each of several colors. A number of balls of different colors are placed on the left pan of a balance while the other balls of the same colors are put on the right one. The balance tips to the left. If you exchange any pair of balls of the same color, however, the balance either tips to the right or stay even. How many balls are there on the balance?

    Solution:
    First, note that every ball on the left is heavier than the ball of the same color on the right. Suppose we have three pairs of balls. Switch the pair whose difference in mass is minimum. Then the nature of the balance is unchanged. Hence we have at most two pairs of balls. With exactly two pairs, the difference in mass between each pair must be the same. We certainly can have only one pair of balls.

Solutions

22. Sixteen baseball teams, ranked 1 through 16, took part in a knock-out tournament. In the first round, eight teams were eliminated; in the second round, four teams were eliminated; in the third round, two teams were eliminated; and the two remaining teams met in the championship game. It turned out that the team with the higher ranking won every game. What is the minimum possible number of uninteresting games, that is, games in which the ranking of the teams differed by at least five?

    Solution:
    We first show that there can be as little as one uninteresting game. In the first round, 1 beats 2, 3 beats 4, 5 beats 6, 7 beats 8, 9 beats 10, 11 beats 12, 13 beats 14 and 15 beats 16. In the second round, 1 beats 3, 5 beats 7, 9 beats 11 and 13 beats 15. In the third round, 1 beats 5 and 9 beats 13. So far, all games are interesting. The only uninteresting game is the championship where 1 beats 9. To avoid even one uninteresting game, the four opponents of 1 must be 2, 3, 4 and 5. Among them, they have eliminated six other teams before being eliminated themselves by 1. One of these six teams must be 11 or lower, resulting in an uninteresting game after all. It follows that the minimum number of uninteresting games is one.

23. A student wrote three positive integers on the blackboard that are consecutive terms of an arithmetic progression. Then he erased the commas between them, creating a seven-digit number. What is the maximum possible value of this number?

    Solution:
    The digits may be distributed among the three terms in the following ways: (1,3,3), (2,2,3), (3,2,2) and (3,3,1). To make the seven-digit number as large as possible, we want to have as many 9s at the beginning as possible. Examining each of the alternatives, we conclude that four 9s at the beginning is impossible, but the last pattern allows for three 9s at the beginning. The first term of the arithmetic progression is 999, the third some one-digit number, and the second the average of the other two terms. Hence we choose the third term to be 9 and the second term is then 504. The maximum value of the seven-digit number is 9995049.

24. Calculate the following quantity to the fifth decimal place without using your calculator: $(\sqrt[3]{2}+1)\sqrt[3]{\frac{1}{3}(\sqrt[3]{2}-1)}$.

Solution:
Let $x^3 = 2$. Then the cube of the given expression is equal to

$$
\begin{aligned}
(x+1)^3 \frac{1}{3}(x-1) &= \frac{1}{3}(x^3 + 3x^2 + 3x + 1)(x-1) \\
&= \frac{1}{3}(x^4 + 3x^3 + 3x^2 + x - x^3 - 3x^2 - 3x - 1) \\
&= \frac{1}{3}(2x + 6 + 3x^2 + x - 2 - 3x^2 - 3x - 1) \\
&= 1.
\end{aligned}
$$

It follows that the given expression is also equal to 1.

25. All sides of triangle ABC are of length 1. D is a point such that $AD = 7$ and both BD and CD are integral. Find these lengths.

    Solution:
    We have $BD \leq AB + AD = 8$ while $BD \geq AD - BD = 6$. Hence $BD = 6$, 7 or 8. Let M be the midpoint of AB. Then $CM^2 = \frac{3}{4}$. If $BD = 6$, then B lies on AD, as shown on the left of the figure below. By Pythagoras' Theorem, $CD^2 = CM^2 + DM^2 = \frac{3}{4} + \frac{169}{4} = 43$, and CD is not integral. If $BD = 8$, then A lies on BD, as shown on the right of the figure below. By Pythagoras' Theorem, we have $CD^2 = CM^2 + DM^2 = \frac{3}{4} + \frac{225}{4} = 57$. Again, CD is not integral. Hence we must have $BD = 74$. Similarly, $CD = 7$ also, and $ABCD$ is a right pyramid with base ABC.

Chapter Six

Past Papers of the Peking University Mathematics Invitational for Youths

Section 1. Problems.

1987

Part A.

1. Correct to two decimal places, $1.4142135 \div 3.14159265$ is equal to _____.

2. The longest side of a $45° - 45° - 90°$ triangle is 12 centimeters. In square centimeters, the area of the triangle is _____.

3. Among 100 positive integers whose sum is 10000, there are more odd numbers than even numbers. The maximum number of even numbers in this group is _____.

4. In the equation $\frac{1}{18} + \frac{1}{\bigcirc} + \frac{1}{\triangle} + \frac{1}{\triangledown} = 1$, each of the symbols $\bigcirc$, $\triangle$ and $\triangledown$ represents a positive integer. The sum of these numbers is _____.

5. In 1980, the age of a father was four times the combined age of his two sons. In 1988, the age of the father will be twice the combined age of his two sons. The father was born in the year _____.

6. In a geometric pattern, there are 100 rectangles, 100 rhombi and 40 squares. The minimum number of parallelograms in this pattern is _____.

7. There are 11112222 chess pieces arranged in a rectangular array. The number of pieces in each row is one greater than the number of pieces in each column. The number of pieces in each row is _____.

8. By using two 1s, one 2 and one 3, the number of different four-digit numbers that can be formed is _____.

9. If A works on a job for 5 hours, it will take B 3 more hours to complete it. If B works on the job for 9 hours, it will take A 3 more hours to complete it. If A works on it for only 1 hour, the number of hours it will take B to complete it is _____.

10. In a game, two players take turns calling out any positive integer from 1 to 8. The total of all numbers called out so far is computed. The player who brings this total to 88 or over is the winner. If you go first, you can have a sure win by calling out _____.

Part B.

1. There are three concentric circular tracks of lengths $\frac{1}{5}$, $\frac{1}{4}$ and $\frac{3}{8}$ kilometers respectively. The starting points are all directly east of the common center. A runs on the innermost track at 3.5 kilometers per hour. B runs on the in-between track at 4 kilometers per hour. C runs on the outermost track at 5 kilometers per hour. All three runners go clockwise. After how many hours will they be back at the starting points simultaneously for the first time?

2. A number has exactly one decimal place, and the digit before the decimal point is 8. A second number has exactly the same property. After rounding off to one decimal place, the product of these two numbers is 76.5. What is its value before rounding off?

3. A boat has a leak and water is entering at a uniform rate. When this is discovered, some water has already accumulated at the bottom of the boat. It takes 3 hours for 12 people to bail the water out, but 10 hours for 5 people to do the same. How many people will be needed if the task is to be accomplished in 2 hours?

4. A toy consists of 100 puppets in a row, plus a yellow button and a red button. A puppet is either sitting down or standing up. When the yellow button is pressed, one of the puppet standing up will sit down. When the red button is pressed, the number of puppet standing up will double. (The yellow button does not work if no puppet is standing up, and the red button does not work if more than fifty puppets are standing up.) Initially, only 3 puppets are standing up. What is the smallest number of times we have to press the buttons in order to have 21 puppets standing up?

5. Each of two yachts covers 7 kilometers per hour going downstream and 5 kilometers going upstream. They leave the same spot at the same time, one going downstream and one going upstream. Each turns around at some point in time and they meet back at their starting spot one hour later. For how many minutes during this hour are the two yachts travelling in the same direction?

6. Each digit of a one-hundred-digit number is 7. When this number is divided by 13, what is

    (a) the remainder;

    (b) the sum of the digits of the quotient?

1988

Part A.

1. The expression $\dfrac{2\frac{2}{3} \times (1\frac{7}{8} - \frac{5}{6})}{3\frac{1}{4} \div (\frac{7}{8} + 1\frac{5}{6})}$ is equal to _____.

2. The large square in Figure 6.1 is formed of 36 small squares, each of area 1 square decimeter. The area of the shaded region, in square decimeters, is _____.

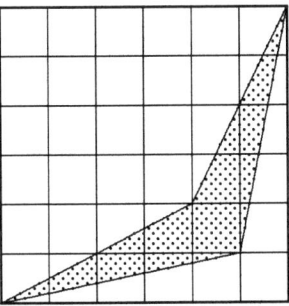

Figure 6.1

3. There were four problems in a mathematics competition. Konrad solved problems 1, 2 and 3 and got 18 points. Michael solved problems 2, 3 and 4 and got 22 points. Wilfred solved problems 1, 3 and 4, and Kenneth solved problems 1, 2 and 4. Each of them got 25 points. Winnie solved all four problems, and the number of points she got was _____.

4. The expression $(123456+234561+345612+456123+561234+612345) \div 6$ is equal to _____.

5. A yacht and a junk were sailing from port A to port B. When the yacht was at the halfway point between the two ports, the junk had covered one-third of the distance. When the yacht arrived at port B, the junk was at the halfway point. The ratio of the speeds of the yacht and the junk is _____.

6. Spinach requires 1 kilogram of fertilizer per 3 acres, and cabbage requires 2 kilograms of fertilizer per 5 acres. The total number of acres of spinach and cabbage is 18, and the total amount of fertilizer used is 7 kilograms. The number of acres of cabbage is _____.

7. The ratio of the costs of 40-watt and 30-watt sun-lamps is 4:3. The total cost of 20 40-watt and 21 30-watt sun-lamps is $715. In dollars, the cost of each 40-watt sun-lamp is _____.

8. The running speeds of A and B were respectively 13 and 11 kilometers per hour. B ran for 20 minutes longer than A, and covered 2 more kilometers. In kilometers, the total distance B had covered was _____.

9. The product of two two-digit numbers is 6975. The smaller of the two numbers is _____.

10. In the equation $\frac{1}{\bigcirc} + \frac{7}{\triangle} = \frac{13}{15}$, each of the symbols $\bigcirc$ and $\triangle$ represents a positive integer. The number represented by $\triangle$ is _____.

Part B.

1. Find the number represented by $\bigcirc$ in
$$\frac{1}{1+\cfrac{1}{2+\cfrac{1}{\bigcirc+\frac{1}{4}}}} = \frac{18}{25}.$$

2. Five hundred students form a long line. The first call out the number 1. Each other student in turn calls out a number according to the following rules. If the preceding student calls out a one-digit number, this student calls out the sum of that one-digit number and 7. If the preceding student calls out a two-digit number, this student calls out the sum of the units digit of that two-digit number and 4. What number does the last student call out?

3. Numbers are to be put into the boxes on the bottom row of Figure 6.2, so that the sum of the numbers in the three boxes on each of the four lines is the same. What is this common sum?

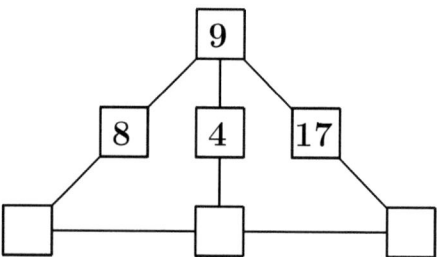

Figure 6.2

4. The ages of four girls and two boys are all different. The oldest is 10 years old and the youngest 4. The oldest girl is older than the younger boy by 4 years, and the older boy is older than the youngest girl by 4 years. Determine the age of the older boy.

5. A circular clock with only one hand is divided into 20 sectors. Every 7 minutes, the hand moves forward 9 sectors. At 8 o'clock this morning, the hand was moving from 0 to 9. Where was the hand pointing at 8 o'clock last evening?

6. Patrick went to the bookstore with 10 dollars in his pocket and bought four books. Half of the money he spent was on the dictionary. In the remaining half, $\frac{12}{25}$ was on a picture book and $\frac{7}{17}$ is on a novel. The book prices did not involve fractions of a cent. How much money did Patrick have left?

1989

Part A.

1. Correct to three decimal places, the value of $0.1\overline{6}+0.\overline{142857}+0.125+0.\overline{1}$ is _____.

2. The expression
$$(1989+1988+1987)-(1986+1985+1984)+(1983+1982+1981)$$
$$-(1980+1979+1978)+\cdots+(9+8+7)-(6+5+4)+(3+2+1)$$
is equal to _____.

3. A and B working together can do a certain job in 10 days. B and C will take 12 days, while C and A will take 15 days. The number of days in which C can do the job alone is _____.

4. The five circles in Figure 6.3 have the same center. The radius of the smallest circle is 1, and the difference between the radii of two consecutive circles is also 1. The ratio of the areas of the shaded part and the unshaded part of the largest circle is _____.

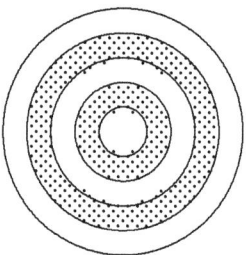

Figure 6.3

5. The total cost of some 4-cent, 8-cent and 10-cent stamps is $1. If the total number of stamps is 15, the maximum number of 10-cent stamps is _____.

6. If we reverse the order of the digits of a 4-digit number, the new number is 7902 more than the original one. The maximum value of the original number is _____.

7. A 3000-meter race was held on a round track of length 400 meters. Mindy's running speed was 5.8 meters per second, while Lana's was $\frac{3}{4}$ of a lap per minute. When Mindy was approaching the finish-line, Lana was running alongside her. At this point, the distance Mindy was from the finish-line, in meters, was _____.

8. The seven circles in Figure 6.4 form a pattern symmetric about the horizontal line. We wish to paint some of them red and the remaining ones blue, so that the coloring pattern is still symmetric about the horizontal line. The number of different ways of painting is _____.

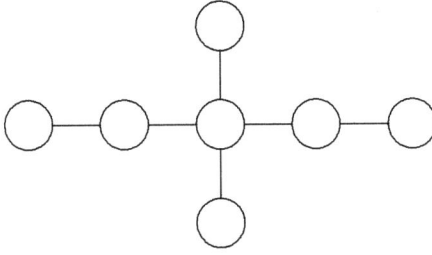

Figure 6.4

9. Nine identical right triangles are arranged around a point as shown in Figure 6.5. The measure, in degrees of the smallest angle of the triangle is _____.

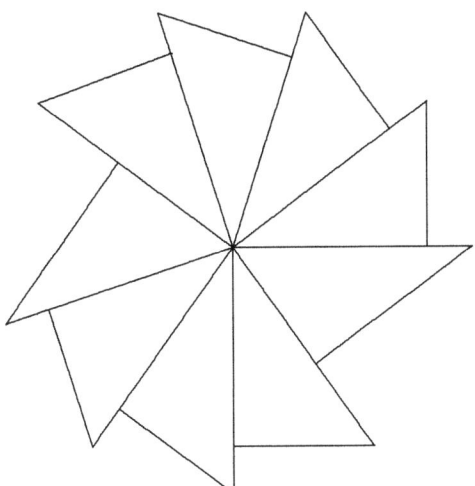

Figure 6.5

10. Michael gave $\frac{1}{3}$ of his apples plus $\frac{2}{3}$ of an apple to A, $\frac{1}{4}$ of the remaining apples and $\frac{1}{2}$ of an apple to B, $\frac{1}{2}$ of the remaining apples to C, and $\frac{1}{2}$ of the remaining apples plus $\frac{1}{2}$ of an apple to D. If 5 apples were left, Michael originally had _____.

Part B.

1. A 10 × 8 × 5 block is formed from 1 × 1 × 1 white cubes. Apart from a 10 × 8 face, the other five faces of the block are painted red. The block is taken apart into 1 × 1 × 1 cubes again. Of these cubes, how many have three red faces, how many have two red faces, how many have one red face and how many have no red faces?

2. In Figure 6.6, the total area of the two small circles is $\frac{3}{5}$ of the large circle. The shaded area inside one of the small circle is $\frac{1}{3}$ of its area while the shaded area inside the other small circle is $\frac{1}{2}$ of its area. On the other hand, the shaded areas inside the large circle are $\frac{1}{4}$ of its area. What is the ratio of the areas of the two small circles?

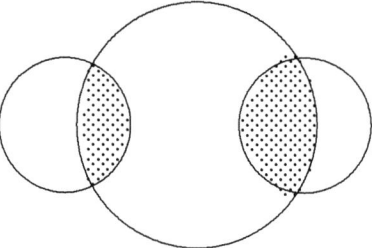

Figure 6.6

3. The top of each of three quadrilaterals is parallel to the base. The ratio of their heights is 1:2:3. The ratio of the lengths of their tops is 6:9:4. The ratio of the lengths of their bases is 12:15:10. The area of the first quadrilateral is 30 square centimeters. What is the total area of other two quadrilaterals?

4. The first number has 9 positive divisors and the second number has 10 positive divisors. What are these two numbers if their least common multiple is 2800?

5. The first number in a sequence is 6 and the second is 3. Starting from the second number, each is 5 less than the sum of the number just before it and the number just after it. What is the sum of the first 398 numbers of this sequence?

6. Lynn is walking from town A to town B, and Mike is riding a bike from town B to town A along the same road. They meet 1 hour after they have started. When Mike reaches town A, he turns around immediately. Forty minutes later, he catches up with Lynn, still on her way to town B. When Mike reaches town B, he turns around immediately. How much time has elapsed between their second and third meeting?

1990

Part A.

1. Expressed as a fraction, $\dfrac{(0.225 + \frac{7}{10}) \times \frac{13}{74}}{10.01 \times \frac{3}{11}}$ is _____.

2. The large equilateral triangle in Figure 6.7 is formed of 36 small equilateral triangles. Each is painted red or blue, so that two triangles sharing a common side have different colors. If there are more red triangles than blue ones, then the number of red triangles exceeds the number of blue ones by _____.

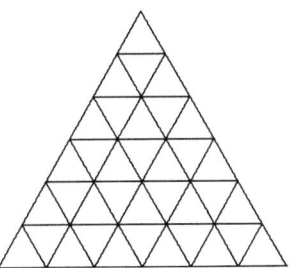

Figure 6.7

3. Consider the following method of computation:

$$1 \times 2 + 2 \times 3 = \frac{1 \times (2 \times 3)}{3} + \frac{3 \times (2 \times 3)}{3} = \frac{2 \times 3 \times 4}{3},$$

$$1 \times 2 + 2 \times 3 + 3 \times 4 = \frac{2 \times (3 \times 4)}{3} + \frac{3 \times (3 \times 4)}{3} = \frac{3 \times 4 \times 5}{3},$$

and so on. Then $1 \times 2 + 2 \times 3 + 3 \times 4 + \cdots + 19 \times 20 + 20 \times 21$ is equal to _____.

4. The average number of reference books A, B, C and D have is more than 30 and less than 40. A's number of books is $\frac{2}{3}$ that of B's, which is in turn $\frac{5}{4}$ that of C's. D has three more reference books than A. The number of reference books C has is _____.

5. Each of 16 balls which look identical weighs either 10 grams or 9 grams. They are divided into 8 groups of 2. When weighed against each other, the 2 balls in the first group do not balance. When weighed against the first group, 3 of the other groups are heavier, 2 lighter and 2 even. In grams, the total weight of these 16 balls is _____.

6. There is a point on each of two adjacent sides of a square of side length 10 meters. They are respectively 4 and 5 meters away from the common corner, as shown in Figure 6.8. A third point is to be chosen on either of the other two sides, so that the triangle formed by these three points is as large as possible. The maximum value of its area, in square meters, is _____.

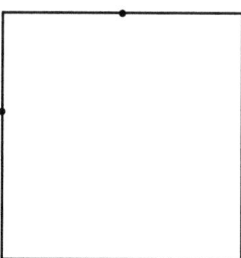

Figure 6.8

7. The large rectangle in Figure 6.9 is formed of ten small rectangles, the area of six of which are shown. The area of the large rectangle is _____.

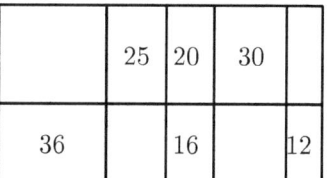

Figure 6.9

8. A perfect score in a science test is 100. Six students have an average of 91, and each has a different score. One of them only has a score of 65. The student in third place among the six has a score of at least _____.

9. If the thousands digit of a 4-digit number is removed, the resulting number is $\frac{1}{15}$ of the original one. This 4-digit number is _____.

10. Some 3-digit number is attached after 789, and we have a 6-digit number which is divisible by 7, 8 and 9. The attached 3-digit number is _____.

Part B.

1. If we round 3.56 up to 4, the rounding error is 4-3.56=0.44. If we round 3.56 down to 3, the rounding error is 3.56-3=0.56. The sum of the five numbers 2.43, 2.53, 2.65, 2.68 and 2.71 is 13. They are to be rounded up or down into five integers whose sum is still 13. What is the smallest possible sum of the five rounding errors?

2. A and C are two diametrically opposite points on a circle. An ant starts crawling from A counterclockwise along the circle at the same time as a cicadas starts crawling from C clockwise along the circle. They meet for the first time at a point B 8 millimeters from A as measured counterclockwise along the circle. They meet for the second time at a point D 6 millimeters from C as measured counterclockwise along the circle.

    (a) What is the length of the circumference of this circle?
    (b) When they meet again at B, how many times will they have met, and how far has the ant crawled up to that point?

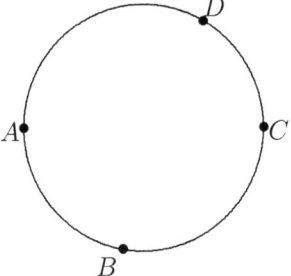

Figure 6.10

3. What is the smallest integral value of the expression $2 \div 3 \div 4 \div 5 \div 6$ when brackets are inserted to make it unambiguous?

4. Compute the product of 3333333333 and 6666666666 and give the details of the calculations.

5. There were 50 students in a class. In a test, the average score of the top 30 students was 12 more than the average score of the bottom 20 students. One of the students added the two average scores and divided the sum by 2, thinking that that was the class average. What was the difference between this and the actual class average?

6. It takes a craftsman and three apprentices to finish a certain job in 4 days. It will take only 3 days if two craftsmen and one apprentice are working on it. How long will it take an apprentice to do the job alone?

1991

Part A.

1. In decimal form, the expression $0.125 + 0.\overline{3} + \frac{5}{12}$ is equal to _____.

2. Let us denote 0.00000011 by 0.0_611. In other words, the subscript 6 counts the number of 0s after the decimal point. In this notation, $0.0_{963}181 \times 0.0_{1028}11$ is _____.

3. The radius of a circle is $123\frac{456}{789}$. Expressed as a fraction in its lowest term, the ratio of the numerical values of its area and its circumference is _____.

4. A train of length 152 meters was travelling at 63.36 kilometers per hour. A man was walking in the opposite direction to the train, and it took 8 seconds for the entire length of the train to go by him. In meters per second, his walking speed was _____.

5. In each of $3.5710\overline{64}$ and $1.6781\overline{89}$, the bar which indicates the decimal digits to be repeated is to be extended to the left over at least one digit, until the difference between the two new numbers is as small as possible. The two new numbers are _____ and _____.

6. The large square in Figure 6.11 is formed of 16 small squares, each of area 1 square centimeter. In square centimeters, the area of the shaded regions is _____.

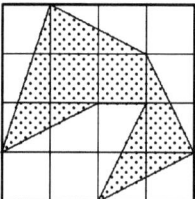

Figure 6.11

7. A test of computation speed consisted of 100 questions. Michael worked out 3 questions every minute. Kenneth took 6 seconds less than Michael to work out 5 questions. When Kenneth had finished the test, the number of questions Michael had worked out was _____.

8. A group of students were discussing their scores in a language arts test. Had Frank's score been higher by 13, their average would have gone up to 90. Had his score been lower by 5, their average score would have gone down to 87. The number of students in this group was _____.

9. When a 3-digit number is divided by 37, the remainder is 17. When it is divided by 36, the remainder is 3. This 3-digit number is _____.

10. Normally, Wallace goes to school by first walking for 5 minutes and then running for 2 minutes. One day, he started running after walking for only 2 minutes, and got to school 1 minute and 40 seconds earlier than usual. The ratio of his running speed to his walking speed is _____.

Part B.

1. A pile of coins is worth 100 cents in total. The ratio of the numbers of 2-cent coins and 1-cent coins is 1:11. The remaining ones are all 5-cent coins. How many 5-cent coins are there?

2. Each battalion consumes the same amount of food each day. A certain regiment consists of several battalions and has a certain amount of food. If one battalion is transferred into the regiment from elsewhere, there is just enough food to last 6 days. If instead, one battalion is transferred out of the regiment to elsewhere, then there will be enough food for 10 days.

    (a) How many battalions are currently in the regiment?
    (b) How many days can the food supply last for a single battalion?

3. List in ascending order all fractions between 0 and 1 such that the numerator and the denominator have no common factors greater than 1, and the product of the numerator and the denominator is 720.

4. Insert nine plus or minus signs between the ten numbers on the left-hand side of the expression 10 9 8 7 6 5 4 3 2 1=37 to make it correct, in such a way that the product of the numbers with minus signs in front of them is as large as possible. What is the largest possible value of this product?

5. In a five-digit number, the number formed of the first three digits is divisible by 9, while the number formed of the last two digits is divisible by 7. If we write down the five digits in reverse order, the sum of the new number and the original number is 67866. What is the original number?

6. Lynn and Wallace were at the entrance of the park. They wished to go to a shop by walking east along the street. Wallace asked, "Would it be faster if we go home by walking west along the street, get our bicycles and then ride to the shop?" Lynn did a bit of calculation and said, "The ratio of our riding speed to our walking speed is 4:1. It would only save time if the distance of the shop from here is at least 2 kilometers." How far, in meters, was their home from the entrance of the park?

1992

Part A.

1. The expression $13.64 \times 0.25 \div 1.1$ is equal to _____.

2. In the addition shown in Figure 6.12, each ◯ represents a different one of the digits 0, 1, 3, 5, 6, 7 and 9. The 4-digit sum is _____.

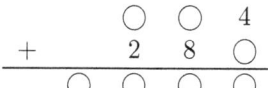

Figure 6.12

3. A track encloses a rectangle plus a semicircle on each end, as shown in Figure 6.13. The total length of the track is 400 meters, and each circular part of the track is 100 meters long. The ratio of the total area of the two semicircles to that of the total area enclosed by the track is _____.

Figure 6.13

4. A shop is selling two items at the same price. One is 25% above its cost while the other is 20% below it cost. The ratio of the total price to the total cost of these two items is _____.

5. The prime numbers 17, 23, 31, 41, 53, 67, 79, 83, 101 and 103 are to be divided into two groups of five, such that the total of each group is the same. In the group containing 101, the second smallest number is _____.

6. Konrad leaves home for school at precisely the same time every morning, walking at 70 meters per second. Mr. Zhang goes out for a stroll every day, also at a fixed time, walking at 40 meters per second. They go in opposite directions along the same road, and meet on the way. One morning, Konrad left early, and met Mr. Zhang 7 minutes earlier than usual. In minutes, the amount of time Konrad had left earlier than usual was _____.

7. In the currency of a certain country, there are only $1, $3, $5, $7 and $9 bills. A stack of bills can be used to pay exactly any integer amount of dollars from $1 to $100. The smallest number of bills in such a stack is _____.

8. In a building, every apartment subscribes to two different ones of the newspapers *China Television Guide*, *Beijing Evening News* and *References and Information*. The numbers of apartments subscribing to these newspapers are 34, 30 and 22 respectively. The number of apartments subscribing to both *Beijing Evening News* and *References and Information* is _____.

9. We use the symbol $\triangledown$ to denote the operation of choosing the larger of two numbers, and $\triangle$ that of choosing the smaller of the two. For example, $5 \triangledown 3 = 3 \triangledown 5 = 5$ and $5 \triangle 3 = 3 \triangle 5 = 3$. The expression $((0.\overline{6} \triangledown \frac{17}{26}) + (0.625 \triangle \frac{23}{33})) \div ((0.\overline{3} \triangle \frac{34}{99}) + (2.25 \triangle \frac{237}{106}))$ is equal to _____.

10. Express 37 as a sum of different prime numbers, so that the product of these prime numbers is as large as possible. The maximum value of the product is _____.

Part B.

1. In the expressions $\frac{1}{2} \odot \frac{1}{9}$, $\frac{1}{3} \odot \frac{1}{8}$, $\frac{1}{4} \odot \frac{1}{7}$ and $\frac{1}{5} \odot \frac{1}{6}$, replace the four $\odot$ by $+$, $-$, $\times$ and $\div$, using each of the four basic operations exactly once, so that the sum of the four answers is as large as possible. What is the largest possible value of this sum?

2. Figure 6.14 shows an eight-sided polygon whose side lengths are 1, 2, 3, 4, 5, 6, 7 and 8 in some order. Segment (b) is longer than segment (c), which is in turn longer than segment (a). Segment (d) is longer than segment (e), which is in turn longer than segment (f), which is in turn longer than segment (g). What is the area of this polygon?

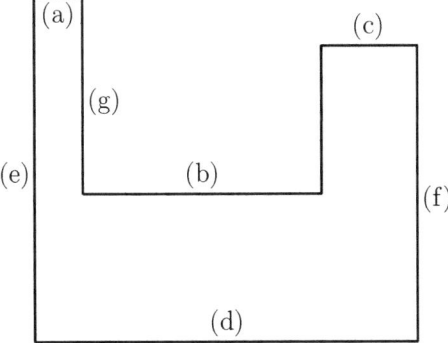

Figure 6.14

3. At most how many of the numbers among 50 consecutive three-digit numbers have the property that the sum of the three digits is divisible by 7?

4. For a project of a single-parent club, one third of the club members brought along one child each. The others came on their own. The club planted 216 trees. Each adult male planted 13 trees, each adult female planted 10 trees and each child planted 6 trees. How many female members came to the project?

5. For a photocopying job, machine A will take 11 hours while machine B will take 13 hours. When both machines are used, the job takes 6 hours and 15 minutes, and it is noticed that some interferences between the machines slow them down so that 28 less pages are processed per hour. How many pages are in this photocopying job?

6. Konrad rides his bike from A to B. The first part of the trip is uphill and the second part on level ground. As soon as he arrives at B, he makes the return journey along the same road, arriving back at A 3 hours after his departure from A. Going uphill, Konrad covers 6 kilometers less per hour than on level ground. Going downhill, he covers 3 kilometers more per hour than on level ground. He notices that he covers 5 kilometers less in the first hour than in the second, and 3 kilometers less in the second hour than in the third.

   (a) How many minutes does Konrad spend riding uphill?
   (b) How many minutes does Konrad spend riding downhill?
   (c) What is the distance between A and B along this road?

Section 2. Solutions.

1987

Part A.

1. The computation below shows that the answer is **0.45**.

$$
\begin{array}{r}
0.450 \\
314159265 \overline{\smash{\big)}\ 141421350.000} \\
125663706.000 \\
\hline
15757644.000 \\
15707963.250 \\
\hline
49680.750
\end{array}
$$

2. **First Solution:**

Put four copies of the triangle together, as shown in Figure 6.15, to form a square of side length 12 centimeters. The area of the square is $12 \times 12 = 144$ square centimeters, and the area of the triangle is $144 \div 4 = $ **36** square centimeters.

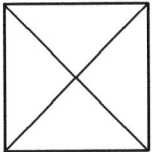

Figure 6.15

Second Solution:

Cut the triangle in halves, each of which is a $45° - 45° - 90°$ triangle with the shortest side 6 centimeters. The two halves may be reassembled, as shown in Figure 6.16, into a square of side length 6 centimeters. The area of the square, and hence the area of the triangle, is $6 \times 6 = $ **36** square centimeters.

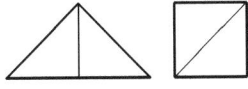

Figure 6.16

3. Since the overall sum is even, the sum of all the odd numbers must be even. Hence the number of odd numbers is also even. Since there are more odd numbers than even numbers, the number of even numbers is at most 48. If we take 48 copies of 100, 26 copies of 99 and 26 copies of 101, the total is 10000 and the number of even numbers is exactly **48**.

4. **First Solution:**
   Let $S = \frac{1}{18} + \frac{1}{\bigcirc} + \frac{1}{\triangle} + \frac{1}{\triangledown}$. We may assume that $\bigcirc \geq \triangle \geq \triangledown$. Suppose $\triangledown \geq 4$. Then $S \leq \frac{1}{18} + \frac{1}{4} + \frac{1}{4} + \frac{1}{4} < 1$. Hence $\triangledown \leq 3$. Suppose $\triangledown = 3$. Then either $S = \frac{1}{18} + \frac{1}{3} + \frac{1}{3} + \frac{1}{3} > 1$ or $S \leq \frac{1}{18} + \frac{1}{4} + \frac{1}{3} + \frac{1}{3} < 1$. Hence $\triangledown = 2$ and we now have $T = \frac{1}{18} + \frac{1}{\bigcirc} + \frac{1}{\triangle} = \frac{1}{2}$. Suppose $\triangle \geq 5$. Then $T \leq \frac{1}{18} + \frac{1}{5} + \frac{1}{5} < 1$. Hence $\triangle \leq 4$. Suppose $\triangle = 4$. Then either $T \geq \frac{1}{18} + \frac{1}{5} + \frac{1}{4} > 1$ or $T \leq \frac{1}{18} + \frac{1}{6} + \frac{1}{4} < 1$. Hence $\triangle = 3$. Now $\frac{1}{\bigcirc} = \frac{1}{6} - \frac{1}{18} = \frac{1}{9}$. Hence $\bigcirc = 9$ and $\bigcirc + \triangle + \triangledown = \mathbf{14}$.

   Second Solution:
   Without $\frac{1}{18}$, there are only three possibilities, namely, $\frac{1}{6} + \frac{1}{3} + \frac{1}{2}$, $\frac{1}{4} + \frac{1}{4} + \frac{1}{2}$ and $\frac{1}{3} + \frac{1}{3} + \frac{1}{3} = 1$. Now $\frac{1}{2} - \frac{1}{18} = \frac{4}{9}$, $\frac{1}{3} - \frac{1}{18} = \frac{5}{18}$, $\frac{1}{4} - \frac{1}{18} = \frac{7}{36}$ while $\frac{1}{6} - \frac{1}{18} = \frac{1}{9}$. Hence we may take $\bigcirc = 9$, $\triangle = 3$ and $\triangledown = 2$, and their sum is **14**.

5. In 1988, the father's age was 8 more than four times his sons' total age in 1980, and $8 \times 4 = 32$ more than twice his sons' total age in 1980, which must therefore be $(32 - 8) \div 2 = 12$. Hence the father's age in 1980 was $12 \times 4 = 48$, so that he was born in $1980 - 48 = \mathbf{1932}$.

6. **First Solution:**
   There are $100 - 40 = 60$ non-square rectangles and $100 - 40 = 60$ non-square rhombi. Hence the minimum number of parallelograms is $40 + 60 + 60 = \mathbf{160}$.

   Second Solution:
   A square is both a rectangle and a rhombus. In the sum $100 + 100 = 200$, we have counted each square twice. Hence the minimum number of parallelograms is $200 - 40 = \mathbf{160}$.

7. **First Solution:**
   From $12 = 3 \times 4$, we guess that $1122 = 33 \times 34$, $111222 = 333 \times 334$ and $11112222 = 3333 \times 3334$. The computation below verifies that the last equation holds, so that the number of pieces in each row is **3334**.

   ```
                    3 3 3 4
         ×          3 3 3 3
         ─────────────────
            1 0 0 0 2
              1 0 0 0 2
                1 0 0 0 2
                  1 0 0 0 2
         ─────────────────
            1 1 1 1 2 2 2 2
   ```

   Second Solution:
   We have $11112222 = 1111 \times 10002 = 1111 \times 3 \times 3334 = 3333 \times 3334$. Hence the number of pieces in each row is **3334**.

Solutions

8. **First Solution:**
   By writing down the possible numbers in increasing order, we have 1123, 1132, 1213, 1231, 1312, 1321, 2113, 2131, 2311, 3112, 3121 and 3211, a total of **12** numbers.

   Second Solution:
   The possible positions for the two 1s are first and second, first and third, first and fourth, second and third, second and fourth, and third and fourth. The 2 can either come before the 3 or after it. Hence the total number of numbers is $6 \times 2 = \mathbf{12}$.

   Third Solution:
   Pretend for now that one of the 1s is a 4. Then we have four choices for the first digit, three choices for the second and two for the third, for a total of $4 \times 3 \times 2 = 24$ numbers. If we now change the 4 back to 1, we can no longer tell whether the 4 comes before the 1 or after it. Hence the total number of numbers is $24 \div 2 = \mathbf{12}$.

9. Suppose one part of the job is completed if each of A and B works for 3 hours. Then the remaining part of the job can be completed if A works for 2 hours or B works for 6 hours. Hence A works 3 times as fast as B, and the whole job would have taken B $5 \times 3 + 3 = 18$ hours to complete. If A works on it for 1 hour, B can complete the job in $18 - 1 \times 3 = \mathbf{15}$ hours.

10. We win if we call out 88. So we say 88 is a safe number. It is not safe to call out any of 87, 86, 85, 84, 83, 82, 81 and 80, because our opponent will win by calling out 1, 2, 3, 4, 5, 6, 7 and 8 respectively. On the other hand, 79 is a safe number. Our opponent is forced to call out one of 80, 81, 82, 83, 84, 85, 86 and 87, all of which are unsafe. Thus each safe number is 9 less than the preceding one. Then our opponent cannot go from one safe position to another, and must go to an unsafe one. From there, we can get back to another safe position. So the safe positions are 88, 79, 70, 61, 52, 43, 34, 25, 16 and **7**, the last one being the unique winning opening move.

Part B.

1. The number of hours A takes to complete a lap is $\frac{1}{5} \div 3.5 = \frac{2}{35}$. The number of hours B takes to complete a lap is $\frac{1}{4} \div 4 = \frac{1}{16}$. The number of hours C takes to complete a lap is $\frac{3}{8} \div 5 = \frac{3}{40}$. Taking common denominators, these numbers become $\frac{32}{560}$, $\frac{35}{560}$ and $\frac{42}{560}$. The least common multiple of 32, 35 and 42 is $2^5 \times 3 \times 5 \times 7 = 3360$. Hence the first time they will be back at the starting points simultaneously is after $3360 \div 560 = \mathbf{6}$ hours.

2. Since the product of the two numbers is less than 100, the integer part of each is 8. Since $9 \times 8.5 = 76.5$, both of them are among 8.6, 8.7, 8.8 and 8.9, probably identical. Now $8.6 \times 8.8 = 75.68$ will only round off to 75.7, as will $8.7 \times 8.7 = 75.69$. On the other hand, $8.7 \times 8.8 = 76.56$ will round off to 76.6. Hence the two numbers must be 8.6 and 8.9, and their product is **76.54**.

3. Compared to the 12 people in 3 hours, the extra amount of water bailed out by the 5 people in 10 hours is what leaks in during 7 hours. Now $5 \times 10 - 12 \times 3 = 14$. Hence it takes $14 \div 7 = 2$ people 1 hours to bail out what leaks in during 1 hour. Since $(5-2) \times 10 = 30 = (12-2) \times 3$, the boat has been leaking for 30 hours. To bail out the accumulated in 2 hours will need $30 \div 2 = 15$ people. Accounting for the amount leaking in during these 2 hours, an extra 2 people will be needed, bringing the total to 15+2=**17**.

4. Working from the back, we see that the only way to get to 21 is from 22. Now 22 can be back-tracked to 23, 24, 12, 6 and 3 or 11, 12, 6 and 3. The latter sequence is faster as we only have to press the buttons **5** times, namely, red, red, yellow, red and yellow.

5. Since the ratio of the speeds downstream and upstream is 7:5, the ratio of the time spent going downstream and upstream is 5:7. Since each yacht has travelled for 1 hour, $60 \times \frac{5}{5+7} = 25$ minutes are downstream and $60 - 25 = 35$ minutes upstream. When the yacht which starts downstream turns around, the other yacht is still going upstream. Hence the amount of time during this hour when they are travelling in the same direction is $2 \times 35 - 60 = 10$ minutes.

6. Note that $1001 = 7 \times 11 \times 13$. Hence $777777 \div 13 = 777 \times 77 = 59829$. When 100 is divided by 6, the quotient is 16 and the remainder is 4. When 7777 is divided by 13, the quotient is 598 and the remainder is 3.

   (a) When the given number is divided by 13, the remainder is **3**.

   (b) When the given number is divided by 13, the sum of the digits of the quotient is given by $16(5+9+8+2+9)+5+9+8=$**550**.

Solutions

1988

Part A.

1. We have

$$\frac{2\frac{2}{3} \times (1\frac{7}{8} - \frac{5}{6})}{3\frac{1}{4} \div (\frac{7}{8} + \frac{11}{6})} = \frac{\frac{8}{3} \times (\frac{45}{24} - \frac{20}{24})}{\frac{13}{4} \div (\frac{21}{24} + \frac{44}{24})}$$

$$= \frac{\frac{8}{3} \times \frac{25}{24}}{\frac{13}{4} \times \frac{24}{65}}$$

$$= \frac{125}{54}$$

$$= 2\frac{17}{54}.$$

2. **First Solution:**
As shown in Figure 6.17 on the left, the shaded region may be divided into four triangles each of which is of base 3 decimeters and height 1 decimeter. Hence their total area is $4 \times (\frac{1}{2} \times 3 \times 1) = $ **6** square decimeters.

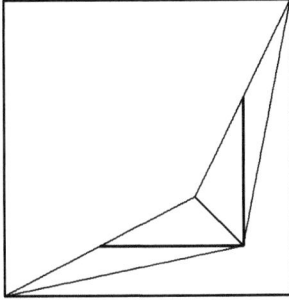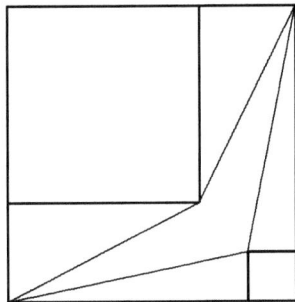

Figure 6.17

Second Solution:
As shown in Figure 6.17 on the right, the unshaded region may be divided into a large square of area 16 square decimeters, a small square of area 1 square decimeter, two large triangles each of area 4 square decimeters and two small triangles each of area $2\frac{1}{2}$ square decimeters. The area of the shaded region is $36 - 16 - 1 - 2(4 + 2\frac{1}{2}) = $ **6** square decimeters.

Third Solution:
As shown in Figure 6.18 on the left, the shaded region may be divided into two triangles each of which is $\frac{1}{3}$ the area of $\frac{1}{4}$ of the large square. Hence their total area is $2 \times \frac{1}{3} \times \frac{1}{4} \times 36 = $ **6** square decimeters.

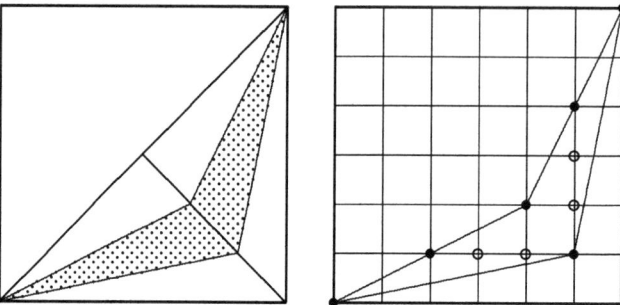

Figure 6.18

Fourth Solution:
There is a result called **Pick's Formula** which states that the area of a polygon whose vertices are all grid points is equal to $I - \frac{1}{2}B - 1$, where I is the number of grid points inside the polygon and B is the number of grid points on the perimeter of the polygon. As shown in Figure 6.18 on the right, $I = 4$ and $B = 6$. Hence the area is $4 + \frac{1}{2} \times 6 - 1 =$ **6** square decimeters.

3. Among them, the four boys solved each problem 3 times and got 18+22+25+25 =90 points. Winnie solved each problem once and got $90 \div 3 =$ **30** points.

4. Since each of the digits 1, 2, 3, 4, 5 and 6 appears exactly once in each place, the sum is equal to $(1 + 2 + 3 + 4 + 5 + 6) \times 111111 = 2333331$ and the quotient is $2333331 \div 6 = 777777 \div 2 =$ **388888.5**.

5. The situation is impossible unless the yacht and the junk did not start at the same time. When the yacht covered the second $\frac{1}{2}$ of the distance, the junk covered $\frac{1}{2} - \frac{1}{3} = \frac{1}{6}$ of the distance. Hence the desired ratio is $\frac{1}{2} : \frac{1}{6} =$ **3:1**.

6. Each acre of spinach requires $\frac{1}{3}$ kilogram of fertilizer, while each acre of cabbage requires $\frac{2}{5}$ kilogram of fertilizer. Hence each acre of cabbage requires $\frac{2}{5} - \frac{1}{3} = \frac{1}{15}$ kilogram of fertilizer. If all 18 acres were spinach, 6 kilograms of fertilizer are needed. The extra $7 - 6 = 1$ kilogram accounts for $1 \div \frac{1}{15} =$ **15** acres of cabbage.

7. The cost for 21 30-watt sun-lamps is equal to the cost for $21 \times \frac{3}{4}$ 40-watt sun-lamps. Hence $20 + \frac{21 \times 3}{4}$ 40-watt sun-lamps cost \$715 so that each costs \$715 $\div \frac{4 \times 20 + 63}{4} =$ **\$20**.

Solutions

8. In 20 minutes, B covered $\frac{11}{3}$ kilometers. A covered $13-11=2$ kilometers more than B per hour. Hence A covered $\frac{11}{3}-2=\frac{5}{3}$ kilometers more than B in $\frac{5}{3} \div 2 = \frac{5}{6}$ hours. During this time, B covered $11 \times \frac{5}{6} = 9\frac{1}{6}$ kilometers. Thus the total distance B had covered is $9\frac{1}{6} + 3\frac{2}{3} =\mathbf{12\frac{5}{6}}$ kilometers.

9. Factoring 6975, we have $3 \times 3 \times 5 \times 5 \times 31$. Since 5×31 is a three-digit number, 25 divides one of the two-digit factors and 31 divides the other. Since 9×25 is a three-digit number, the smaller factor must be $3 \times 25=\mathbf{75}$.

10. We have $\frac{7}{\triangle} = \frac{13}{15} - \frac{1}{\bigcirc} = \frac{13 \times \bigcirc - 15}{15 \times \bigcirc}$. Hence $13 \times \bigcirc - 15$ is divisible by 7. Now $-15+13 = -2$, $-2+13 = 11$, $11+13 = 24$, $24+13 = 37$, $37+13 = 50$ and $50+13 = 63$. We finally hit a multiple of 7. So, if $\bigcirc = 6$, then $\frac{7}{\triangle} = \frac{13 \times 6 - 15}{15 \times 6} = \frac{7}{10}$ so that $\triangle=\mathbf{10}$. If $\bigcirc > 6$, then $\triangle < 10$ but $\frac{13}{15} - \frac{7}{9} = \frac{4}{45}$ does not have 1 as its numerator, and $\frac{13}{15} - \frac{7}{\triangle} < 0$ if $\triangle < 9$.

Part B.

1. We have

$$1 + \cfrac{1}{2 + \cfrac{1}{\bigcirc + \frac{1}{4}}} = \frac{25}{18},$$

$$2 + \cfrac{1}{\bigcirc + \frac{1}{4}} = \frac{18}{7},$$

$$\bigcirc + \frac{1}{4} = \frac{7}{4},$$

so that $\bigcirc=\mathbf{1\frac{1}{2}}$.

2. The first few numbers called out are 1, 8, 15, 9, 16, 10, 4, 11, 5, 12, 6, 13, 7, 14 and 8. Thus the numbers called out, apart from the first, form a cycle of length 13. We may pretend that the first number called is 14 since this has no effect on subsequent values. When 500 is divided by 13, the remainder is 6. This means that the numbers move round the cycle a number of times, and there are still 6 more to go. Hence the last number called is the same as the sixth, namely, **10**.

3. The sum of the numbers on the three non-horizontal lines, with the top number counted three times, is three times the common sum, which is $3 \times 9 + 8 + 4 + 17 = 56$ plus the numbers on the horizontal line. It follows that the common sum is $56 \div 2 = \mathbf{28}$. This can be realized. In fact, the missing numbers from left to right must be $28 - 9 - 8=11$, $28 - 9 - 4=15$ and $28 - 9 - 17=2$.

4. Suppose the older boy is 10. Then the youngest girl is 6, the younger boy 4 and the oldest girl 8. This is impossible since there are 4 girls of different ages. Hence the oldest girl must be 10, the younger boy 6, the youngest girl 4 and the older boy **8**. This is not impossible as the other two girls may be any two of 5, 7 and 9.

5. Every 140 minutes, the hand moves forward 180 sectors and returns to its starting point. A 12-hour interval may be divided into five 140-minute blocks with 20 minutes left. Hence at 8:20 last evening, the hand was moving from 0 to 9. At 8:13, it was moving from 11 to 0, and at 8:06, it was moving from 2 to 11. Thus at 8 o'clock last evening, the hand was pointing at **2**.

6. Since $\frac{12}{25} + \frac{7}{17} = \frac{379}{425}$ and 379 is relatively prime to 425, half of what Patrick spent must be a multiple of 425 cents so that the total amount must be a multiple of 850 cents. Since he had only 10 dollars, the amount was exactly 850 cents. What he had left was $1000 - 850 = 150$ cents or **1** dollar and **50** cents.

1989

Part A.

1. We have $0.1\bar{6} = 1.\bar{6} \div 10 = 1\frac{6}{9} \div 10 = \frac{1}{6}$, $0.\overline{142857} = \frac{142857}{999999} = \frac{1}{7}$, $0.125 = \frac{125}{1000} = \frac{1}{8}$ and $0.\bar{1} = \frac{1}{9}$. Hence the given expression is equal to $\frac{1}{6} + \frac{1}{7} + \frac{1}{8} + \frac{1}{9} = \frac{84+72+63+56}{504} = \frac{275}{504} = 0.5456\ldots$. Correct to three decimal places, the answer is **0.546**.

2. The value of the part of the expression consisting of two brackets linked by a minus sign is 9 since each number in the first bracket is 3 more than the corresponding number in the second bracket. When 1989 is divided by 6, the quotient is 331 and the remainder is 3. The sum of the three numbers left over is 6. Hence the value of the expression is $331 \times 9 + 6 = $ **2985**.

3. In one day, A and B can do $\frac{1}{10}$ of the job, B and C can do $\frac{1}{12}$ while C and A can do $\frac{1}{15}$. Hence C alone can do $(\frac{1}{12} + \frac{1}{15} - \frac{1}{10}) \div 2 = \frac{1}{40}$. Hence C can do the job alone in **40** days.

4. The area of the shaded part is $\pi(4^2 - 3^2 + 2^2 - 1^2) = 10\pi$ while the area of the unshaded part is $\pi(5^2 - 10) = 15\pi$. Hence the desired ratio is 10:15 or **2:3**.

5. **First Solution:**
We may have as many as six 10-cent stamps if we also have one 8-cent stamp and eight 4-cent stamps. Since 4, 8 and 100 are all multiples of 4 but 10 is only a multiple of 2, the number of 10-cent stamps must be even. If we have eight of them for a total cost of 80 cents, the remaining seven stamps will have a total cost of only 20 cents. This is impossible even if they are all 4-cent stamps. Hence the maximum number of 10-cent stamps is **6**.

    Second Solution:
If all fifteen stamps are 4-cent stamps, the total cost is 60 cents, 40 cents short of \$1. Replacing a 4-cent stamp by an 8-cent stamp makes up for 4 cents, and replacing a 4-cent stamp by a 10-cent stamp makes up for 6 cents. Dividing 40 by 6, we have a quotient of 6 and a remainder of 4. This means that we must replace one 4-cent stamp by an 8-cent stamp and eight 4-cent stamps by **6** 10-cent stamps.

6. In the original number, the thousands digit must be 1 or 2, and the units digits 8 or 9. Since the units digit of the difference between the new number and the old number is 2, the thousands digit of the original number must be 1 and the units digit must be 9. Since the tens digit of the difference is 0, the tens digit of the original number must be 1 less than the hundreds digit. Thus the maximum value of the original number is **1989**.

7. Lana's running speed was $400 \times \frac{3}{4} \div 60 = 5$ meters per second, which was less than Mindy's. Hence when Mindy was approaching the finish-line and Lana was running alongside her, Mindy has run a number of laps more than Lana. Mindy laps Lana in $400 \div (5.8-5) = 500$ seconds. During this time, Mindy had run $500 \times 5.8 = 2900$ meters of the 3000-meter race. Hence she could not have lapped Lana a second time. The distance Mindy was from the finish-line was therefore $3000-2900=\mathbf{100}$ meters.

8. Each of the five circles on the horizontal line may be red or blue. The other two circles may either be both red or both blue. Hence the total number of ways is $2 \times 2 \times 2 \times 2 \times 2 \times 2 = \mathbf{64}$.

9. At the point where the nine triangles meet, the larger acute angle appears 2 times while the smaller acute angle appears 7 times. The sum of 2 angles of each size is $2 \times 90° = 180°$. Hence the sum of 5 times the smaller acute angle is also $180°$, so that each is $180° \div 5 = 36°$. It follows that the measure of the smallest angle of the triangle is $\mathbf{36°}$.

10. Working backwards, we find that before giving apples to D, Michael still had $(5+\frac{1}{2}) \div (1-\frac{1}{2}) = 11$ applies. Before giving apples to C, he still had $11 \div (1-\frac{1}{2}) = 22$ apples. Before giving apples to B, Michael still had $(22+\frac{1}{2}) \div (1-\frac{1}{4}) = 30$ apples. Before giving apples to A, he had $(30+\frac{2}{3}) \div (1-\frac{1}{3}) = \mathbf{46}$ apples.

Part B.

1. Call the unpainted 10×8 face be the top. The only cubes with three red faces are the corners of the bottom, and there are **4** of them. The cubes with two red faces are those along the edges of the bottom, as well as along the four vertical edges. Their total number is

$$2(10-2) + 2(8-2) + 4(5-1) = \mathbf{44}.$$

The cubes with one red face are those in the interior of the bottom and the four vertical faces. Their total number is

$$(10-2)(8-2) + 2(10-2)(5-1) + 2(8-2)(5-1) = \mathbf{160}.$$

The cubes with no red faces are those in the interior of the cube. Their total number is
$$(10-2)(8-2)(5-1) = \mathbf{192}.$$

Solutions

2. Suppose the area of the large circle is 20. Then the sum of the areas of the two shaded regions is $20 \times \frac{1}{4} = 5$. The sum of the areas of the two small circles is $20 \times \frac{3}{5} = 12$. The sum of the areas of their unshaded parts is $12 - 5 = 7$, and the difference in total area between the unshaded parts and the shaded parts is $7 - 5 = 2$. In the first small circle, the ratio of the areas of the unshaded part to the shaded part is 2:1, while it is 1:1 in the second small circle. Therefore the area of the shaded part of the first small circle is 2 and the area of the unshaded parts is 4. In the second small circle, both the shaded and unshaded parts have area 3. Therefore, the ratio of the areas of the two small circles is **1:1**.

3. The area of such a quadrilateral will not change if we reduce its height to $\frac{1}{k}$ of its value, and increase both its top and its base to k times their respective values. Applying this transformation with $k = 2$ to the second quadrilateral and $k = 3$ to the third, the new ratio of their heights, tops and bases are 1:1:1, 6:18:12 and 12:30:30. The total length of the tops of the second and the third quadrilaterals is $(18 + 12) \div 6 = 5$ times the top of the first quadrilateral, and the total length of the bottoms of the second and the third quadrilaterals is $(30 + 30) \div 12 = 5$ times the top of the first quadrilateral. It follows that the total area of the second and the third quadrilateral is also 5 times that of the first quadrilateral, that is, $30 \times 5 =$ **150** square centimeters.

4. The divisors of a number come in pairs whose products are equal to the number itself. When the number of divisors is odd, the two divisors in one of the pairs must be the same, so that the number is a square. Now $2800 = 2^4 \times 5^2 \times 7$ is divisible by the following squares: 2^2, $(2^2)^2$, 5^2, $(2 \times 5)^2$ and $(2^2 \times 5)^2$. Among them, only $100 = 2^2 \times 5^2$ has exactly 9 divisors, namely, 1, 2, 4, 5, 10, 20, 25, 50 and 100. The other number must be divisible by $2^4 \times 7 = 112$, and 112 itself happens to have exactly 10 divisors, namely, 1, 2, 4, 7, 8, 14, 16, 28, 56 and 112. Hence the two numbers are **100** and **112**.

5. The third number is $5 + 3 - 6 = 2$, the fourth $5 + 2 - 3 = 4$, the fifth $5 + 4 - 2 = 7$, the sixth $5 + 7 - 4 = 8$, the seventh $5 + 8 - 7 = 6$ and the eighth $5 + 6 - 8 = 3$. Hence the sequence repeats in a cycle (6,3,2,4,7,8) of length six. When 398 is divided by 6, the quotient is 66 and the remainder is 2. Hence the sum of the first 398 numbers in this sequence is $(6 + 3 + 2 + 4 + 7 + 8) \times 66 + 6 + 3 =$ **1989**.

6. At the third meeting, Lynn and Mike between them have covered the distance between A and B three times. Since it takes them 60 minutes to cover the distance once, they meet for the third time 180 minutes after they have started. Since the first meeting occurs after 60 minutes, and the second meeting occurs 40 minutes after that, the third meeting occurs 80 minutes after that. So the time elapsed is **1** hour and **20** minutes.

Solutions

1990

Part A.

1. The numerator is equal to $(\frac{9}{40} + \frac{28}{40}) \times \frac{13}{74} = \frac{13}{80}$. The denominator is equal to $\frac{1001}{100} \times \frac{3}{11} = \frac{273}{100}$. It follows that the expression is equal to $\frac{13}{80} \div \frac{273}{100} = \frac{5}{84}$.

2. There are two kinds of small triangles, those pointing up and those pointing down. Two triangles sharing a common side must be of different kinds. Hence all triangles of each kind are in the same color. Since there are more red triangles than blue ones, and more triangles pointing up than those pointing down, the triangles pointing up are red and those pointing down are blue. The triangles in each row alternate in color, and each starts and ends with a red one. Hence there is an extra red triangle in each row. Since there are 6 rows, there are **6** extra red triangles.

3. The pattern suggests that the expression is equal to $\frac{20 \times 21 \times 22}{3} = \mathbf{3080}$.

   Remark:
   This result can be justified by a method known as *mathematical induction*.

4. A's number of books is $\frac{2}{3} \times \frac{5}{4} = \frac{5}{6}$ of C's. If D puts aside three of his books, the ratio of the numbers of books they have is $\frac{1}{6} : \frac{1}{4} : \frac{1}{5} : \frac{1}{6}$ or $\frac{10}{47} : \frac{15}{47} : \frac{12}{47} : \frac{10}{47}$. Now the total number of books they have, minus the three D puts aside, lies strictly between $4 \times 30 - 3 = 117$ and $4 \times 40 - 3 = 157$. The only multiple of 47 in this range is 141. It follows that C's number of books is $141 \times \frac{12}{47} = \mathbf{36}$.

5. Since the 2 balls in the first group do not balance each other, one weighs 10 grams and the other weighs 9 grams. The total weight of each of the 2 groups which balance the first group is also 19 grams. A heavier group than the first has a total weight of 20 grams while a lighter group than the first has a total weight of 18 grams. Hence the total weight of the 16 balls is $3 \times 19 + 3 \times 20 + 2 \times 18 = \mathbf{153}$ grams.

6. If the side of the triangle connecting the two given points is considered to be its base, clearly the height is maximum when the third vertex is at the point of intersection of the other two sides of the square. Subtracting off the areas of the three outside triangles in Figure 6.19 from the area of the square, the area of the inside triangle is given by $10 \times 10 - \frac{1}{2}(4 \times 5 + 5 \times 10 + 10 \times 6) = \mathbf{35}$ square meters.

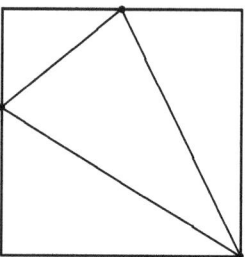

Figure 6.19

7. From the third column, the ratio of the height of the two rows is 5:4. Hence the missing areas of the rectangles in the first and fifth columns are respectively $36 \times \frac{5}{4} = 45$ and $12 \times \frac{5}{4} = 15$. Hence the total area of the first row is $45 + 25 + 20 + 30 + 15 = 135$, and the area of the whole rectangle is $135 \times \frac{5+4}{5} =$ **243**.

8. The total score of the six students is $6 \times 91 = 546$. Hence the total score of the top five is $546 - 65 = 481$. To calculate the minimum score of the student in third place, we first maximize the scores of the top two students at 100 and 99, leaving a total score of $481 - 100 - 99 = 282$ for the other three. The average here is $282 \div 3 = 94$. Hence the minimum score for the student in third place is 94+1=**95**.

9. The thousands digit times 1000 is $1 - \frac{1}{15} = \frac{14}{15}$ of the original number. Hence 14 times the original number is a multiple of 1000. It follows that this multiple must be 7000, and the original number is given by $7000 \div \frac{14}{15} =$ **7500**.

10. Since no two of 7, 8 and 9 have any common divisors greater than one, a common multiple of 7, 8 and 9 must be a multiple of $7 \times 8 \times 9 = 504$. Dividing 790000 by 504, the remainder is 232. It follows that 504 divides $790000 - 232 = 789768$ and $789768 - 504 = 789264$. Thus we may attach either **264** or **768**.

Part B.

1. The sum of the integer parts of the five numbers is 2+2+2+2+2=10. Hence $13 - 10 = 3$ of them must be rounded up and $5 - 3 = 2$ of them rounded down. Arrange the number in descending order of the size of their decimal parts. Clearly, to minimize the sum of the rounding errors, we should round up the top 3 and round down the bottom 2. Hence the sum of the rounding errors is

$$3 - 0.71 - 0.68 - 0.65 + 0.53 + 0.43 = \mathbf{1.92}.$$

2. (a) At their first meeting at B, the ant and the cicadas have crawled a semicircle between them, and the ant had crawled 8 millimeters. At their second meeting at D, they have crawled three semicircles between them, so that the ant must have crawled $8 \times 3 = 24$ millimeters. Since the distance between C and D is 6 millimeters, the distance between A and C is $24 - 6 = 18$ millimeters. Hence the circumference of the circle is $18 \times 2 =$ **36** millimeters.

   (b) The distance between B and C is $18 - 8 = 10$ millimeters. Hence the crawling speeds of the ant and the cicadas are in the ratio 4:5. Between consecutive meetings at the same point, the ant must have crawled around the circle 4 times and the cicadas 5 times. The ant must have crawled a total of $8 + 4 \times 36 =$ **152** millimeters, counting from its starting point at A.

3. When the expression is expressed in the form of a simple fraction, 2 must be in the numerator and 3 must be in the denominator. Each of 4, 5 and 6 may either be in the numerator or the denominator. In order to have an integral value, we must have both 5 and 6 in the numerator. To minimize the value of the expression, we put 4 in the denominator. This can be achieved as $2 \div 3 \div (4 \div 5 \div 6) =$ **5**.

4. We have

   $$\begin{aligned}
   3333333333 \times 6666666666 &= 3333333333 \times 3 \times 2222222222 \\
   &= 9999999999 \times 2222222222 \\
   &= (10000000000 - 1) \times 2222222222 \\
   &= 22222222220000000000 - 2222222222 \\
   &= \mathbf{22222222217777777778}.
   \end{aligned}$$

5. The calculated class average is 6 points more than the average of the bottom 20 students. The total score of the class is 50 times the average of the bottom 20 students plus $30 \times 12 = 360$. Hence the actual class average is $360 \div 50 = 7.2$ higher than the average of the bottom 20 students. It follows that the calculated average is $7.2 - 6 =$ **1.2** points lower than the actual class average.

6. It will take two craftsmen and six apprentices to finish the job in 2 days. Thus five extra apprentices save 1 day out of 3, meaning that they can do in 2 days one-third of the job. To do the job alone, an apprentice would need $5 \times 2 \div \frac{1}{3} =$ **30** days.

1991

Part A.

1. We have $0.125 = \frac{1}{8}$ and $0.\overline{3} = \frac{1}{3}$. Hence the sum is $\frac{3+8+10}{24} = \frac{7}{8} = \mathbf{0.875}$.

2. The first factor is to 966 decimal places while the second factor is to 1030 places. Hence the product is to 966+1030=1996 decimal places. Now $181 \times 11 = 1991$. Hence the product is $\mathbf{0.0_{1992}1991}$.

3. The area of a circle is π times the square of its radius while the circumference is π times twice its radius. Hence the desired ratio is equal to half of its radius. Now $123\frac{456}{789} = 123\frac{152}{263} = 122\frac{415}{263}$, and half of this is $\mathbf{61\frac{415}{526}}$.

4. The speed of the train was $63360 \div 60^2 = 17.6$ meters per second. The total speed of the train and the man was $152 \div 8 = 19$ meters per second. Hence the speed of the man was $19 - 17.6 = \mathbf{1.4}$ meters per second.

5. To minimize the difference, we should make the first number as large as possible and the second as small as possible. In each number, the first six decimal places are fixed. The smallest possible value of the next digit in the first number is 0. Hence it should become $\mathbf{3.572\overline{064}}$. The largest possible value of the next digit in the second number cannot be 9 since we must extend the bar over at least one digit. The next best choice is 8, but there are two of them. The first 8 is followed by a 1 while the second is by a 9. Hence the new number should be $\mathbf{1.678\overline{189}}$.

6. **First Solution:**
As shown in Figure 6.20 on the right, the unshaded region may be divided into a rectangle of area 2 square centimeters, a small square of area 1 square centimeter, a large triangle of area 1.5 square centimeters and five small triangles each of area 1 square centimeters. Hence the area of the shaded region is $16-2-1-1.5-5=\mathbf{6.5}$ square centimeters.

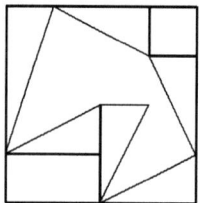 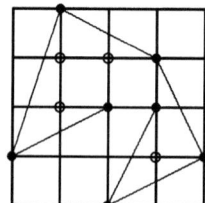

Figure 6.20

Second Solution:
There is a result called **Pick's Formula** which states that the area of a polygon whose vertices are all grid points is equal to $I - \frac{1}{2}B - 1$, where I is the number of grid points inside the polygon and B is the number of grid points on the perimeter of the polygon. As shown in Figure 6.20 on the right, $I = 4$ and $B = 7$. Hence the area is $4 + \frac{1}{2} \times 7 - 1 = \mathbf{6.5}$ square centimeters.

7. Kenneth finished the test $6 \times (100 \div 5) = 120$ seconds or 2 minutes less than Michael. In these 2 minutes, Michael could work out $3 \times 2 = 6$ questions. Hence Michael had worked out $100 - 6 = \mathbf{94}$ questions.

8. An $18 = 13 + 5$ point swing in the total score results in a $3 = 90 - 87$ point swing in the average score. Hence the number of students is $18 \div 3 = \mathbf{6}$.

9. Clearly, the quotient in the first division is smaller than that in the second. Suppose the difference is just 1. If we reduce the quotient in the second division by 1, we must increase the remainder to 3+36=39. The difference between the remainders in the two divisions will then be $39 - 17 = 22$. This must be the quotient of the first division, so that the number is $37 \times 22 + 17 = \mathbf{831}$. If the difference between the two quotients is greater than 1, the resulting number would have more than 3 digits.

10. Wallace usually takes 7 minutes to get to school, but only $7 - 1\frac{2}{3} = 5\frac{1}{3}$ minutes on that day. He walked for 2 minutes and ran for $5\frac{1}{3} - 2 = 3\frac{1}{3}$ minutes. Thus on that day, he walked for 3 minutes less than usual but ran for $3\frac{1}{2} - 2 = \frac{4}{3}$ minutes more than usual. Since he covered the same distance, the ratio of his running speed to his walking speed is $3 : \frac{4}{3} = \mathbf{9:4}$.

Part B.

1. For each 2-cent coin, we have 11 1-cent coins, for a total value of 13 cents. Since the remaining ones are all 5-cent coins, the total value of all 2-cent and 1-cent coins must be a multiple of 5. The only common multiple of 5 and 13 under 100 is 65. Hence there are $(100 - 65) \div 5 = \mathbf{7}$ 5-cent coins.

2. (a) A swing of 2 battalions results in a swing of 4 days. This means that what these 2 battalions eat in 6 days can support the remaining battalions for 4 days. Hence there are $2 \times 6 \div 4 = 3$ battalions besides these two, and there are currently $3 + 1 = \mathbf{4}$ battalions in the regiment.

(b) The food supply lasts 6 days for 4+1=5 battalions. Hence it lasts 6 × 5=**30** days for a single battalion.

3. Note that $720 = 2^4 \times 3^2 \times 5$. Since the numerator and the denominator have no common factors greater than 1, we can only factor 720 as a product of two such numbers in four ways, namely,

$$\begin{aligned} 720 &= 1 \times 720 \\ &= 5 \times 144 \\ &= 9 \times 80 \\ &= 16 \times 45. \end{aligned}$$

Hence the desired list of fractions is $\frac{1}{720} < \frac{5}{144} < \frac{9}{80} < \frac{16}{45}$.

4. The sum of the ten numbers is 55. Hence the sum of the numbers with minus signs in front of them is $(55 - 37) \div 2 = 9$. As long as we do not use the number 1, the product should have as many factors as possible. Since the sum of the smallest four numbers is already 10, we can only partition 9 as a sum of three different numbers, and in only one way, as 2+3+4. Hence the largest possible value of the product is $2 \times 3 \times 4=$**24**.

5. Since the first and the last digits of 67866 are the same, there is no carrying from the units digits in the addition. Since the second digit of 67866 exceeds the fourth digit by 1, there must be a carrying from the hundreds digits. Since the hundreds digit of 67866 is 8, there is no carrying from the tens digits, and the hundreds digit of the original number must be 9. Since the number formed of the first three digits of the original number is divisible by 9, so is the number formed of its first two digits. This two-digit number can only be 54 or 45. Then the number formed of the last two digits is 21 or 12, but only 12 is not divisible by 7. It follows that the original number is **54921**.

6. In Wallace's plan, they will be riding instead of walking from the entrance of the park to the shop. However, they must both walk and ride between the entrance of the park to their home. The ratio of the sum of their riding and walking speeds to the difference of these two speeds is $4+1 : 4-1 = 5 : 3$. The extra time taken by walking home and riding back to the entrance of the park must be equal to time saved by riding over walking a distance of 2 kilometers. It follows that their home was $2000 \times \frac{3}{5} =$**1200** meters from the entrance of the park.

1992

Part A.

1. We have $13.64 \times 0.25 \div 1.1 = 3.41 \div 1.1 =$ **31**.

2. Clearly, the thousands digit in the sum must be 1, and the hundreds digit of the first summand is at least 7. We have already used 8, and if we use 9 here, then the hundreds digit of the sum will have to be 1 or 2, both of which have been used. Hence the hundreds digit of the first summand is 7, and the hundreds digit of the sum is 0. The two missing units digits cannot be 5 and 9 since in the tens digit, 3 is odd and 6 and 8 are even. Hence the two missing units digits are 9 and 3 with a carrying, and the two missing tens digits are 6 and 5. The sum is therefore **1053**.

3. The radius of each semicircle is $\frac{100}{\pi}$ meters so that the total area of the two semicircles is $\pi(\frac{100}{\pi})^2 = \frac{10000}{\pi}$ square meters. The area of the rectangle is $100 \times \frac{100}{\pi} = \frac{10000}{\pi}$ also. Hence the desired ration is **1:2**.

4. We take the price of each item to be $100. Then their costs are given by $100 \div (1 + 25\%) = \$80$ and $100 \div (1 - 20\%) = \$125$ respectively. Hence the desired ratio is given by $200 : (80 + 125)=$**40:41**.

5. The sum of all ten numbers is 598 so that the sum of the five numbers in each group is 299. Dividing the numbers by 6, the respective remainders are 5, 5, 1, 5, 5, 1, 1, 5, 5 and 1. Hence exactly two of 31, 67, 79 and 103 are in one group. The other numbers are 41, 101, 23, 53, 83 and 17, arranged according to their units digits. Suppose 103 goes with 31. The other three numbers must add up to $299 - 103 - 31 = 165$. To get the correct units digit, we must take 41, 101 and one of 23, 53 and 83. The smallest one, 23, is exactly what is needed. Now this group contains 101, and the second smallest number in it is **31**. Suppose 103 goes with 67. The other three numbers must add up to $299 - 103 - 67 = 129$. To get the correct units digit, they are either 41, 101 and 17, or 23, 53 or 83. The sum of either alternative is too large. Suppose 103 goes with 79. The other three numbers must add up to $299 - 103 - 79 = 117$. To get the correct last digit, we must take one of 41 and 101, and two of 23, 53 and 83. The smallest sum among such choices is 41+23+53=117, which is exactly what is needed. The other group contains 101, and the second smallest number there is also **31**.

6. In 1 minutes, Konrad and Mr. Zhang can cover a total distance of 70+40=110 meters. In 7 minutes, they can cover a total distance of $110 \times 7 = 770$ meters. On that particular day, Konrad had to cover this distance all by himself, which takes $770 \div 70=$**11** minutes, the time he had to leave earlier than usual.

7. In order to pay exactly $2, we need at least two $1 bills. To pay exactly $4, it is better to add a $3 bill than two more $1 bills. To pay exactly $6, it is best to add a $5 bill. With these four bills, we can pay any integer amount of dollars from $1 to $9, and no smaller number of bills will do. Since the sum of these four bills is $10, we can add ten $9 bills to bring the total to $100. Hence the smallest number of bills in the stack is **14**.

8. **First Solution:**
Of the 30+22=52 copies of *Beijing Evening News* and *References and Information*, 34 copies are delivered to apartments which also subscribe to *China Television Guide*. The remaining $52 - 34 = 18$ copies are delivered to apartments subscribing to both. Hence the number of such apartments is $18 \div 2 = $ **9**.

Second Solution:
The number of apartments in the building is $(34+30+22) \div 2 = 43$. The number of apartments subscribing to both *Beijing Evening News* and *References and Information* is the same as the number of apartments not subscribing to *China Television Guide*, which is $43 - 34 =$ **9**.

9. Note that we have $0.\overline{6} = \frac{2}{3} > \frac{17}{26}$, $0.625 = \frac{5}{8} < \frac{23}{33}$, $0.\overline{3} = \frac{3}{10} < \frac{34}{99}$ and $2.25 = \frac{9}{4} > \frac{237}{106}$. The given expression is equal to $(\frac{2}{3}+\frac{5}{8}) \div (\frac{1}{3}+\frac{9}{4}) = \mathbf{\frac{1}{2}}$.

10. Since 1 is not a prime number anyway, we should have as many factors in the product as possible. Since 3+5+7+11+13=39 is already too big, we use four prime numbers, one of which must be 2. Now 35 can be expressed as a sum of three prime numbers in five ways, namely, 3+13+19=5+7+23=5+11+19=5+13+17 =7+11+17. It is easy to verify that the product of 7, 11 and 17 is the largest, so that the maximum value of the overall product is $2 \times 7 \times 11 \times 19 = $ **2618**.

Part B.

1. Of the four expressions, the largest one comes from $\div$ and the smallest from $\times$. So we place $\div$ in the first expression and $\times$ in the fourth. Since we wish to subtract off as smallest a number as possible, we place $-$ in the second expression and $+$ in the third. The maximum sum is $\frac{1}{2} \div \frac{1}{9} + \frac{1}{3} - \frac{1}{8} + \frac{1}{4} + \frac{1}{7} + \frac{1}{5} \times \frac{1}{6} = \mathbf{5\frac{113}{840}}$.

2. The longest horizontal segment is (d) and the longest vertical segment is (e). Since segment (d) is longer than segment (e), segment (d) has length 8. The total length of the segments (a), (b) and (c) is 8. Hence segment (a) has length 1, and segments (b) and (c) have lengths 2 and 5 or 3 and 4, respectively. Now segment (e) must have length 7 and segment (f) 6. Hence the lengths of segment (g) and the eighth segment are consecutive integers. It follows that they must be 4 and 3 respectively, so that the lengths of segments (b) and (c) are 2 and 5 respectively. The area of the polygon is $1 \times 7 + 5 \times 3 + 2 \times 6 = $ **34**.

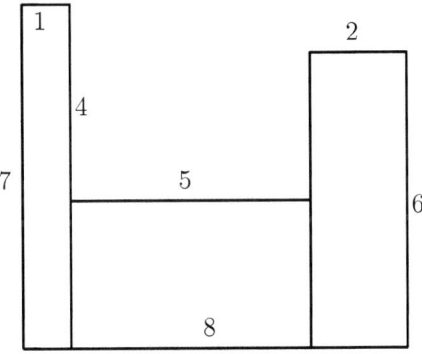

Figure 6.21

3. In each block of 10 consecutive numbers with their units digits running from 0 to 9, at most 2 can have the sum of their digits divisible by 7. In order for this to happen, the sum of the other digits must leave a remainder of 0, 5 or 6 when divided by 7. A set of consecutive blocks of this type is called a good run. Among three-digit numbers, all good runs consist of at most three blocks except for two, each of which consists of five blocks, namely, from 480 to 529 and from 570 to 619. The numbers are 482, 489, 491, 498, 502, 509, 511, 518, 520 and 527 in the first case, and 572, 579, 581, 588, 590, 597, 601, 608, 610 and 617 in the second case. Hence among 50 consecutive three-digit number, the maximum number of those the sum of whose digits is divisible by 7 is **10**.

4. Since there were three times as many adults as children and each child planted 6 trees, the adults could pick up the slack by planting 2 more trees each if they just let the children play. Then each adult male would have planted 15 trees and each female adults 12. Since 12 and 216 are both multiples of 4, the number of adult males must also be a multiple of 4. This number could only be 4, 8 or 12, as otherwise more than 216 trees would have been planted.

The corresponding numbers of female adults were $(216-4\times 15)\div 12=13$, $(216-8\times 15)\div 12=8$ and $(216-12\times 15)\div 12=3$ respectively. Now $4+13=17$, $8+8=16$ and $12+3=15$. Since the total number of adults must be a multiple of 3, the number of female members who came was **3**.

5. In one hour, machine A can do $\frac{1}{11}$ of the job and machine B can do $\frac{1}{13}$ of the job. With interferences, they can do $\frac{1}{6.25} = \frac{4}{25}$ of the job. The loss of 28 pages per hour is $\frac{1}{11} + \frac{1}{13} - \frac{4}{25} = \frac{28}{3575}$ of the job. Hence the job has $28 \div \frac{28}{3575} =$ **3575** pages.

6. If Konrad rides downhill for the entire third hour, he must ride uphill for the entire first hour. The difference in the speeds is 9 kilometers per hour, but yet he covers only 8 kilometers more in the third hour than in the first. Hence Konrad must be riding on level ground for part of the third hour.

    (a) Konrad covers only 5 kilometers more in the second hour than in the first hour, while his speed on level ground is 6 kilometers per hour more than his speed riding uphill. Hence he spends the entire first hour riding uphill and $\frac{5}{6}$ of the second hour riding on level ground. It follows that he spends 60+10=**70** minutes riding uphill.

    (b) By riding on level ground for part of the third hour, Konrad loses 1 kilometer than if he rides downhill for the entire third hour. Since his speed on level ground is 3 kilometers per hour less than his speed going downhill, he rides on the level ground for $\frac{1}{3}$ of the third hour, so that he spends **40** minutes riding downhill.

    (c) In one hour, Konrad covers $\frac{6}{7}$-th of the first part of the road riding uphill and $\frac{6}{4}$-th of the first part of the road riding downhill. The difference is $\frac{6}{4} - \frac{6}{7} = \frac{9}{14}$. Since he rides 9 kilometers more per hour going downhill than uphill, the length of the first part of the road is $9 \div \frac{9}{14} = 14$ kilometers. Hence his riding speed uphill is $14 \times \frac{6}{7} = 12$ kilometers per hour, and his riding speed on level ground is 18 kilometers per hour. Since Konrad spends $3\times 60-70-40 = 70$ minutes covering the second part of the road twice, the length of the second part of the road is $18\times \frac{7}{6} \div 2 = 10.5$ kilometers. It follows that the distance between A and B along this road is 14+10.5=**24.5** kilometers.

Part IV: Mathematics Celebration

In 1997, I was contacted by **Mike Dumanski**, then vice-principal of Our Lady of Victory Catholic Elementary School in Edmonton. He lamented that he was deeply dissatisfied with the school's Science Fair. Most of the projects were beyond the understanding of the children, and were apparently the work of the parents on behalf of their offspring.

Mike wondered if there was something called a Math Fair. I put my foot in my mouth and said, "Of course, this would be a celebration of mathematics." I committed myself to serve as adviser on this matter for the school. This was already in late May, so I promised to go to the school in late September.

I had the whole summer to think about this Math Fair that I knew nothing about. After much thought, I decided that it should be similar to a Science Fair in that it also consisted of projects by the students, and every student, not just the elite, should participate. Thus the first principle was laid down, summarized in the word **all-inclusive**.

However, there is one significant difference. There will be no formal judging and no prizes in the Math Fair. The proof of the pudding is in the eating, to see which project attracted the most attention from the audience. This was the second principle, summarized in the word **non-competitive**, cutting out any incentive for direct parent intervention. Children are competitive by nature in any case. I remember one boy who said to another after this principle was spelt out, "I am more non-competitive than you are." There is no need to fuel this fire. Students should participate just for the fun of it.

As for the subject matters for the projects, we do not want the projects to turn into teaching stations. They should present attractive interactive puzzles to the audience. This suggested the third and the fourth principles, summarized in the words **student-oriented** and **problem-based**.

The inaugural Math Fair was held on November 12, 1997 in Meadowlark Shopping Mall, in the neighborhood of the school. The movement has spread like wild fire, across Canada and into the United States, as well as overseas. In 2004, the S.N.A.P. Mathematics Foundation was chartered in Edmonton to oversee the development of the Math Fair, embodying the four basic principles. An annual conference is held at the Banff International Research Stations in the scenic Canadian Rockies, renewing existing relationship and introducing new friends.

My colleague **Ted Lewis** then worked the S.N.A.P. Math Fair into the curriculum for students training to become elementary school teachers at the University of Alberta.

Each class had to go to a city school and present a S.N.A.P. Math Fair to students there. Typically, they would set things up in the school gymnasium, and various classes would take turns coming in to try the puzzles. The program proved so successful that our waiting list became unmanageable. Well, if Mohammed cannot go to the mountain, the mountain must come to Mohammed. The S.N.A.P. Math Fair moves to the university campus, and schools from within the city and surrounding rural area send classes to it. This is held three times each academic year, and is the most popular community event on campus. All participating education students agree that this is the most valuable experience in their university career. For further details about the S.N.A.P. Math Fair, see the S.N.A.P. Mathematics Foundation's website http://www.mathfair.com.

The attention span of school children is relatively short, but for logistic reasons, the visiting group must come for a whole morning or a whole afternoon. G.A.M.E, an acronym for Graduates of Alberta Mathematics Etcetera, comes to the rescue. It is a graduate student organization founded by my former student **Tom Holloway**. While half of the visiting students are engaged in the S.N.A.P. Math Fair, the other half will learn to play mathematical games with the graduate students. The children have christened the event the G.A.M.E. Math Unfair, because the graduate students usually cream them. Nevertheless, the children have a lot of fun, especially when they play against one another.

The S.M.A.R.T. Circle members gather and develop mathematical games for the G.A.M.E. Math Unfair. They also organized a S.N.A.P. Math Fair during the 2004 CMS/AITF Camp. Although this had not been repeated in later camps, it has been organized on other occasions. For instance, during a Faculty of Science Open House at the University of Alberta, Circle members were invited to present a S.N.A.P. Math Fair for the visitors.

The true celebration came in 2010, with the passing of *Martin Gardner*, the great American writer of popular mathematics. As he did not want any memorials, an international event called the Celebration of Mind was initiated to highlight his lifetime achievements. Each October, Circle members present a S.N.A.P. Math Fair to the public at the Telus World of Science, Edmonton, our local science museum, under the direction of Grant MacEwan University professor **Tiina Hohn** and museum staff **Samantha Marion**, a former president of G.A.M.E..

Another part of the Mathematics Celebrations is the participation of select Circle members in the Gathering for Gardner. The late *Robert Barrington Leigh* went in 2006, David Rhee went in 2010, 2012 and 2014. Also participating in 2010 were Mariya Sardarli, along with Brian Chen of Chiu Chang Mathematical Circle. In 2014 and 2016, Ryan Morrill attended the event.

The Gathering, fondly referred to as **G4G**, is the brain-child of the late *Tom Rodgers*. This most prestigious conference is held in the spring once every two years, always in Tom's hometown Atlanta. The first one was held in 1994. Martin attended the first two, but not since. Tom was terminally ill during the Gathering in 2012, but his strength of character enabled him to still oversee the event, passing away on the day after it was over.

Participation is by invitation only, and mainly for people whose life has been influenced by the writing of Martin. In some cases, the relation to Martin is much more personal. Martin's circle is extremely wide, but the following three are the main groups, mathematicians, magicians and people who expose pseudo-sciences.

The Gathering typically starts on a Wednesday evening and lasts until Sunday evening. Most of the time is devoted to presentations of things that would be of interest to Martin, which may be anything under the sun. Nevertheless, people are not going to embarrass themselves on such an auspicious occasion by not having some really worthy material. Thus the standards are uniformly high.

On one of the days, depending on weather conditions, the conference moves in the afternoon and the evening to Tom's Japanese-style house on top of a small hill. About two hundred people gather around outdoor toys for adults, including a life-size "Impossible Cube" inside which people can have their photographs taken. It is a very nice break from the talks, and enables people to meet more intimately and leisurely.

In Chapter 7, we provide some sample projects. Most of them here are extracted from the conference proceedings of the Gatherings. Papers presented at the Gatherings can be found in the following volumes. They were originally published by A. K. Peters, founded by **Klaus** and **Alice Peters**.

[1] Elwyn Berlekamp and Tom Rogers, *The Mathemagician and the Pied Puzzler*, CRC Press, 1999.

[2] David Wolfe and Tom Rodgers, *Puzzlers' Tribute*, CRC Press, 2001.

[3] Barry Cipra, Erik Demaine, Martin Demaine and Tom Rodgers, *Tribute to a Mathemagician*, CRC Press, 2004.

[4] Erik Demaine, Martin Demaine and Tom Rodgers, *A Lifetime of Puzzles*, CRC Press, 2008.

[5] Ed Pegg Jr., Tom Rodgers and Alan Schoen, *Homage to a Pied Puzzler*, CRC Press, 2009.

[6] Ed Pegg Jr., Tom Rodgers and Alan Schoen, *Mathematical Wizardry for a Gardner*, CRC Press, 2009.

Sample G.A.M.E. Math Unfair games are provided in Chapter 8. These are largely extracted from Martin's famed monthly column *Mathematical Games* which graced the pages of the magazine *Scientific American* for over a quarter of a century. Later, the columns were anthologized and became chapters, in fifteen volumes.

Unfortunately, since these books were put out by several publishers, they varied greatly in format. The Mathematical Association of America has packaged all of them into a compact disc, but maintaining the original format. A new edition, published jointly by the Association and Cambridge University Press, will have uniform format and updated material. Three volumes were edited by Martin himself before his passing in 2010, but other volumes will continue to appear. Here are the titles.

1. *Hexaflexagons, Probability Paradoxes, and the Tower of Hanoi.*
2. *Origami, Eleusis, and the Soma Cube.*
3. *Sphere Packing, Lewis Carroll, and Reversi.*
4. *Knots and Borromean Rings, Rep-Tiles, and Eight Queens.*
5. *Klein Bottles, Op-Art, and Sliding-Block Puzzles.*
6. *Sprouts, Hypercubes, and Superellipses.*
7. *Nothing and Everything, Polyominoes, and Game Theory.*
8. *Random Walks, Hyperspheres, and Palindromes.*
9. *Words, Numbers, and Combinatorics.*
10. *Wheels, Life, and Knotted Molecules.*
11. *Knotted Doughnuts, Napier Bones, and Gray Codes.*
12. *Tangrams, Tilings, and Time Travel.*
13. *Penrose Tiles, Trapdoor Ciphers, and the Oulipo.*
14. *Fractal Music, Hypercards, and Chaitin's Omega.*
15. *Hydras, Eggs and Other Mathematical Mystifications.*

Chapter Seven
Sample SNAP Math Fair Projects

Section 1. Problems.

Geometric Puzzles

Project One: Six-Piece Tangram

Equipment:
The six pieces in Figure 7.1.

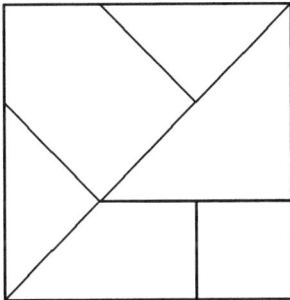

Figure 7.1

Example:
Each half of the square in Figure 7.1 is a right isosceles triangle constructed with three of the pieces. Find two other solutions.

Solution to Example:
The constructions are shown in Figure 7.2. The two figures are of different sizes.

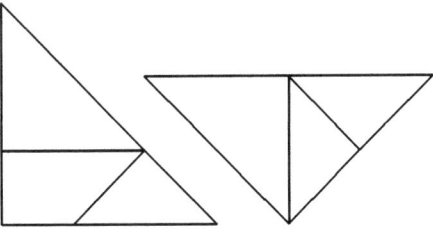

Figure 7.2

Problem 1:
Use five of the pieces to construct a square.

Problem 2:
Use five of the pieces to construct a right isosceles triangle.

Project Two: In Our Own Images

Equipment:
Four playing boards in Figure 7.3; four pieces, one of each of the shapes in the playing boards, but half in linear dimensions.

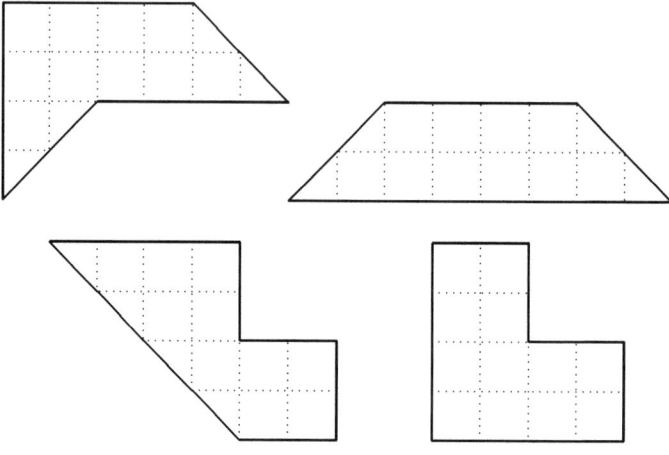

Figure 7.3

Problem:
Construct the figure on each of the playing boards using all four pieces. The pieces may be rotated and reflected.

Project Three: U and I

Equipment:
Three copies of each of the three U shapes and seven copies of the I shape in Figure 7.4.

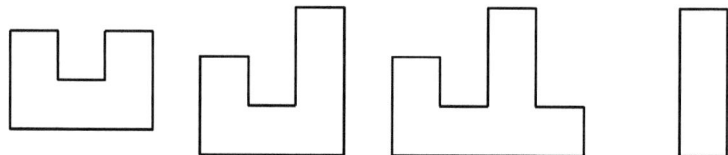

Figure 7.4

Example:
Construct a figure with two copies of the second U shape, and then construct the same shape with four copies of the I shape. Shapes may be rotated or reflected but may not overlap.

Solution to Example:
A construction is shown in Figure 7.5.

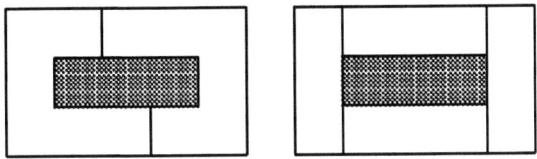

Figure 7.5

Problem 1:
Construct a figure with all three copies of the first U shape, and then construct the same shape with five copies of the I shape. Shapes may be rotated but may not overlap.

Problem 2:
Construct a figure with all three copies of the second U shape, and then construct the same shape with six copies of the I shape. Shapes may be rotated or reflected but may not overlap.

Problem 3:
Construct a figure with all three copies of the third U shape, and then construct the same shape with all seven copies of the I shape. Shapes may be rotated or reflected but may not overlap.

Project Four: Tetromino Symmetries

Equipment:
The five pieces in Figure 7.6.

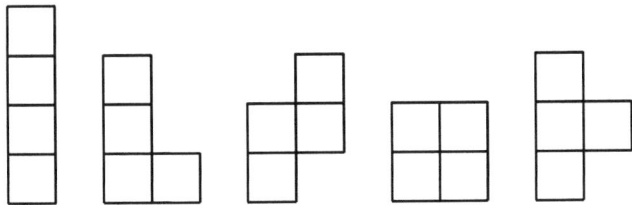

Figure 7.6

Elaboration:
A tetromino is a shape formed of four unit squares joined edge-to-edge. There are five tetrominoes which are the pieces in this project. They are called the I-tetromino, the L-tetromino, the N-tetromino, the O-tetromino and the T-tetromino, respectively.

Example:
Find all combinations of two tetrominoes which can form figures with reflectional symmetry.

Solution to Example:
There are seven possible combinations, namely, IL, IO, IT, LN, LO, LT and NT. One construction for each combination is shown in Figure 7.7.

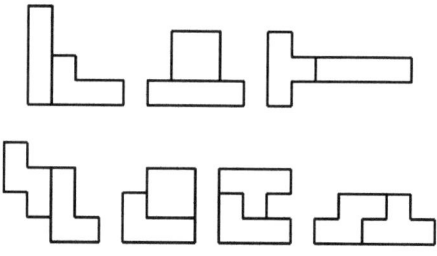

Figure 7.7

Problem 1:
Find all combinations of three tetrominoes which can form figures with reflectional symmetry.

Problem 2:
Find all combinations of four tetrominoes which can form figures with reflectional symmetry.

Counter-Moving Puzzles

Project Five: Magical Swap

Equipment:
The playing board in Figure 7.8; a counter and a small marker.

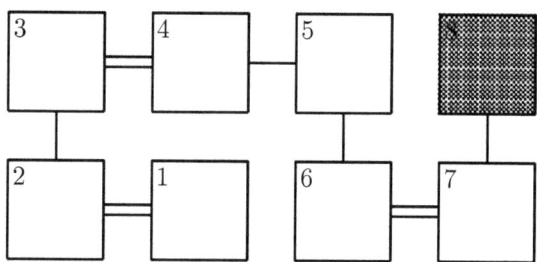

Figure 7.8

Elaboration:
The playing board portrays a magical archipelago consisting of eight islands linked by seven bridges. The bridges marked by single lines are made of wood while those marked by double lines are made of stone. A mouse can cross any bridge but an elephant can only cross stone bridges.

If the mouse and the elephant are on adjacent islands (such as 1 and 4 or 5 and 8 but not 1 and 5), they can magically swap places, regardless of whether there is a bridge between these two islands.

Problem:
The mouse lives on island 1 and the elephant on island 5. After the mouse has pulled the splinter out of the sole of the elephant's foot, the elephant still find it painful and must visit a vet's office on island 8. How can the mouse help the elephant get there?

Project Six: Safe Haven

Equipment:
The playing board shown in Figure 7.9; two counters R and P of different colors and three small markers a, b and c of the same color as the counter R.

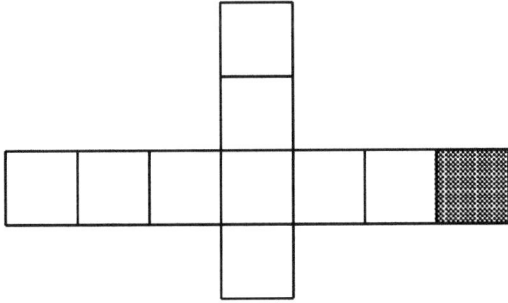

Figure 7.9

Elaboration:
The counter R represents a robber and the counter P represents a corrupt policeman. The markers represents minor crooks in cohort with the robber. The playing board portrays a small town with one long avenue and one short street. The shaded square at the right end of the avenue is a safe haven for the robber. Each square may be occupied by one counter or marker. For the sake of appearance, the robber may not occupy a square sharing a common side with the square occupied by the policeman. A counter or marker can jump over a block of occupied squares of any length, to the vacant square immediately beyond. The robber can dictate the movement of the policeman as well as the small fries.

Example:
The initial position is shown in Figure 7.10, with the street closed for repair. How can the robber get to the safe haven?

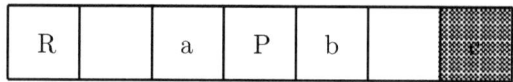

Figure 7.10

Solution to Example:
The task can be accomplished in seven jumps, as shown in Figure 7.11.

R		a	P	b		c
R		a		b	P	c
R		a	c	b	P	
R	c	a		b	P	
	c	a	R	b	P	
P	c	a	R	b		
P		a	R	b	c	
P		a		b	c	R

Figure 7.11

Problem 1:
The initial position is shown in Figure 7.12. How can the robber get to the safe haven?

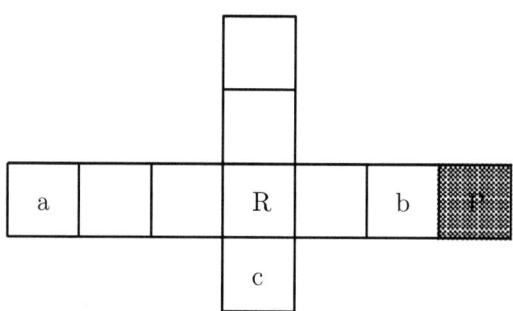

Figure 7.12

Problem 2:
The initial position is shown in Figure 7.13. How can the robber get to the safe haven?

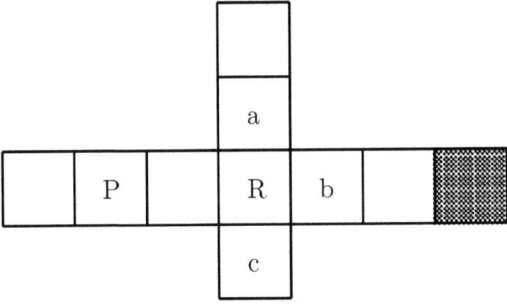

Figure 7.13

Project Seven: Knight Mares

Equipment:
The playing board in Figure 7.14; two counters of one color and two counters of another color.

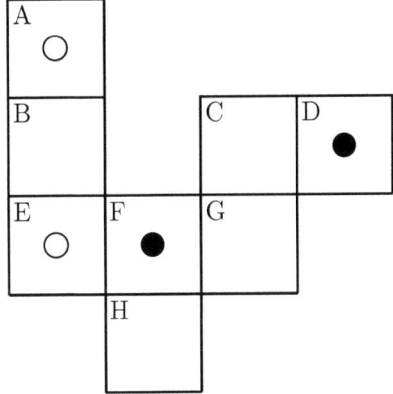

Figure 7.14

Elaboration:
The counters of one color represent knights while the counters of the other color represent their mares. The knights are trying to mount their mares, but the mares have other ideas. Any of the counters may move, but one at a time. It makes a Knight's move in chess but must land on a vacant square.

Problem:
Place the two knights on squares A and E and the two mares on squares D and F. Show how to move the knights to D and F and the mares to A and E.

Project Eight: Pentomino Transformations

Equipment:
A 3 × 3 playing board; 5 counters.

Elaboration:
A pentomino is a shape formed of five unit squares joined edge-to-edge. There are twelve pentominoes, five of which are shown In Figure 7.15. They are called the F-pentomino, T-pentomino, V-pentomino, W-pentomino and Z-pentomino, respectively.

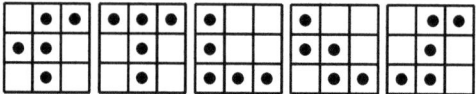

Figure 7.15

The task is to transform one pentomino into another, not necessarily in the given orientation. In each move, you may move any coin to a vacant square, provided that this square is adjacent to an occupied square in the same row as well as an occupied square in the same column. A stationary coin is represented by a black circle. A coin which is being moved is represented by a white circle. A coin which has just arrived is represented by a target circle.

Example 1:

(a) Transform the V- and the W-pentominoes into each other, using only 1 move in each direction.

(b) Transform the W- and the Z-pentominoes into each other, using only 1 moves in each direction.

(c) Transform the V- and the Z-pentominoes into each other, using 2 moves in each direction.

Solution to Example 1:

(a) The reversible transformation is shown in Figure 7.16 on the left.

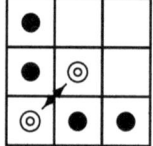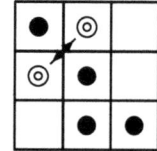

Figure 7.16

Problems

(b) The reversible transformation is shown in Figure 7.16 on the right.

(c) This can be accomplished by combining (a) and (b).

Example 2:

(a) Transform the F-pentomino into the W-pentomino in 2 moves.

(b) Transform the W-pentomino into the F-pentomino in 2 moves.

Solution to Example 2:

(a) We can transform the F-pentomino into the W-pentomino in 2 moves, as shown in Figure 7.17.

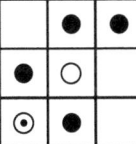 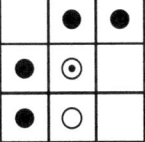

Figure 7.17

(b) We can transform the W-pentomino into the F-pentomino in 2 moves, as shown in Fiure 7.18.

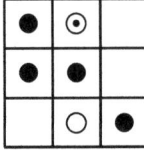 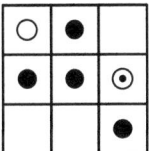

Figure 7.18

Problem 1:
Transform the F- and the Z-pentominoes into each other, using 2 moves in each direction.

Problem 2:

(a) Transform the T-pentomino into the F-pentomino in 2 moves.

(b) Transform the F-pentomino into the T-pentomino in 3 moves.

Problem 3:

(a) Transform the F-pentomino into the V-pentomino in 2 moves.

(b) Transform the V-pentomino into the F-pentomino in 3 moves.

Problem 4:

(a) Transform the T-pentomino into the W-pentomino in 3 moves.

(b) Transform the W-pentomino into the T-pentomino in 5 moves.

Problem 5:

(a) Transform the T-pentomino into the Z-pentomino in 3 moves.

(b) Transform the Z-pentomino into the T-pentomino in 5 moves.

Problem 6:

(a) Transform the T-pentomino into the V-pentomino in 4 moves.

(b) Transform the V-pentomino into the T-pentomino in 6 moves.

Counter-Sliding Puzzles

Project Nine: Space Colonization:

Equipment:
The playing board in Figure 7.19; six counters.

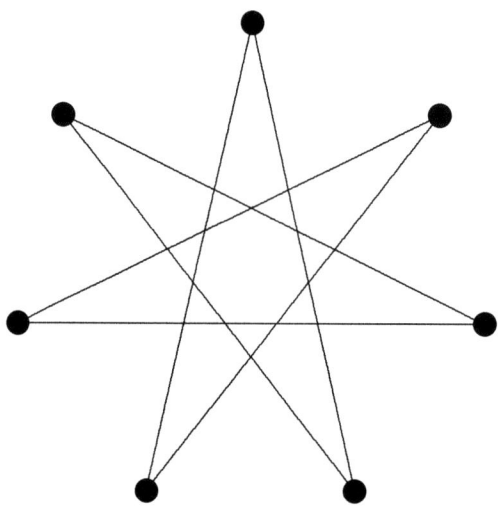

Figure 7.19

Elaboration:
The playing board portrays a planet with seven inhabitable sites represented by black dots. They are linked by seven hyperpaths represented by lines. They pass over and under one another without intersecting.

Six alien races wish to colonize this planet, arriving one race at a time in their flying saucers and landing on one of the seven sites. However, they are never satisfied with anything. They will move along a hyperpath to another site and then settle down. Landing on or moving to a site already settled is prohibited.

Problem:
How can all six alien races settle in this planet?

Project Ten: Martin to Gardner

Equipment:
The playing board in Figure 7.20; four counters.

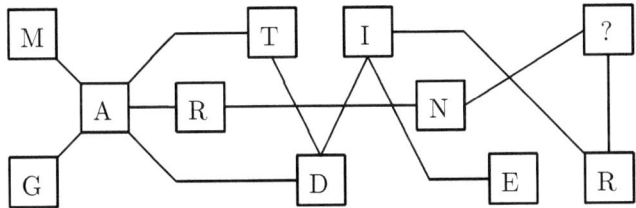

Figure 7.20

Elaboration:
The letters in the boxes of the top two rows spell "MARTIN" from left to right, with a question mark on the last box because Martin does not have a middle initial. The letters in the boxes of the bottom two rows spell "GARDNER" from left to right. Counters move one at a time between two boxes connected by a line, but may not moved to boxes connected by lines to boxes occupied by other counters.

Problem:
Place the counters on the four boxes on the bottom row so that the visible letters spell "MARTIN". Move them according to the rules to the boxes in the top row, so that the visible letters spell "GARDNER".

Project Eleven: Limited Visibility

Equipment:
The playing board in Figure 7.21; two large counters labeled A and B; five small counters labeled c, d, e, f and g.

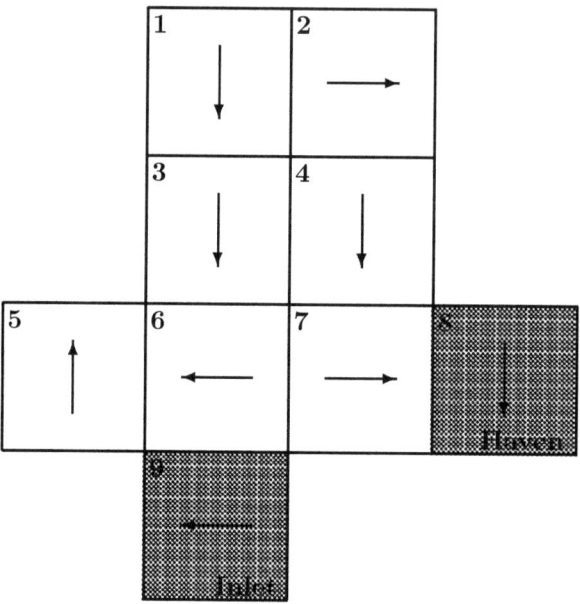

Figure 7.21

Elaboration:
The playing board consists of nine numbered squares representing an ocean. Square 8 is a Haven and square 9 is an Inlet. Each square may be occupied by one ship. A fleet consists of an aircraft carrier, a battle cruiser, a corvette, a destroyer, an electronic surveilance ship, a frigate and a gunboat. Their starting positions are squares 7, 6, 5, 4, 3, 2 and 1, respectively. The ships move one at a time, to an adjacent vacant square. Each square has an arrow which indicates an allowable direction in which ships may move. However, the ocean is fogged, so that the arrow of an occupied square is hidden from view. This means that at any time, only two arrows are visible, and these are the only options available. While one option is carried over to the next move, the other option usually changes.

A move is described by two numbers (m, n), meaning that the ship currently in square m moves to square n. From the opening position, as shown in the diagram below on the left, the only possible first move is (6,9). If the second move is (3,6), then the third move must be (1,3), and we are stuck. Hence the next three moves must be (7,6), (4,7) and (2,4). If the fifth move is (1,2), we are stuck. Hence it must be (7,8), resulting in the position as shown in the diagram below on the right. These five moves from a position where squares 8 and 9 are vacant form what we call Operation X.

Problems

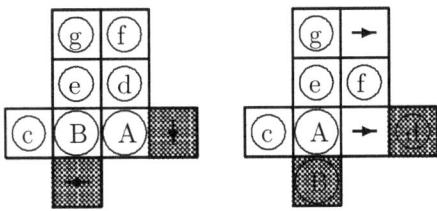

Figure 7.22

Example:
We start from a position where squares 8 and 9 are vacant.

(a) Find a sequence of moves which results in a double-switch between the ships in squares 4 and 7, and between the ships in squares 5 and 6.

(b) Find a sequence of moves which results in a cyclic permutation of all seven ships in squares 7, 6, 5, 3, 1, 2 and 4.

Solution to Example:

(a) We first perform Operation X. The next nine moves are (6,7), (5,6), (4,2), (7,4), (6,7), (9,6), (6,5), (7,6) and (8,7). From the opening position, we obtain the position as shown in Figure 7.23 on the left.

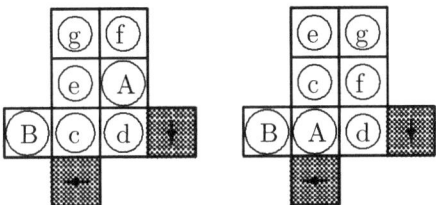

Figure 7.23

(b) We first perform Operation X. The next nine moves are (6,7), (5,6), (1,2), (3,1), (6,3), (9,6), (6,5), (7,6) and (8,7). From the opening position, we obtain the position as shown in Figure 7.23 on the right.

For future reference, we call these two sequences Operations Y and Z respectively. They are useful, along with Operation X, in solving the following problems.

Problem 1.
Get the aircraft carrier to the Inlet and the battle cruiser to the Haven.

Problem 2.
Get the aircraft carrier to the Haven and the battle cruiser to the Inlet.

314   7 Sample SNAP Math Fair Projects

Project Twelve: Solar Sortie

Equipment:
A 3×7 playing board; a counter R and five small markers a, b, c, d and e of a different color from R.

Elaboration:
The playing board portrays a solar panel in which Robbie the Robot and some of his fellow robots are trapped. Robbie is represented by the counter and the others by small markers. To escape, they must get Robbie to the shaded central square to open the trapdoor there. The other robots may occupy that square temporary, but they cannot open the trapdoor. Robbie cannot open the trapdoor either if he only passes through that square. At any time, any of the robots can move, but one at a time. It moves from along a row or a column, but it must move as far as it can in the same direction, before stopping in the square in front of the first robot it encounters. If there are no other robots in line, it may not move in that direction.

Example:
In each of the following scenarios, the initial position is shown in Figure 7.24. How can the other robots help Robbie get to the trapdoor?

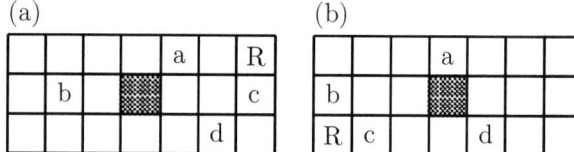

Figure 7.24

Solution to Example:

(a) First, c moves left towards b. Then R moves left towards a, down towards d and left towards c.

(b) First c moves right towards d, up towards a and left towards b. Then R moves right towards d, up towards a and left towards c.

Problem:
In each of the following scenarios, the initial position is shown in Figure 7.25. How can the other robots help Robbie get to the trapdoor?

(a)

a			b		c	
					d	
	e				R	

(b)

	a		b		c	
	R					
		d			e	

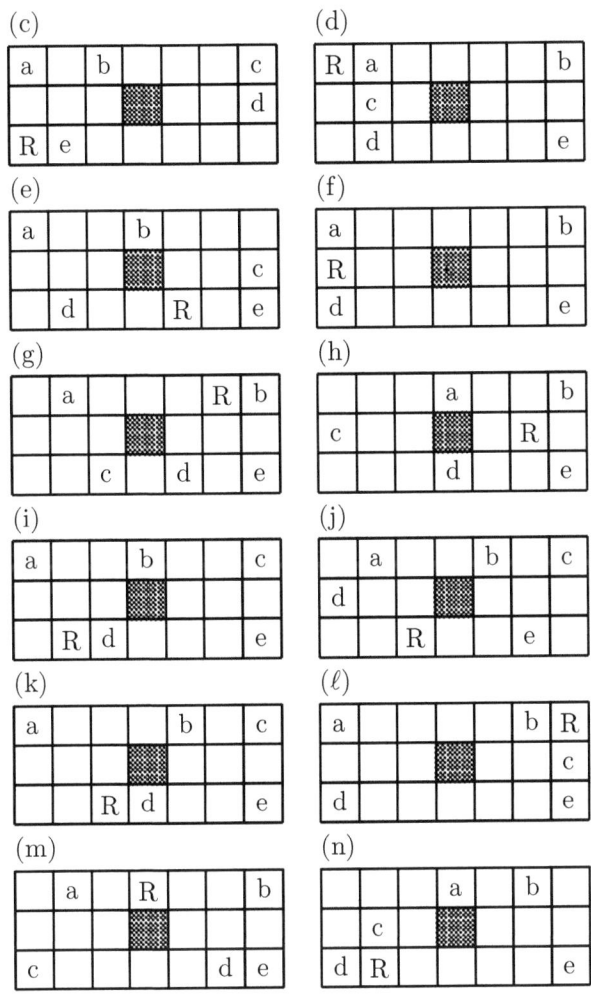

Figure 7.25

Maze Puzzles

Project Thirteen: No Going Straight

Equipment:
The playing board in Figure 7.26; a counter, preferably in the shape of a miniature car.

Elaboration:
The playing board portrays the street map of a town. As one approaches an intersections, there are three options, turn right, turn left or go straight. In this town, you may not go straight.

316   7 Sample SNAP Math Fair Projects

Problem:
Enter and exit the town along the given arrows.

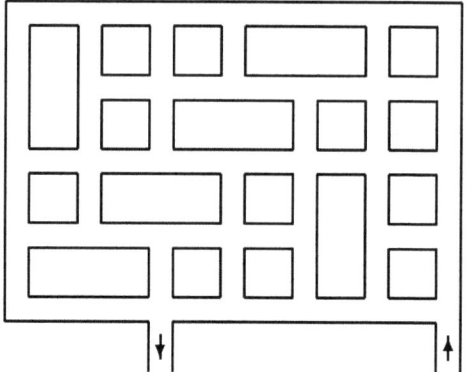

Figure 7.26

Project Fourteen: No Turning Left

Equipment:
The playing board in Figure 7.27; a counter, preferably in the shape of a miniature car.

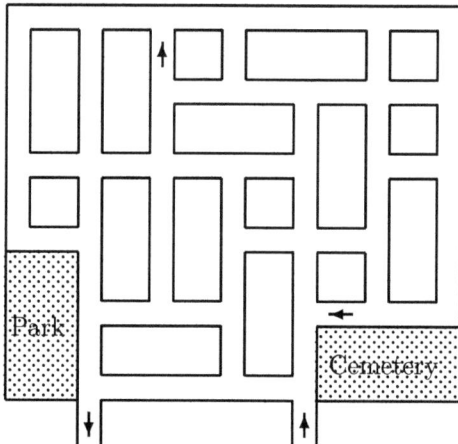

Figure 7.27

Elaboration:
The playing board portrays the street map of a town. As one approaches an intersections, there are three options, turn right, turn left or go straight. In this town, you may not turn left, and may not go against the two arrows inside the town.

Problems

Problem:
Enter and exit the town along the given arrows.

Project Fifteen: Pentagon Heist

Equipment:
The playing board in Figure 7.28; a counter.

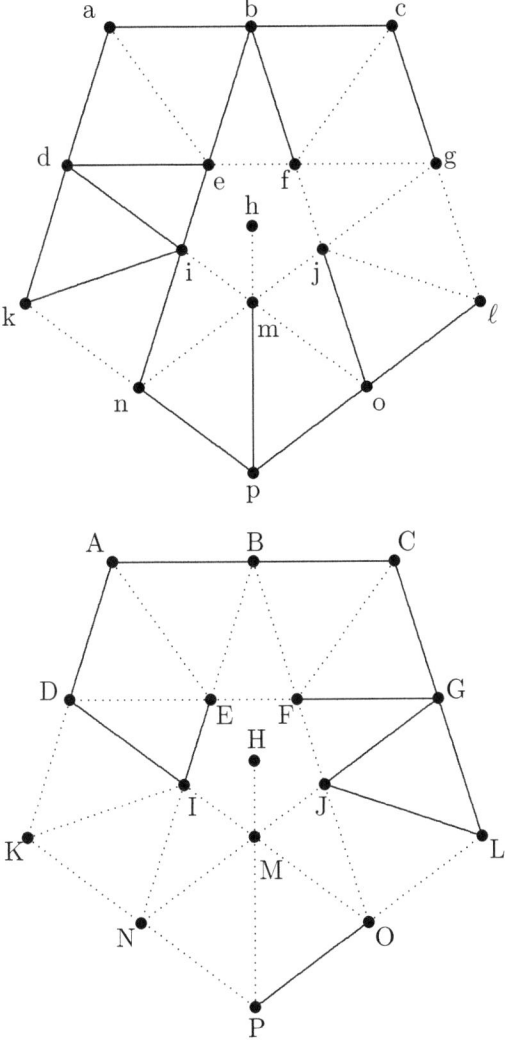

Figure 7.28

Elaboration:

The playing board portrays the ground level and the underground level of the Pentagon. Between the two levels are 16 pairs of matching rooms (identified by the same letter in lower and upper cases respectively) linked by 15 vertical elevators. The exception is between room h (the hiding place) and room H (the Headquarters), because there is no record for the existence of room h. Various pairs of rooms on the same level are connected by sealed passages. The air inside each passage is either acidic or alkaline. Acidic passages are indicated by solid lines while alkaline passages are indicated by dotted lines. The dosage is not lethal, but if one takes two successive passages containing the same kind of air, the result is fatal. Resting in the rooms or the elevator in between does not help.

Problem:
Start from the hiding place and go to the Headquarters safely.

Project Sixteen: Turn-Tiles

Equipment:
A 4×4 playing board; four copies of each of the four types of tiles shown in Figure 7.29, preferably with a different color for each type: **T** for Turn to the left, **U** for U-turn, **R** for Right-turn and **N** for No-turns.

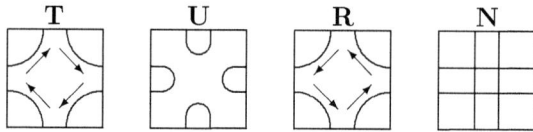

Figure 7.29

Elaboration:
Before entering a maze, you must first prepare it by placing turn-tiles on its squares. Squares with the same label are covered by tiles of the same kind. Squares with the same label in different mazes need not be covered by tiles of the same kind.

Example:
Enter the 3×3 maze shown in Figure 7.30 along the given arrow, follow the turn-tiles and exit the maze along the given arrow.

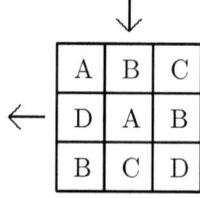

Figure 7.30

Solution to Example:

Obviously, neither the B squares nor the D squares are to be covered with the U-tiles. Suppose the U-tiles cover the A squares. Then the B squares cannot be covered by the N-tiles. If they are covered by the R-tiles, then whether the C squares are covered by the T-tiles or the N-tiles, the path will exit at the wrong place. If they are covered by the T-tiles, then whether the C squares are covered by the R-tiles or the N-tiles, the path will exit at the wrong place. Hence the U-tiles cover the C squares. If the B squares are covered by the N-tiles, then whether the A squares are covered by the T-tiles or the R-tiles, the path will exit at the wrong place. If the B squares are covered by the R-tiles, then the A squares must be covered by the T-tiles and the D squares by the N-tiles. Again the path will exit at the wrong place. It follows that the B squares must be covered by the T-tiles, the A squares by the R-tiles and the D squares by the N-tiles, as shown in Figure 7.31.

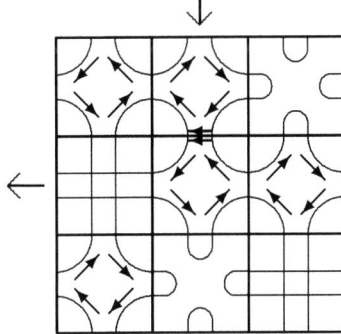

Figure 7.31

Problem 1:

For each of the two 4 × 4 mazes shown in Figure 7.32, enter along the single arrow, follow the turn-tiles and exit along the single arrow. Also enter along the double arrow, follow the turn-tiles and exit along the double arrow.

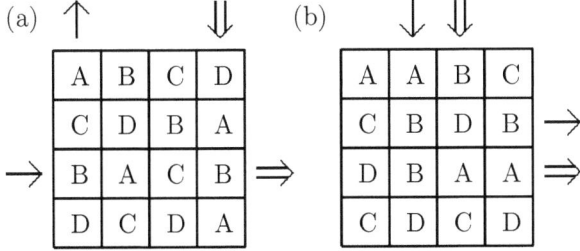

Figure 7.32

Problem 2:

For each of the two mazes shown in Figure 7.33, enter along the arrow, follow the turn-tiles and along the arrow.

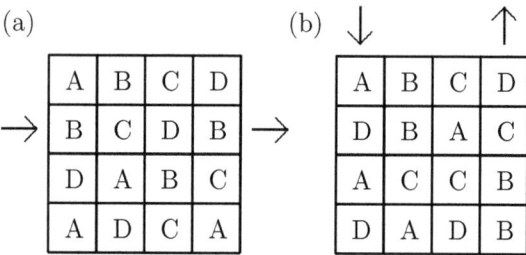

Figure 7.33

Solutions

Section 2. Solutions

Problem One-1:
The construction is shown as the unshaded part in Figure 7.34.

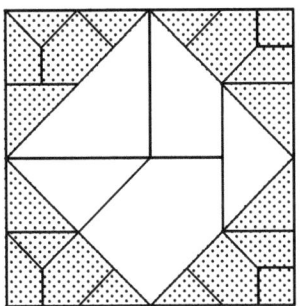

Figure 7.34

Problem One-2:
The four possible constructions are shown as the shaded parts in Figure 7.34.

Problem Two:
The four pieces form two pairs, and the two pieces in each pair can be used to construct a chisel, as shown in Figure 7.35. The figure in each playing board can be constructed with the two chisels.

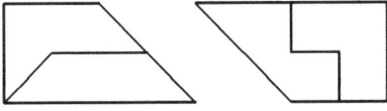

Figure 7.35

Problem Three-1:
A construction is shown in Figure 7.36.

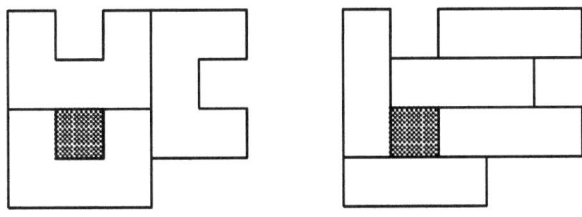

Figure 7.36

Problem Three-2:
A construction is shown in Figure 7.37.

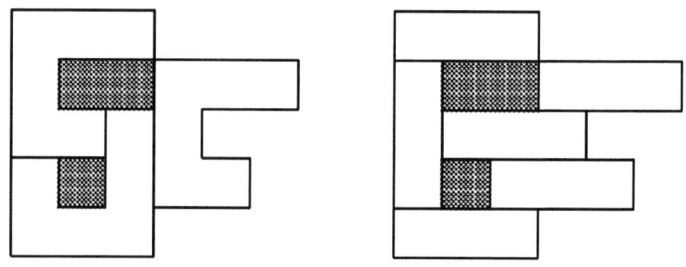

Figure 7.37

Problem Three-3:
A construction is shown in Figure 7.38.

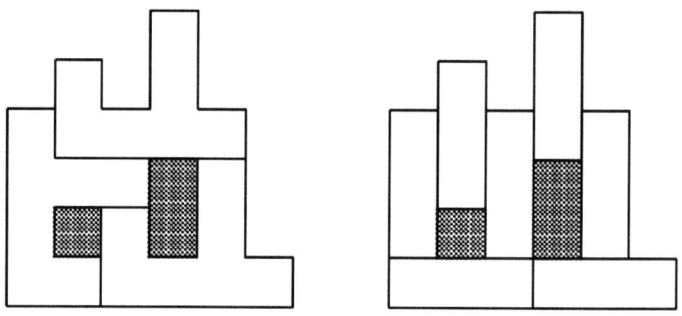

Figure 7.38

Problem Four-1:
There are six possible combinations, namely, ILN, ILT, INO, INT, LNO and LNT. One construction for each combination is shown in Figure 7.39.

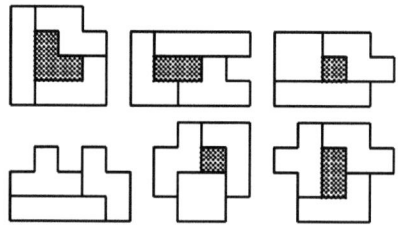

Figure 7.39

Solutions

Problem Four-2:
All five combinations are possible. One construction for each combination is shown in Figure 7.40.

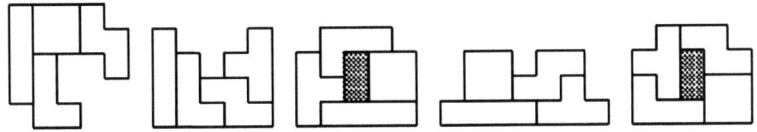

Figure 7.40

Problem Five:
The mouse goes to island 4 and do a magical swap. The elephant then goes to island 3. The mouse goes to island 4, do a magical swap, go to island 1, do another magical swap, go to island 6 and do yet another magical sway. The elephant then goes to island 7. The mouse goes to island 6, do a magical swap and goes to island 8. Then the elephant goes to island 7 and do a magical swap.

Problem Six-1:
The task can be accomplished in the following fifteen jumps by: **c**, **R**, P, b, P, b, a, b, P, a, **R**, a, P, a and R. The entries in boldface represent jumps along the street, and the other entries represent jumps along the avenue.

Problem Six-2:
The task can be accomplished in the following fifteen jumps by: **c**, R, **c**, c, R, b, **R**, P, b, P, **R**, c, P, c and R. The entries in boldface represent jumps along the street, and the other entries represent jumps along the avenue.

Problem Seven:
The task can be accomplished in 40 moves grouped into twelve sequences of moves made by the same counter. They parallel the moves of counters in the playing board in Fiure 7.41. Here, the counters can only move to adjacent squares in the same row or the same column.

A → C → H → B → G;
F → A → C → H → B;
D → F → A → C → H;
E → C → A → F → D;
H → C → A → F;
B → H → C → A;
G → B → H → C → E;
A → C → H → B;
F → A → C → H;
E → C → A → F;
H → C → A;
B → H → C → E.

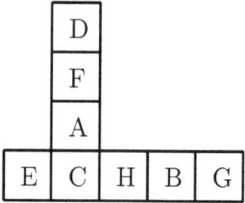

Figure 7.41

Problem Eight-1:
The reversible transformation is shown in Figure 7.42.

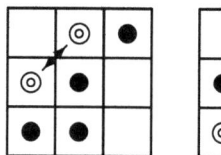

Figure 7.42

Problem Eight-2:

(a) We can transform the T-pentomino into the F-pentomino in 2 moves, as shown in Figure 7.43.

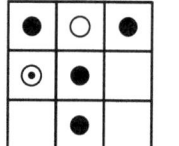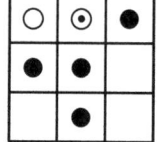

Figure 7.43

(b) We can transform the F-pentomino into the T-pentomino in 3 moves, as shown in Figure 7.44.

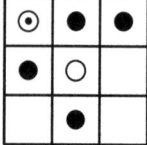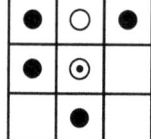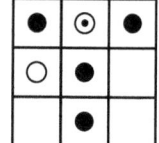

Figure 7.44

Solutions

Problem Eight-3:

(a) We can transform the F-pentomino into the V-pentomino in 2 moves, as shown in Figure 7.45.

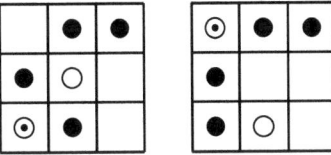

Figure 7.45

(b) We can transform the V-pentomino into the W-pentomino in only 1 move, and then to the F-pentomino in 2 more moves.

Problem Eight-4:

(a) We can transform the T-pentomino into the W-pentomino in 3 moves, as shown in Figure 7.46.

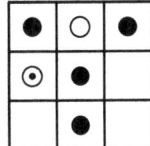 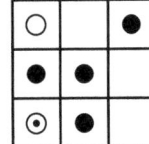 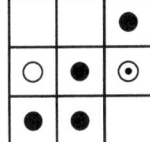

Figure 7.46

(b) We can transform the W-pentomino into the F-pentomino in 2 moves, and then to the T-pentomino in 3 more moves.

Problem Eight-5:

(a) We can transform the T-pentomino into the Z-pentomino in 3 moves, as shown in Figure 7.47.

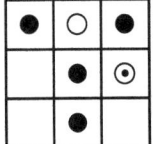 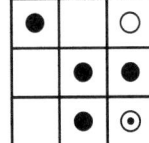 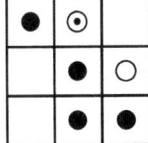

Figure 7.47

(b) We can transform the Z-pentomino into the F-pentomino in 2 moves, and then to the T-pentomino in 3 more moves.

Problem Eight-6:

(a) We can transform the T-pentomino into the F-pentomino in 3 moves, and then to the V-pentomino in 2 more moves.

(b) We can transform the V-pentomino into the F-pentomino in 3 moves, and then to the T-pentomino in 3 more moves.

Problem Nine:
By symmetry, we may let the first alien race land on any site and move to either adjacent site. The next alien race should land on a site from which they can move to the site just vacated by the preceding alien race. Continue in this manner and all six alien race can be settled in this planet.

Problem Ten:
Let the counters which start on the boxes G, D, E and R be denoted by g, d, e and r respectively. The task can be accomplished via the following sequence of moves.

$$
\begin{array}{ll|ll}
r & R \to ? \to N \to R, & g & G \to A \to M, \\
d & D \to T, & r & N \to R, \\
e & E \to I \to R \to ?, & e & R \to ?, \\
d & T \to D \to I \to E, & d & E \to I \to D \to T, \\
e & ? \to R, & e & ? \to R \to I, \\
r & R \to N, & r & R \to N \to ?.
\end{array}
$$

Problem Eleven-1:
Performing Operations Y, Z, Y, Z, Y and Z results in the position as shown in Figure 7.48 on the left. The task is completed by performing Operations Z and X, resulting in the final position as shown in Figure 7.48 on the right.

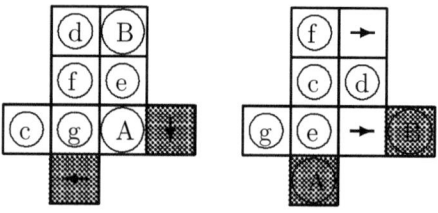

Figure 7.48

Solutions

Problem Eleven-2:
Performing Operations Z, Z, Z, Z and Z results in the position as shown in Figure 7.49 on the left. The task is completed by performing Operations Y, Z and X, resulting in the final position as shown in Figure 7.49 on the right.

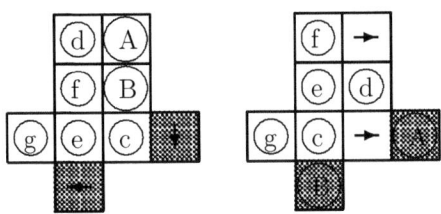

Figure 7.49

Problem Twelve:

(a) First, b goes left and down. Then d goes left. Finally, R goes up and left.

(b) First, c goes down. Then a goes right, down and right. Finally, R goes right.

(c) First, b goes left, down and right. Then c goes left, down and right. Finally, R goes up and right.

(d) First, d goes right. Then a goes right down, and left. Finally, R goes right, down and left.

(e) First, a goes right. Then R goes left, up and right. Next, e goes left and up. Finally, R goes left.

(f) First, d goes right. Then a goes right and down. Next, c goes right. Finally, R goes right.

(g) First, d goes right. Then R goes down. Next, b goes left and down. Finally, R goes left.

(h) First, e goes left. Then d goes up and left. Next, b does left, down and left. Finally, R goes left.

(i) First, a goes right and down. Then c goes left. Next, d goes right. Finally, R goes right, up and left.

(j) First, c goes left and down. Then d goes right. Next, b goes left. Finally, R goes up and right.

(k) First, e goes left and up. Then b goes left. Next, c goes left. Finally, R goes up and right.

(ℓ) First, a goes down. Then d goes right, up and left. Next, c goes left. Finally, R goes down and left.

(m) First, d goes left and up. Then c goes right. Next, e goes up and left. Finally, R goes right, down and left.

(n) First, R goes right. Then d goes right. Next, a goes right and down. Finally, R goes up and left.

Problem Thirteen:
Follow the dotted path in Figure 7.50.

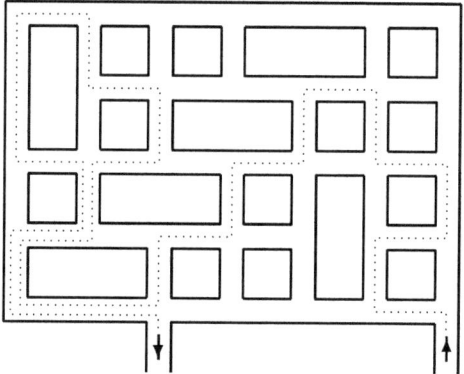

Figure 7.50

Problem Fourteen:
Follow the dotted path in Figure 7.51.

Solutions 329

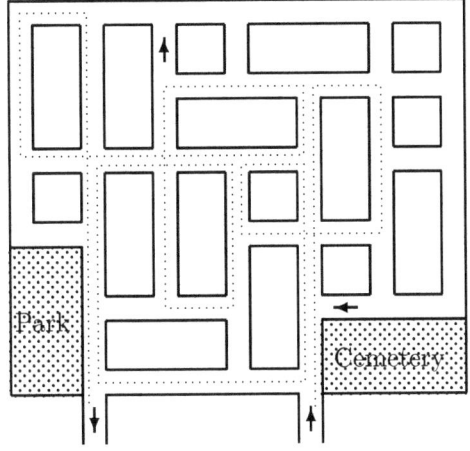

Figure 7.51

Problem Fifteen

First stage: h, m, p/P, N/n, i/I, K/k, d/D, E/e, b/B, F, G.
Second stage: G/g, ℓ/L, J/j and g (or G/g, j/J, L/ℓ, g).
Third stage: g/G, F, B/b, e/E, D/d, k/K, I/i, n/N, P/p, m/M, H.
The moves are marked in numerical order in Figure 7.52.

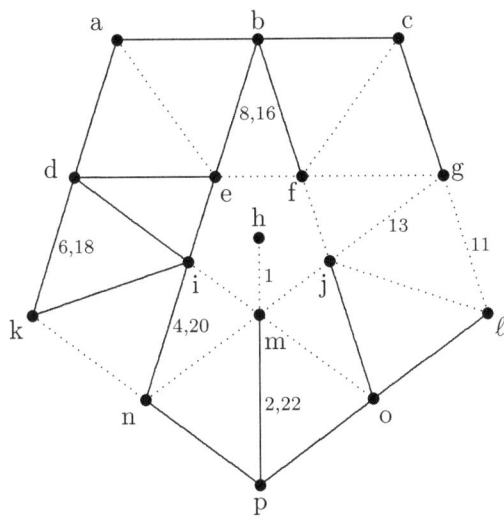

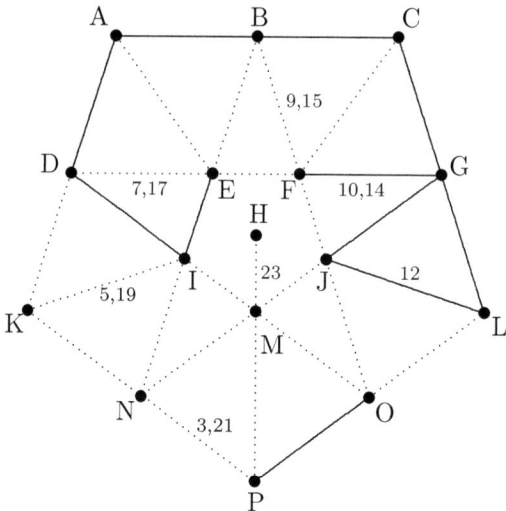

Figure 7.52

Problem Sixteen-1:

(a) The A squares are covered by the R-tiles, the B squares by the T-tiles, the C squares by the U-tiles and the D squares by the N-tiles, as shown in Figure 7.53.

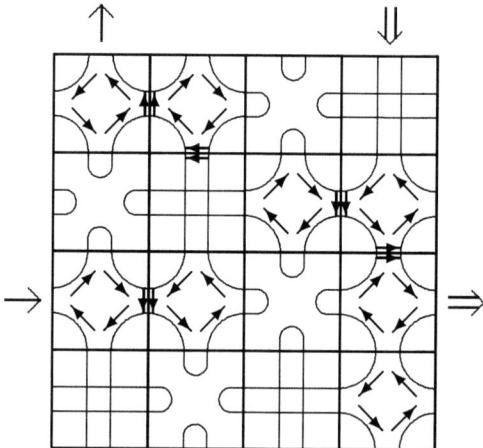

Figure 7.53

Solutions

(b) The A squares are covered by the N-tiles, the B squares by the T-tiles, the C squares by the R-tiles and the D squares by the N-tiles, as shown in Figure 7.54.

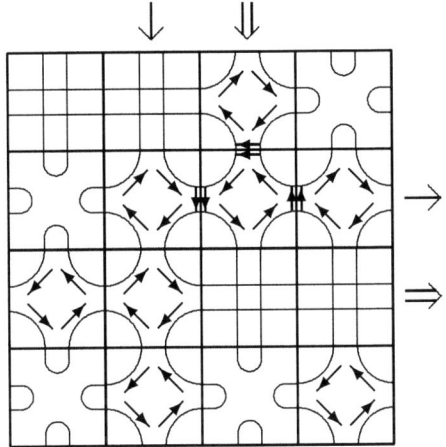

Figure 7.54

Problem Sixteen-2:

(a) The A squares are covered by the N-tiles, the B squares by the R-tiles, the C squares by the U-tiles and the D squares by the T-tiles, as shown in Figure 7.55.

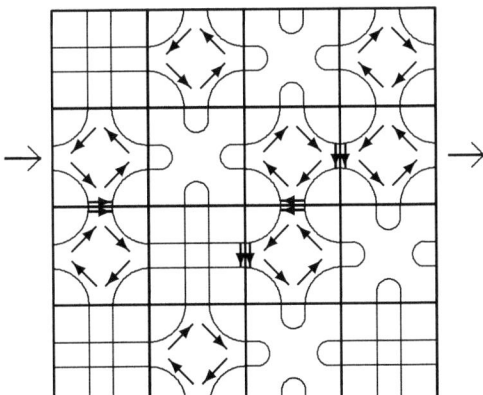

Figure 7.55

(b) The A squares are covered by the N-tiles, the B squares by the U-tiles, the C squares by the R-tiles and the D squares by the T-tiles, as shown in Figure 7.56.

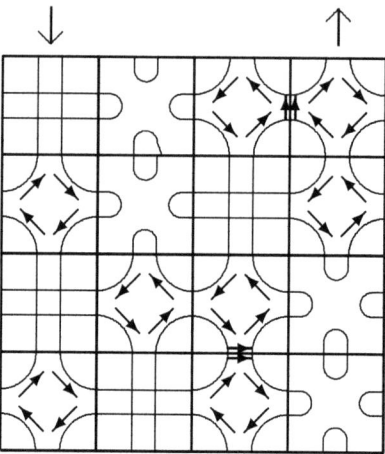

Figure 7.56

Chapter Eight
Sample GAME Math Unfair Games

Section 1. Problems.

Ticktacktoe-type Games

Game One: Ticktacktoe with Motion

Equipment:
A 3×3 board; three white counters and three black counters.

Rules:
The two players take turns, each placing one counter of her color on any vacant square on the board. The objective is for each player to have her three counters in a row, a column or a diagonal. If victory has not been achieved after all six counters have been placed, the players continue their turns. Instead of placing an additional counter of her color, she moves a counter of her color to a vacant square sharing a common side with its current square. The game is a draw if victory is not achieved after each player has made three moves.

Game Two: Wild Ticktacktoe

Equipment:
A 3×3 board; nine counters which are black on one side and white on the other.

Rules:
The two players take turns, each placing one counter with either side up on any vacant square on the board. The objective for either player is to have three counters of the same color in a row, a column or a diagonal. If victory has not been achieved after all nine counters have been placed, the game is a draw.

Game Three: Extended Ticktacktoe

Equipment:
The playing board shown in Figure 8.1; four white counters and four black counters.

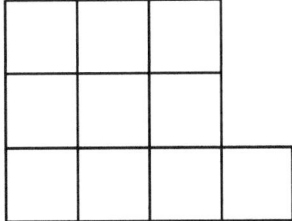

Figure 8.1

Rules:

The two players take turns, each placing one counter of her colors on any vacant square on the board. The objective is for each player to have three counters of her color in a row, a column or a diagonal, except that in the bottom row, four counters of her color are required. If victory has not been achieved after all eight counters have been placed, the game is a draw.

Game Four: Tri-Hex

Equipment:

The playing board shown in Figure 8.2; four white counters and four black counters.

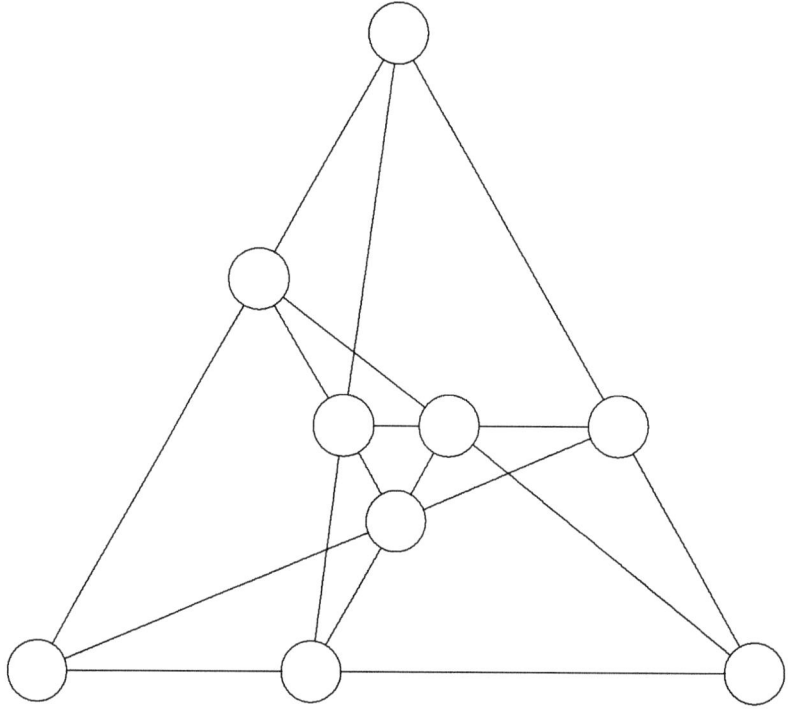

Figure 8.2

Rules:

The two players take turns, each placing one counter of her colors on any vacant circle on the board. The objective is for each player to have her three counters in a line. If victory has not been achieved after all eight counters have been placed, the game is a draw.

Problems

Game Five: West

Equipment:
Nine cards with the words ASIA, BROW, FREE, HAWK, HEEL, MOTH, PART, SOON and WEST printed on them.

Rules:
The two players take turns, each taking one card which has not already been taken. The objective is to have three cards with a common letter. If victory has not been achieved after all nine cards have been taken, the game is a draw.

Game Six: Fifteen

Equipment:
Nine cards with the numbers 1, 2, 3, 4, 5, 6, 7, 8 and 9 printed on them.

Rules:
The two players take turns, each taking one card which has not already been taken. The objective is to have three cards with numbers totalling 15. If victory has not been achieved after all nine cards have been taken, the game is a draw.

Nim-type Games

Game Seven: Nim

Equipment:
Twelve counters arranged in three piles, with 3, 4 and 5 counters respectively.

Rules:
The two players take turns, each taking at least one counter from any pile. The objective is to take the last counter.

Game Eight: Kayles

Equipment:
Twelve counters arranged in three rows, with 3, 4 and 5 counters respectively.

Rules:
The two players take turns, each taking one counter or two adjacent counters from any row. A row is broken up into two if counters are taken from the middle part of it. The objective is to take the last counter.

Game Nine: Tac-Tix

Equipment:
A 3×3 board; eight counters.

Rules:
The counters are placed on eight of the nine squares of the board, except for a corner square. The two players take turns, each taking one counter or several adjacent counters from any row or column. The objective is to take the last counter.

Game Ten: Cram

Equipment:
The playing board shown in Figure 8.3; fifteen counters. one on each square of the playing board.

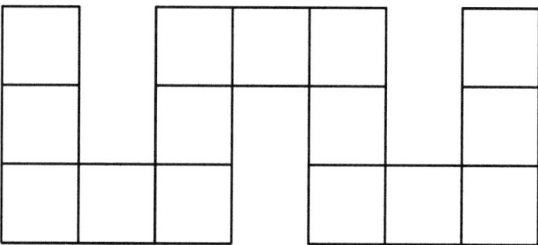

Figure 8.3

Rules:
The counters are placed the squares of the playing board. The two players take turns, each taking two adjacent counters. The objective is to make the last move.

Game Eleven: Chomp

Equipment:
A 3×5 board; fifteen counters.

Rules:
The counters are placed on the squares of the board. The two players take turns, each choosing a counter and taking all counters in squares above or to the right (or both) of the square of the chosen counter. The objective is *not* to take the last counter.

Game Twelve: Bynum's Game

Equipment:
A 3×5 board; fifteen counters.

Rules:
The counters are placed on the squares of the board. The two players take turns. The first player chooses a column and takes every counter in that column. As the game progress, rows and columns are broken into separate groups of adjacent counters. Each group is then considered a separate row or column. The second player chooses a row and takes every counter in that row, the first player then chooses a column and takes every counter in that column, and so on. The objective is to take the last counter.

Miscellaneous Games

Game Thirteen: Cop and Robber

Equipment:
The playing board shown in Figure 8.4; one black counter and one white counter.

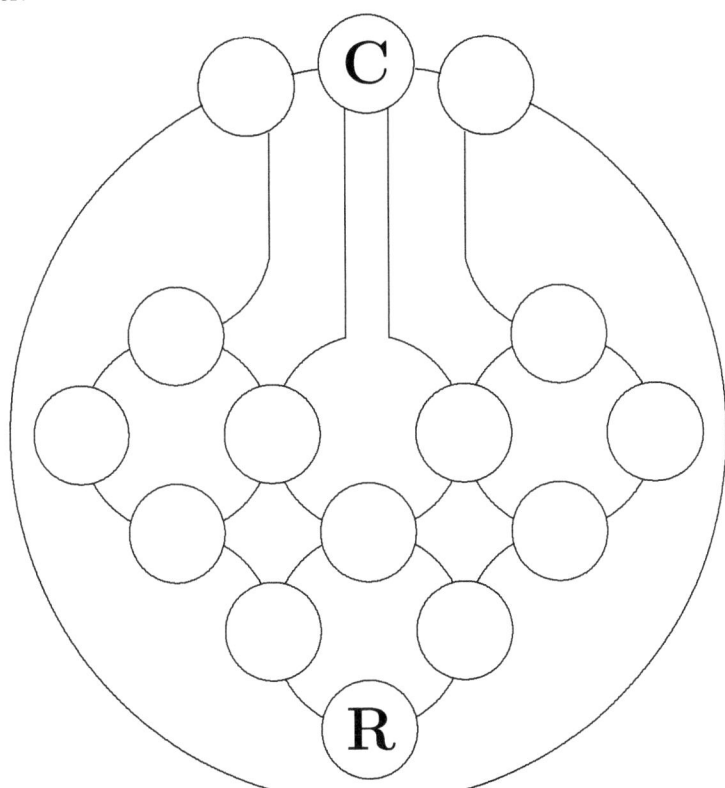

Figure 8.4

Rules:
The black counter represents the Cop and starts on the circle marked C. The white counter represents the Robber and starts on the circle marked R. One of the two players controls the Cop while the other controls the Robber. They take turns moving their counters, with the Cop moving first. Movement is between two circles connected by an arc, four of which are extended by line segments. A Cop victory is achieved when the Cop moves onto the circle currently occupied by the Robber. If a Cop victory is not achieved after the Cop has moved eight times, the game is a Robber victory.

Game Fourteen: Lewthwaite's Game

Equipment:
A 5×5 board; twelve white counters and twelve black counters.

Rules:
The first player White controls the twelve white counters and the second player Black controls the twelve black counters. The counters are placed on the board except for the center square, and squares sharing a common side contain counters of different color. The four squares surrounding the vacant center square contain white counters. The players take turns moving a counter of her color into the vacant square from a square sharing common side with it. The player without a move loses the game.

Game Fifteen: Trax

Equipment:
The playing board shown in Figure 8.5 on the left; fourteen square cards marked front and back as shown in Figure 8.5 on the right.

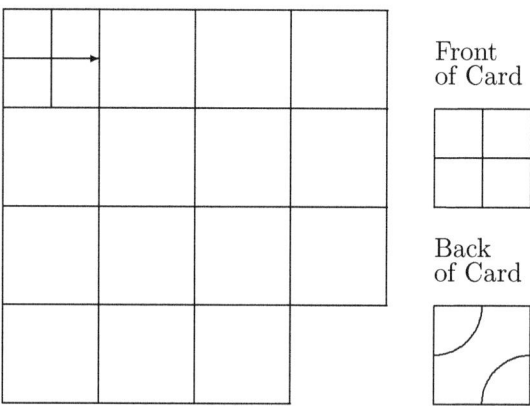

Figure 8.5

Rules:
The two players take turns placing a card on the board, continuing the path which starts with the marked arrow on the playing board. The player may choose either side of the card to be up. The player who runs the path to the boundary of the playing board loses the game.

Game Sixteen: Non-Attacking Rooks

Equipment:
A standard 8×8 chessboard in which the rows are numbers from 1 to 8 and the columns are labeled from a to h; a White Rook and a Black Rook.

Rules:
The first player places the White Rook on square b2 and the second player places the Black Rook on square c4. The players then take turns moving their Rooks to another square in the same row or the same column which is not under attack by the other Rook, and has not been landed on before by either Rook. The player without a move loses the game.

Game Seventeen: Increasing Distances

Equipment:
A standard 8×8 chessboard; a counter.

Rules:
The second player places the counter on any square. Then the players take turns moving the counter to another square such that the distance covered by the counter exceeds the distance in the preceding move. Distance is measured from the center of the initial square to the center of the final square. The player without a move loses the game.

Game Eighteen: Cats and Mouse

Equipment:
The playing board shown in Figure 8.6; three black counters and one white counter.

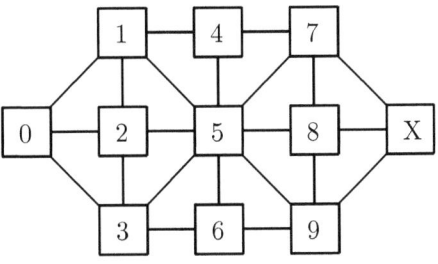

Figure 8.6

Rules:
The black counters represent Cats and start in the squares 0, 1 and 3. The white counter represents the Mouse. One of the two players controls the Cats while the other controls the Mouse. The Mouse player starts the game by placing the Mouse on any circle not occupied by a Cat. Then the Cat player moves one of her Cats, and the Mouse player moves the Mouse. Turns alternate thereafter. Movement is between two circles connected by a line segment. Only one Cat can move at a time, and a Cat may never move from right to left, orthogonally or diagonally. A Cat victory is achieved when the Cats surround the Mouse so that it cannot move. This may happen if the Mouse is in square 4, 6 or most likely X. If a Cat victory is not achieved after the Cats among them have moved a total of twelve times, the game is a Mouse victory. A Mouse victory is also achieved when the Mouse has broken through to the left of the Cats.

Solutions

Section 2. Solutions

Game One: Ticktacktoe with Motion

The first player has a sure win by taking the center square. The second player can either take a corner square or a side square. In Figure 8.7, these responses are marked by large black dots. The first player can force the response of the second player in the next two moves, both marked by small black dots. After placement has been completed, the first player wins in two moves, indicated by arrows.

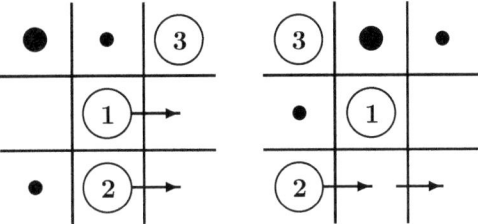

Figure 8.7

Game Two: Wild Ticktacktoe

The first player has a sure win by placing a counter in the center square, say black side up. The second player cannot place a counter black side up anywhere. If she places a counter white side up in a corner square, the first player places a counter white side up in the opposite corner square and wins. If the second player places a counter white side up in a side square, the first player places a counter white side up in the opposite side square. The only move available to the second player without losing immediately is to place a counter white side up on another side square. The first player places a counter white side up in the last side square and wins.

Game Three: Extended Ticktacktoe

The first player has a sure win by taking the square adjacent to the extra square diagonally, labeled 1 in Figure 8.8. The second player's nine responses are shown by large black dots. In each case, the first player can make a second move, labeled 2. This is forcing, and create an unstoppable double threat on her third move, labeled 3. She then wins on her fourth and last move. The small black dots represent the second player's forced second move.

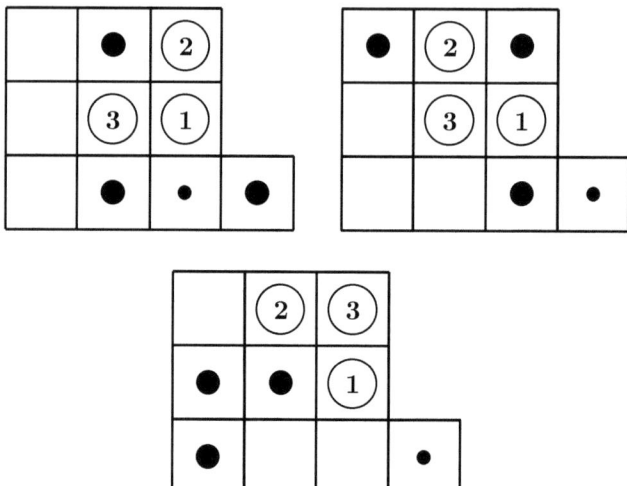

Figure 8.8

Game Four: Tri-Hex

The nine circles fall into three groups of three, the outer circles, the inner circles and the intermediate circles. The outer circles are equivalent to the inner circles by symmetry.

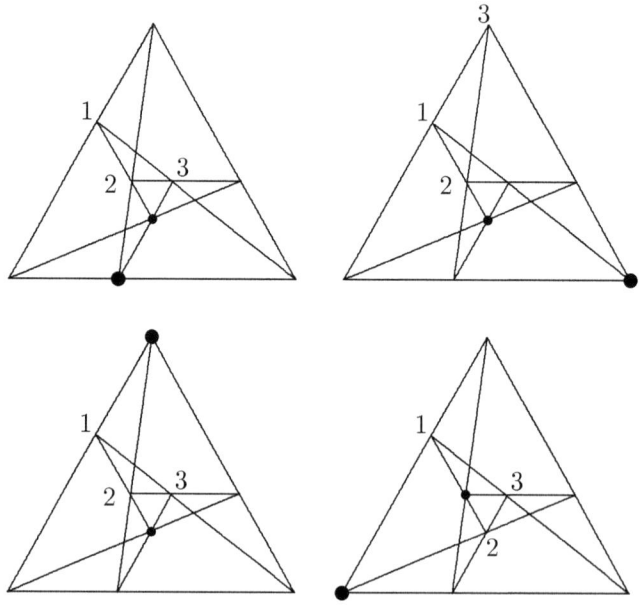

Figure 8.9

Solutions

The first player has a sure win by taking any of the intermediate circles, labeled 1 in Figure 8.9. The second player's nine responses are shown by large black dots. In each case, the first player can make a second move, labeled 2. This is forcing, and create an unstoppable double threat on her third move, labeled 3. She then wins on her fourth and last move. The small black dots represent the second player's forced second move.

Game Five: West

There are exactly eight winning lines in the standard Ticktacktoe game. Here, we have exactly eight letters which are shared by three words. They are A (ASIA-HAWK-PART), E (FREE-HEEL-WEST), H (HAWK-HEEL-MOTH), O (BROW-MOTH-SOON), R (BROW-FREE-PART), S (ASIA-SOON-WEST), T (MOTH-PART-WEST) and W (BROW-HAWK-WEST).

Thus this game is isomorphic to standard Ticktacktoe game. The link between them is provided by the table below.

MOTH	HEEL	HAWK
SOON	WEST	ASIA
BROW	FREE	PART

Note that the letters B, F, I, K, M, L, N and P play no part in the game, and the letters A (in ASIA), E (in FREE and HEEL) and O (in SOON) have been doubled. Only in WEST are all four letters essential.

Game Six: Fifteen

This game is also isomorphic to standard Ticktacktoe game. There are exactly eight ways in which three of the numbers 1, 2, 3, 4, 5, 6, 7, 8 and 9 add up to 15. They are 1+5+9, 1+6+8, 2+4+9, 2+5+8, 2+6+7, 3+4+8, 3+5+7 and 4+5+6. The link between the two games is provided by the magic square below.

6	1	8
7	5	3
2	9	4

Game Seven: Nim

We represent the starting position by (3,4,5). The first player has a sure win by moving to the position (1,4,5), that is, taking 2 counters from the pile with 3 counters. Figure 8.10 shows the first player's winning strategy. Victory is achieved by leaving behind the position (0,0,0), but if a player can leave behind any of (0,1,1), (0,2,2), (0,3,3) or (0,4,4), she can win by mimicking the moves of the other player. These positions are called *safe* positions. Other safe positions are (1,2,3) and (1,4,5). These are shaded.

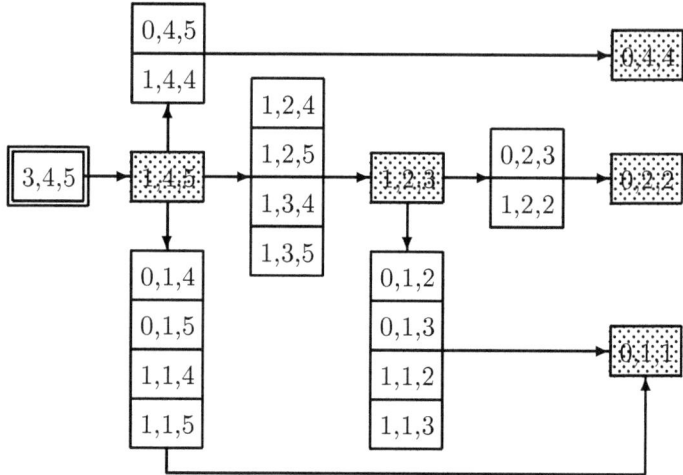

Figure 8.10

An *unsafe* position is simply one that is not safe. Whether a position is safe or unsafe can be worked out by backtracking from the ultimate winning position using the following general definition of safe and unsafe positions.

(1) The ultimate winning position is a safe position.

(2) If from a position, there is a move which leads to a safe position, then this position is unsafe.

(3) If from a position, every move leads to an unsafe position, then this position is safe.

However, for Nim, there is a direct way of doing so by means of binary numbers. Take the starting position (3,4,5). These three numbers become 11, 100 and 101 in base 2, as shown in the diagram below on the left. The total number of 1s in each column is computed. A Nim position is defined to be safe if and only if among the digits in any column, the number of 1s is always even. Since the number of 1s in the second column is 1, (3,4,5) is an unsafe position.

$$
\begin{array}{c|ccc}
3 & 0 & 1 & 1 \\
4 & 1 & 0 & 0 \\
5 & 1 & 0 & 1 \\
\hline
  & 2 & 1 & 2
\end{array}
\qquad
\begin{array}{c|ccc}
? & 0 & 0 & ? \\
4 & 1 & 0 & 0 \\
5 & 1 & 0 & 1 \\
\hline
  & 2 & 0 & ?
\end{array}
\qquad
\begin{array}{c|ccc}
1 & 0 & 0 & 1 \\
4 & 1 & 0 & 0 \\
5 & 1 & 0 & 1 \\
\hline
  & 2 & 0 & 2
\end{array}
$$

Solutions

We now show that this definition agrees with the general definition given earlier. Taking the unsafe position (3,4,5) as an example. We must show that there is a move which leads to a safe position. We can leave the first column alone, but we must work on the second column. Since 0s cannot be changed into 1s, we must change the lone 1 into a 0. This means that we are taking counters from the pile which starts with 3. As shown in the diagram above in the middle, the number of counters is this pile is reduced, regardless of what the digits in the remaining columns on this row may be. In this case, there is no need to make further changes, and we have arrived at the safe position (1,4,5), as shown in the diagram above on the right.

From the safe position (1,4,5), all moves lead to unsafe positions. This is because we must take counters from only one pile, so that at most one digit is changed in each column. At least one digit must change since we must take at least one counter, and this digit can only change from 1 to 0. This means that the column containing this 1, which has an even number of 1s before, will now have an odd number of 1s. Thus the resulting position must be unsafe.

Game Eight: Kayles

We represent the starting position by (3,4,5). The first player has a sure win by moving to the position (1,3,3,4), that is, taking 1 counter from the pile with 5 counters to split it into a pile with 1 counter and a pile with 3 counters. Figure 8.11 shows the first player's winning strategy.

Note that the number of piles may increase as well as decrease. A pile which has been depleted will no longer be represented in the position. Thus victory is achieved by leaving behind the empty set. However, if a player can leave behind any symmetric position such as (1,1), (2,2), (3,3), (1,1,1,1), (1,1,2,2), (1,1,3,3) or (1,1,1,1,1,1), she can win by mimicking the moves of the other player. Thus these positions are safe. Other safe positions are (1,3,3,4), (1,2,3), (2,3,4) and (1,1,1,1,1,4). These are shaded in the diagram above.

In all safe positions containing 4, if we replace the 4 by a 1, we still have a safe position. It appears that a pile with 4 counters acts just like a pile with only 1 counter. This is in fact correct, and we will give a more formal justification later.

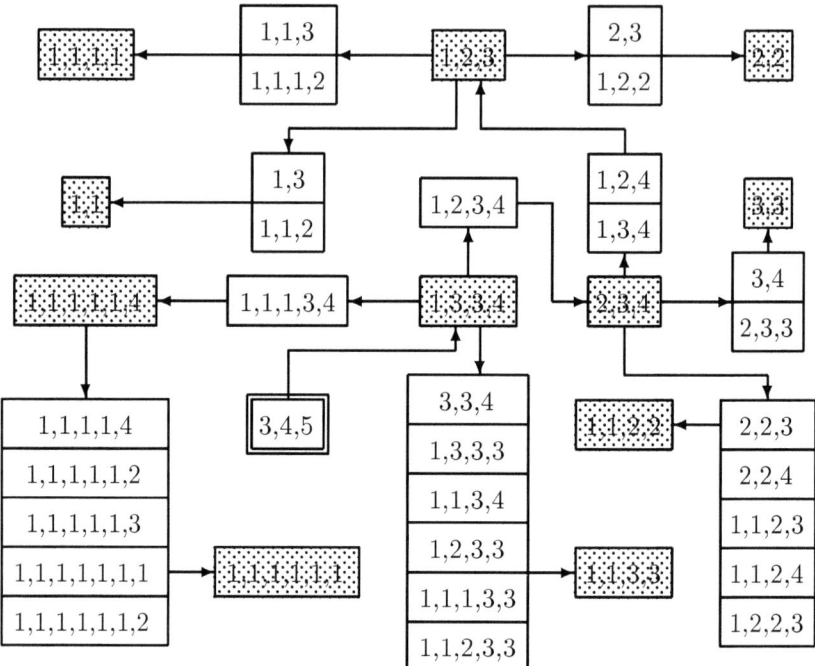

Figure 8.11

As usual, we assign the value 0 to the final position, the empty set. The values of non-empty piles are computed based on the Principle of Minimum Excluded Value. We give some simple examples.

From a pile with 1 counter, we can only leave behind the empty set with value 0. The smallest value not represented is 1, and that is the value assigned to a pile with 1 counter.

From a pile with 2 counters, we can either leave behind the empty set of a pile with 1 counter. Their respective values are 0 and 1. The smallest value not represented is 2, which is the value assigned to a pile with 2 counters. The assigned value of a pile of 3 counters is 3. While we cannot leave behind the empty set, we can split it into two piles each with 1 counter.

From a pile with 4 counters, we can leave behind a pile with 3 counters, a pile with 2 counters, two piles each with 1 counter, or two piles with 1 and 2 counters. Their respective values are 3, 2, 1+1=0 and 1+2=3 using arithmetic modulo 2. The smallest value not represented is 1, and this value is assigned to a pile with 4 counters. Note that the same principle is used to compute the values of Nim positions, with a different result for a pile with 4 counters.

Solutions

To see why a pile with 4 counters acts as if it is a pile with only 1 counter, consider the position (1,4). We prove that it is safe. If your opponent takes the only counter from the smaller pile, you can convert the larger pile into something with value 0. Suppose he takes counters from the larger pile. He cannot leave behind something with value 1 because that value is excluded. If he leave behind something with value 0, you win by taking the only counter from the smaller pile. If he leaves behind something with value greater than 1, then you can convert it into something with value 1 since 1 is no longer excluded.

Game Nine: Tac-Tix

This game is a sure win for the second player. There are twelve non-equivalent opening moves by the first player, as shown in Figure 8.12.

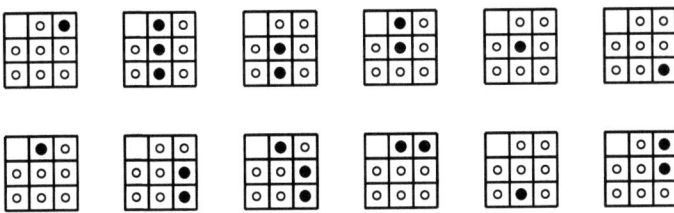

Figure 8.12

The second player can reduce each of the first four positions into two groups of 1×2, and each of the next two positions into two groups of 2×2 with a corner missing. Thereafter, the second player plays symmetrically between the two groups. Each of the next four positions can be reduced to a 2×2. If the first player takes two counters, the second player wins immediately. If the first player takes one counter, the second player wins by reducing the position to two groups of 1×1.

The second player can reduce each of the positions resulting from the first player's last two opening moves to the position shown in Figure 8.13, with five non-equivalent responses from the first player. The second player wins by reducing each position into two groups of 1×1.

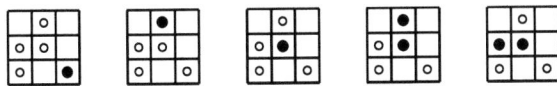

Figure 8.13

The Principle of Minimum Excluded Value may be applied to analyze the game of Tac-Tix, especially for larger boards for which a position by position strategy may be difficult to construct.

Game Ten: Cram

The second player has a sure win. For convenience, we change the playing board to its equivalent form, a 1×15 board, with the squares numbered 1 to 15 in their natural order. Instead of removing counters, the first player will place two black counters on adjacent squares and the second player will place two white counters on adjacent squares.

We shall first analyze the game for a $1 \times n$ board for small values of n. On a 1×1 board, the second player wins as the first player cannot move. On a 1×2 or a 1×3 board, the first player wins. On a 1×4 board, the first player wins by placing two counters on squares 2 and 3, but she can also play to lose by placing them on squares 1 and 2. It is desirable to do so under certain circumstances.

On a 1×5 board, the second player wins. On a 1×6 board, the first player can play to either win or lose. On a 1×7 or a 1×8 board, the first player can win. However, if she does not play to win, then the second player can decide who wins. We now show that on a 1×9 board, the second player has a sure win. Figure 8.14 shows the four non-equivalent opening moves by the first player and the winning responses by the second player.

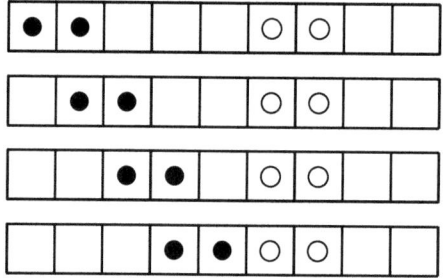

Figure 8.14

On the 1×15 board, the first player has seven non-equivalent opening moves as shown in Figure 8.15, along with the winning responses by the second player.

In the second and the fifth cases, the second player wins on the 1×9 board. In the first case, if the first player plays on the 1×3 board, the second player plays to win on the 1×8 board. If the first player plays to win on the 1×8 board, the second player plays on the 1×3 board. If the first player plays on the 1×8 board but not to win, she must leave behind a 1×4 board on which the second player also play not to win. The analysis of the remaining cases are analogous to that for the first case.

Solutions

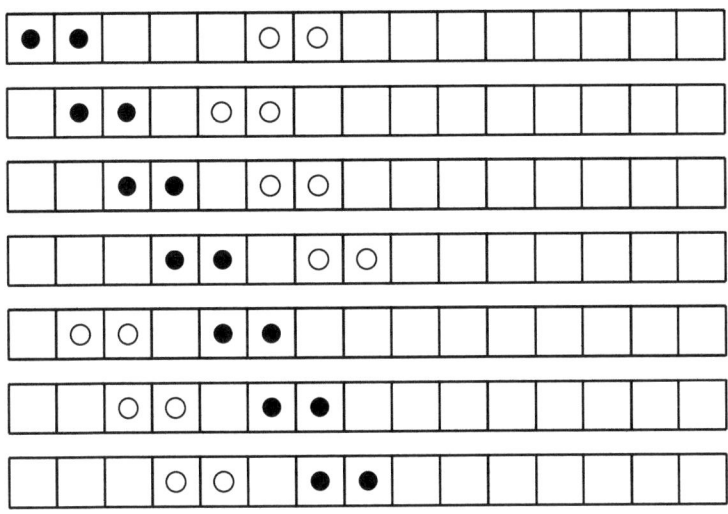

Figure 8.15

Game Eleven: Chomp

The first player has a sure win. Label the rows of the board 1 to 3 from bottom to top, and the columns of the board a to e from left to right. The winning opening move is to choose the counter on d3 and take it along with the counter on e3. There are twelve possible responses by the second player, grouped into six pairs, namely (a2,b1), (a3,e2), (b2,d1), (b3,c1), (c2,e1) and (c3,d2). In each case, the winning continuation is the other move in the same pair. These are shown in Figure 8.16.

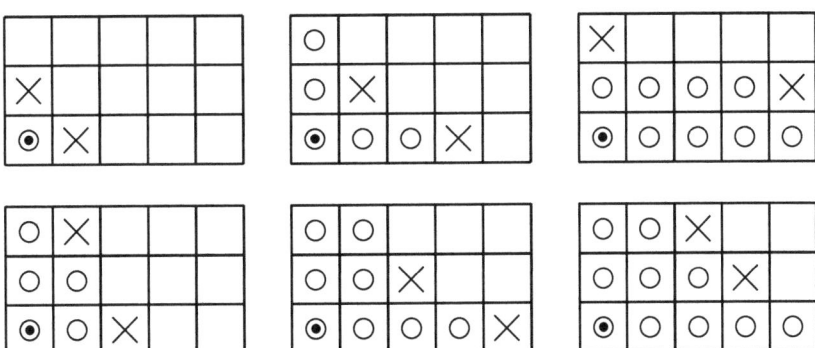

Figure 8.16

In the first case, the first player has already won. In the second case, the first player plays symmetrically on row 1 and column a. In the third case, there is one more counter on row 1 than in row 2. In the fourth case, there is one more counter on column a than in column b. The first player can keep this up and win.

In the fifth case, there seven possible responses by the second player are grouped into four pairs, namely (a2,b1), (a3,d2), (b2,d2) and (b3,c1), with d2 appearing twice. In each case, the winning continuation is the other move in the same pair, leading to an earlier case.

In the sixth case, there nine possible responses by the second player are grouped into five pairs, namely (a2,b1), (a3,e2), (b2,12), (b3,c1) and (c2,e2), with e2 appearing twice. In each case, the winning continuation is the other move in the same pair, again leading to an earlier case.

Game Twelve: Bynum's Game

The first player has a sure win, by taking every counter in the second column. There are four non-equivalent responses by the second player, as shown in Figure 8.17. The moves are represented by vertical and horizontal arrows.

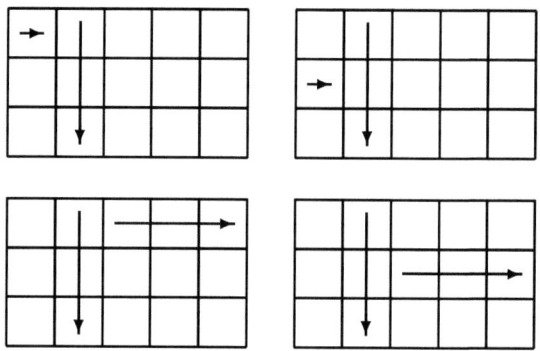

Figure 8.17

In the first two cases, the first player takes all the counters in the fourth column, leaving behind 8 counters. In the last two cases, the first player takes all the counters in the first column, leaving behind 6 counters. Both numbers are even. In all subsequent moves, the second player takes exactly one counter. The first player can take either 1 or 3 counters, and wins.

Solutions

Game Thirteen: Cops and Robber

The Cop player has a sure win. The playing board is equivalent to the one shown in Figure 8.18, where the circles are painted black and white in the standard chessboard pattern. All moves are between a black circle and a white circle except for the segment at the top, which corresponds to the large circle in the original playing board.

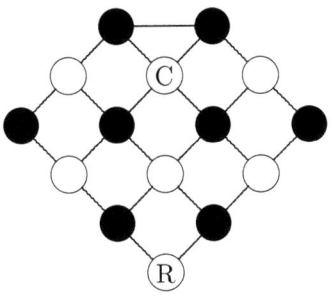

Figure 8.18

Initially, the Cop and the Robber are on circles of the same color. In order for the Cop to be able to move onto the circle currently occupied by the Robber, they must be on circles of opposite colors. Hence the Cop must make the only possible move between two black circles, and wins since the Robber can be prevented from doing the same.

Game Fourteen: Lewthwaite's Game

Paint the squares of the board black and white in the standard chessboard pattern, with black squares at the corners. Initially, all white counters are on white squares and all black counters are on black squares. When it is White's turn to move, the vacant square is black, and when it is Black's turn to move, the vacant square is white. Since every move is between a white square and a black square, each counter can move at most once, so that the players cannot be making moves forever.

Figure 8.19

The game is a sure win for Black. The winning strategy is to partition the board except the center square into dominoes, as shown in Figure 8.19. At every turn, Black just moves her counter on the same domino as the square which White has just vacated.

Game Fifteen: Trax

The game is a sure win for the first player. The winning opening move is to use the back of the card to curve the path towards the center of the playing board. There are three possible responses from the second player, as shown in Figure 8.20. In each case, the remaining part of the playing board is partitioned into dominoes, except for the two squares connected by the unused track of the square which the second player has just placed. These two squares constitute a split domino.

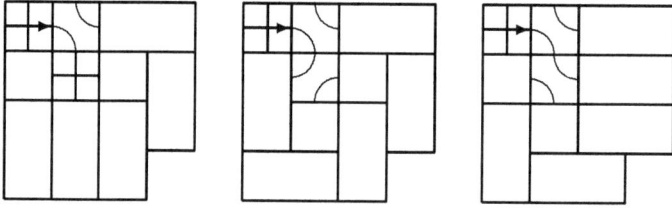

Figure 8.20

The winning strategy for the first player is simple. Continue the path so that it remains within the domino (including the split domino), thereby not running it to the boundary of the playing board. The second player must then complete the domino. If the path is not already run to the boundary, it will point to an uncovered domino, allowing the first player to continue her strategy. The second player must lose by the time no uncovered dominoes remain, if not before.

Game Sixteen: Non-Attacking Rooks

The second player has a winning strategy. Divide the eight rows into four pairs $(1,3)$, $(2,4)$, $(5,7)$ and $(6,8)$, and the eight columns also into four pairs (b, c), (d, e), (f, g) and (h, a). Then divide the sixty-four squares into thirty-two pairs. Two squares are in the same pair if and only if they are on two different rows which form a pair, and on two different columns which also form a pair. Thus the starting squares of the two rooks form a pair. The second player's strategy is to move the Black Rook to the square which forms a pair with the square where the White Rook has just landed.

Solutions

First, this can always be done, because if the White Rook stays on its current row, the Black Rook will do the same, and if the White Rook stays on its current column, the Black Rook will do the same. Second, since the square on which the White Rook has just landed cannot have been landed on before, the square to which the Black Rook is moving has never been landed on before, since the squares are occupied by the two rooks in pairs. Third, the Black Rook will not be under attack by the White Rook since the two squares in the same pair are on different rows and on different columns. Hence the second player always has a move, and can simply wait for the first player to run out of moves.

Game Seventeen: Increasing Distances

The first player has a winning strategy. Label the cells of the board as shown in Figure 8.21. The maximum distance between two squares with the same label occurs when the two squares are symmetric about the center of the board, and is at least the distance between two squares with smaller or equal labels. In each move, the first player takes the counter to the square symmetric about the center of the chessboard. The second player is forced to take the counter to a square with a higher label. Eventually, the first player takes the counter from one corner square to the opposite corner square and wins.

9	8	7	6	6	7	8	9
8	6	5	4	4	5	6	8
7	5	3	2	2	3	5	7
6	4	2	1	1	2	4	6
6	4	2	1	1	2	4	6
7	5	3	2	2	3	5	7
8	6	5	4	4	5	6	8
9	8	7	6	6	7	8	9

Figure 8.21

Game Eighteen: Cats and Mouse

This being the last item in the book, we give a full analysis in the multi-page Figure 8.22.

The Cats player has a sure win, regardless of where the Mouse starts. There are 53 non-equivalent positions at the moment when it is the turn for the Cat player to move. For each position, the best move for the Cats is given, followed by all possible responses by the Mouse.

The starting position is any of 48 to 53 inclusive. The Cat victory is achieved in position 1 or 2, to which there are no responses. In all other cases, each response leads to an earlier position given in boldface.

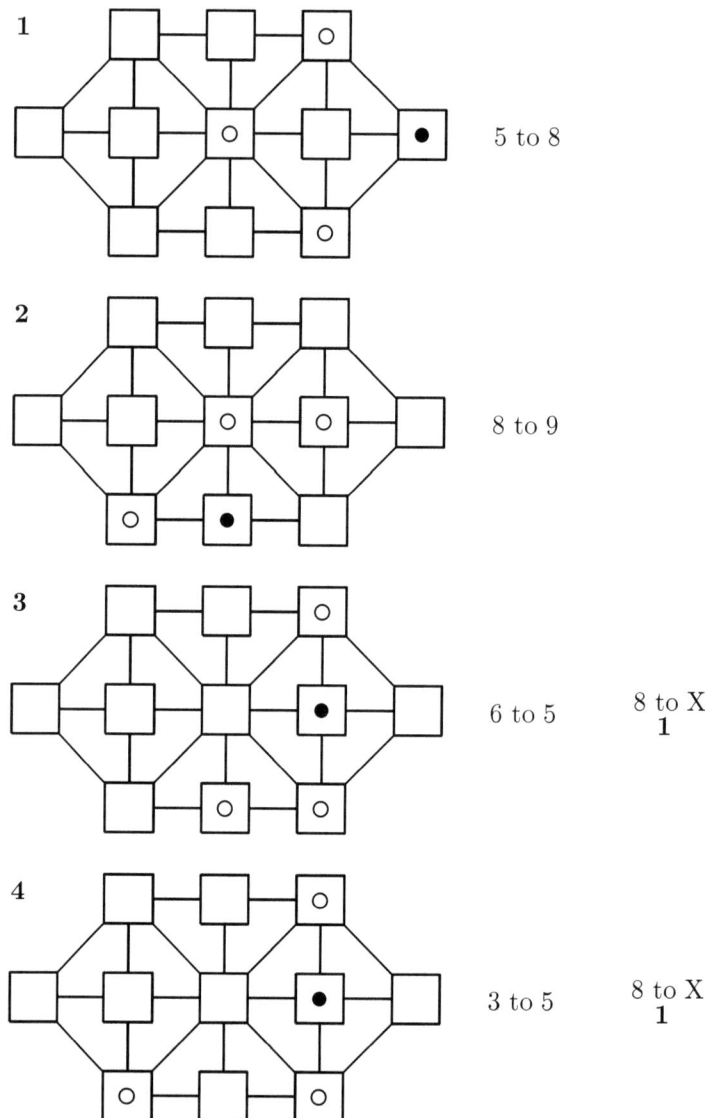

Solutions

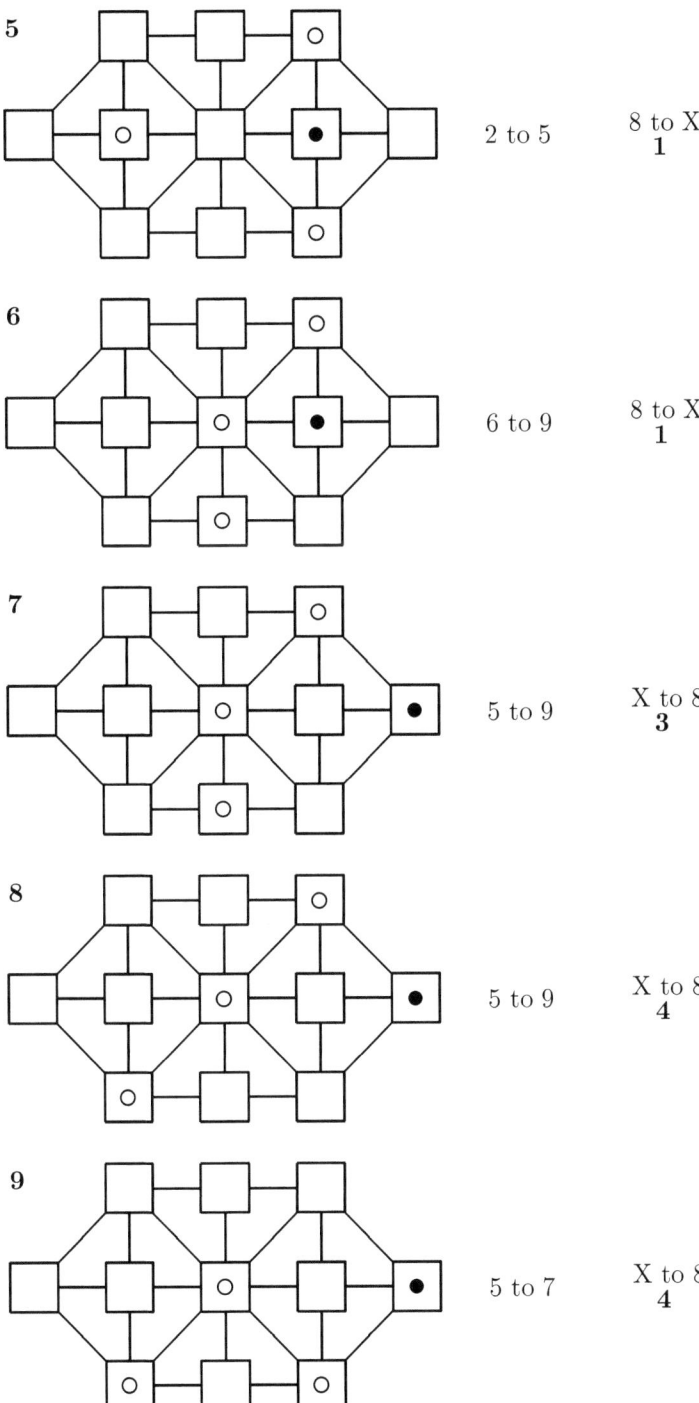

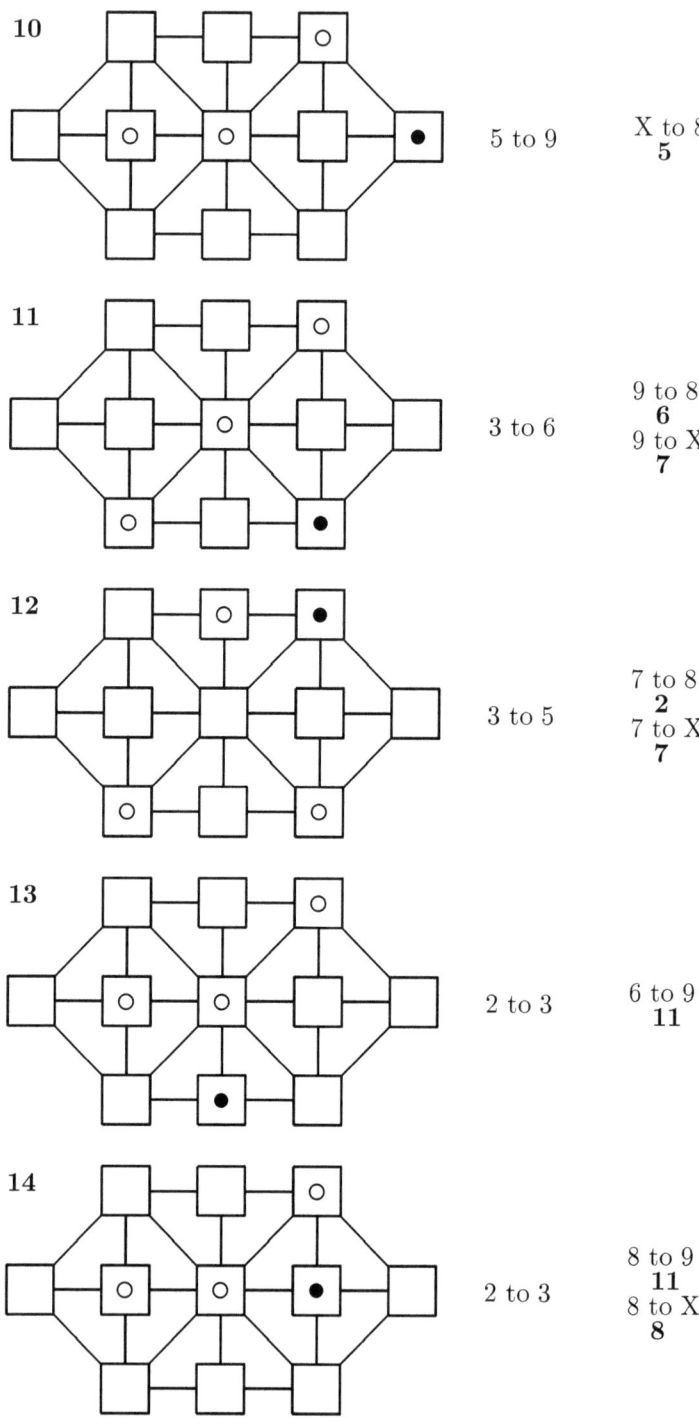

Solutions

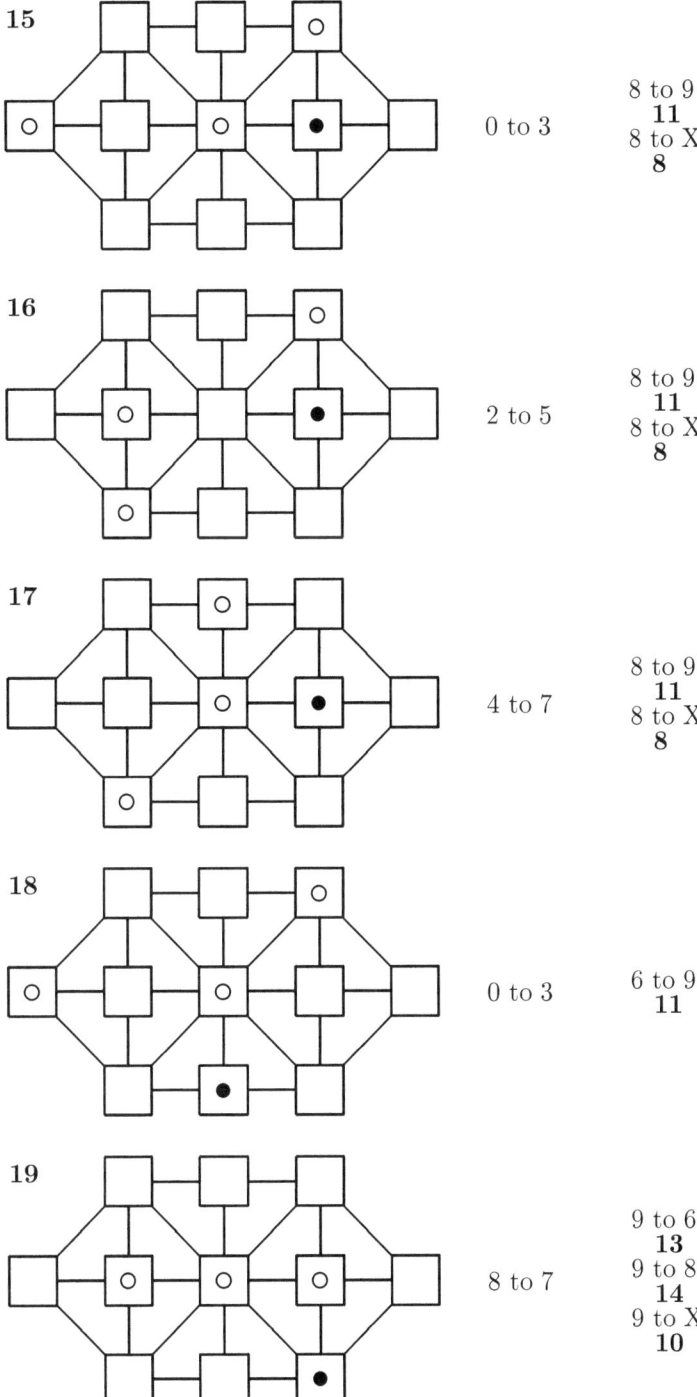

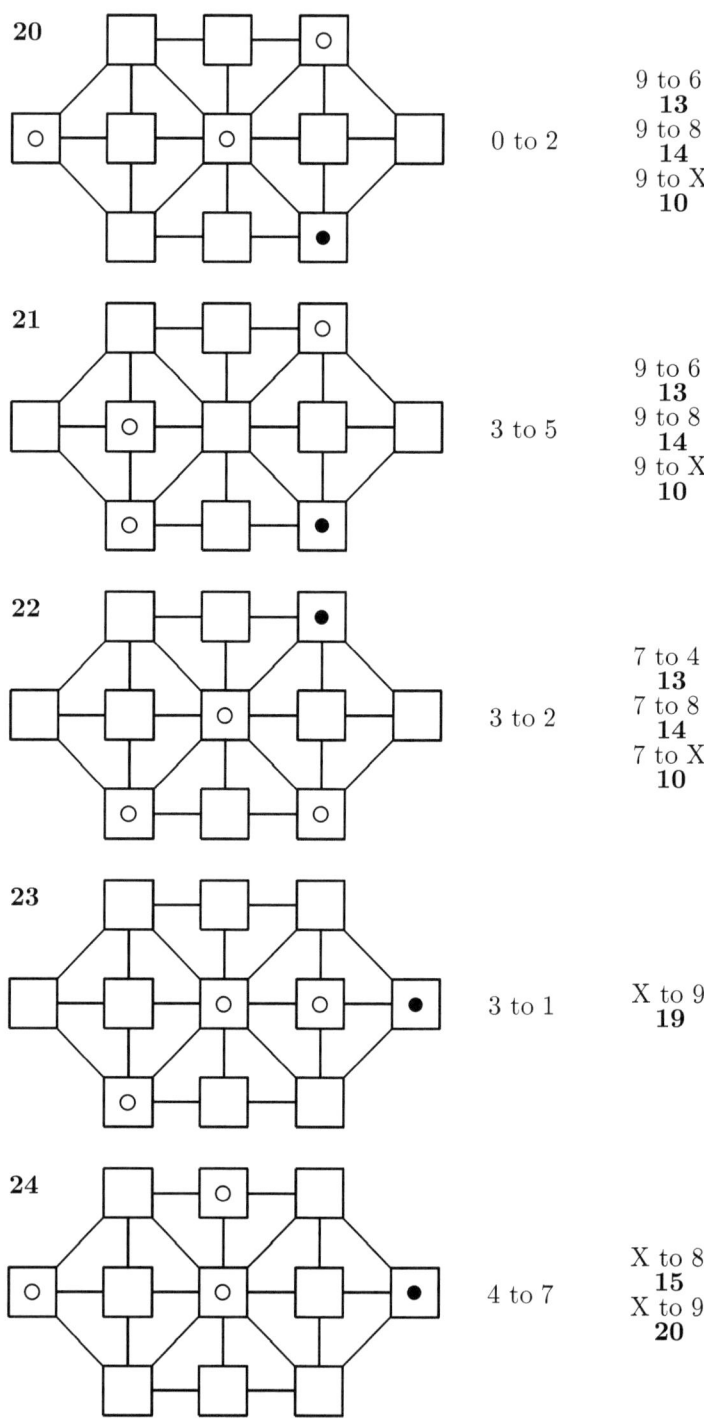

Solutions

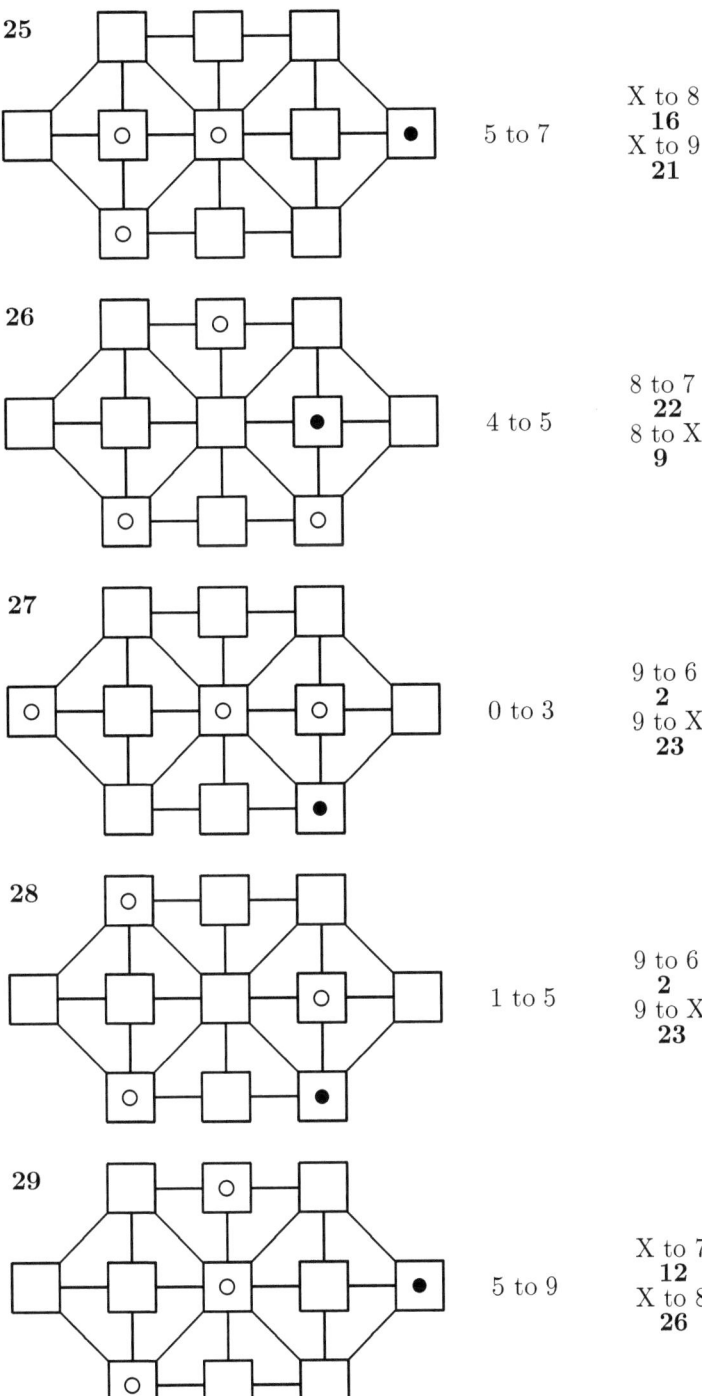

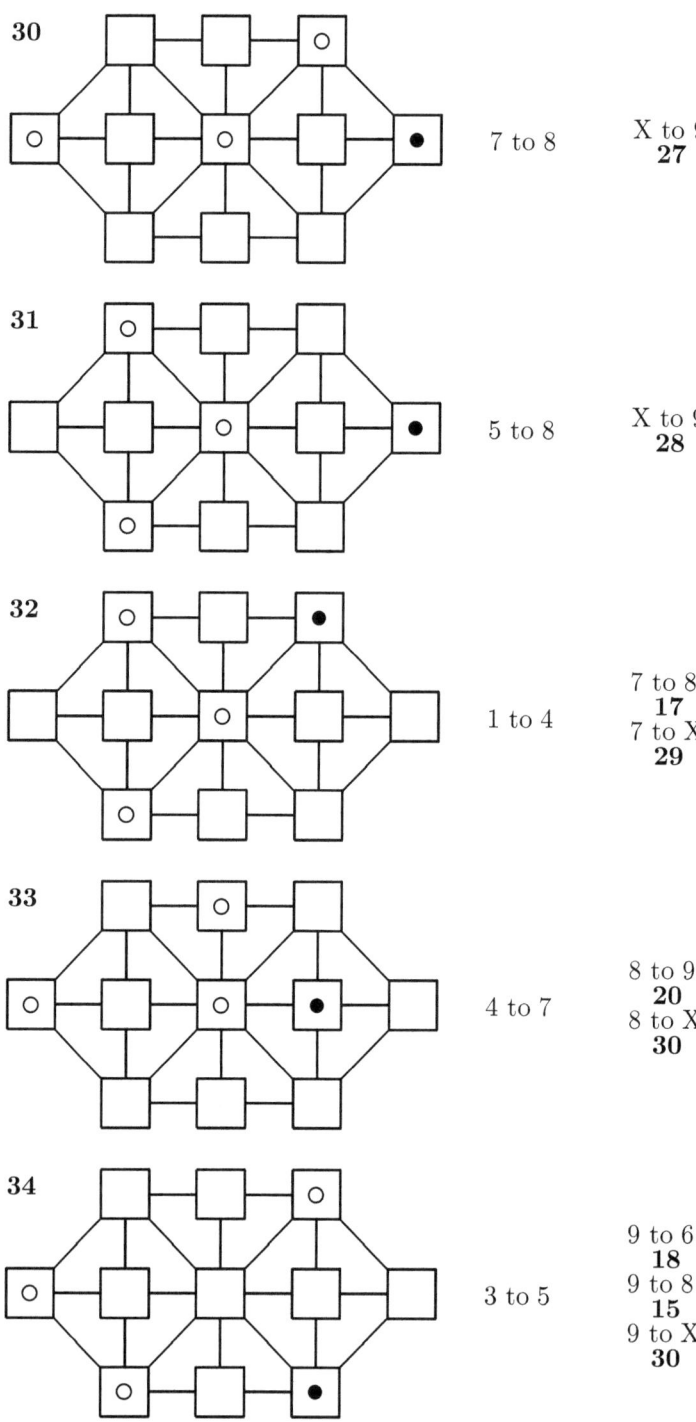

Solutions 361

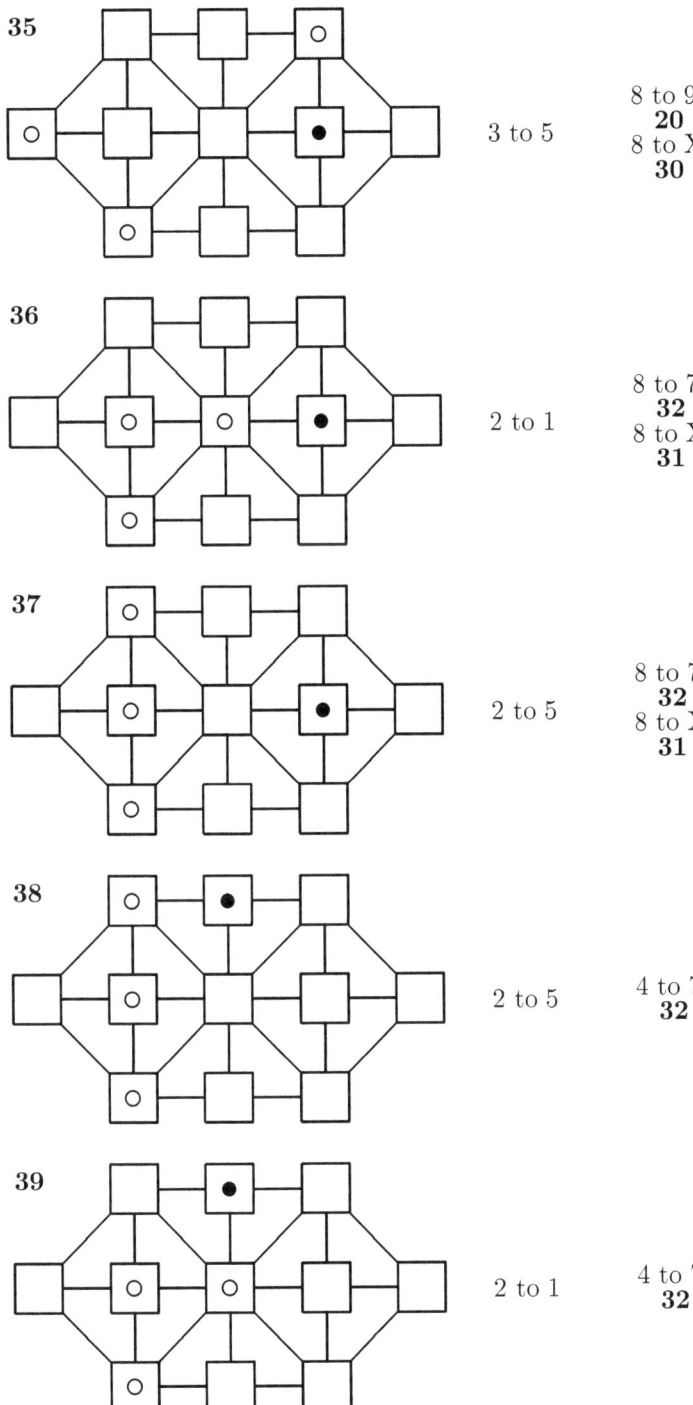

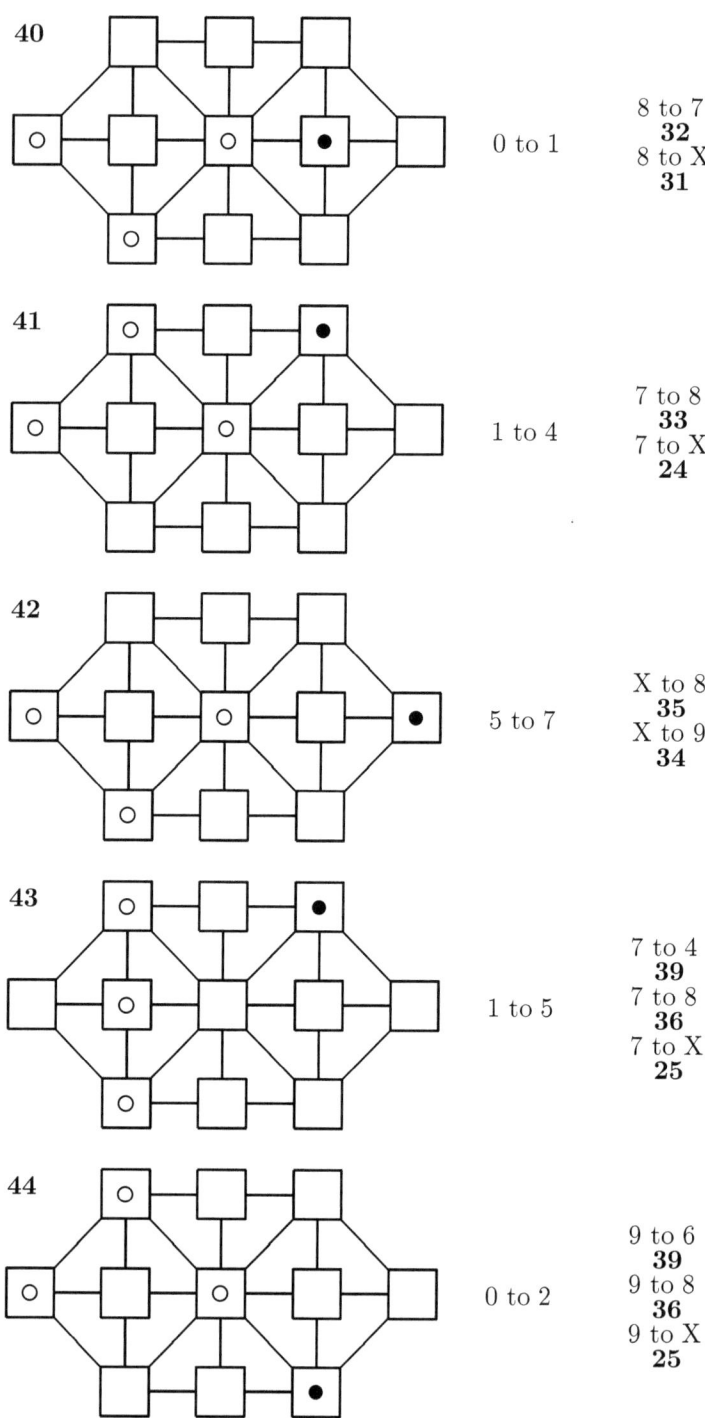

Solutions 363

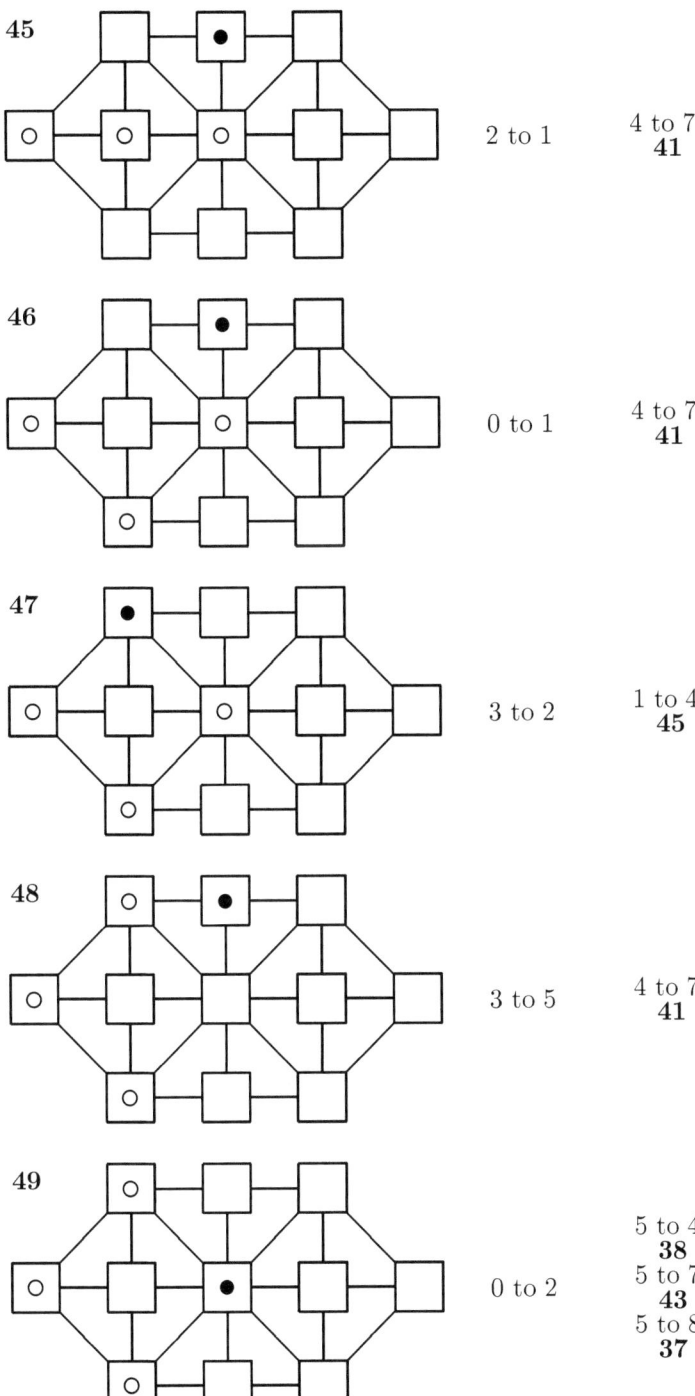

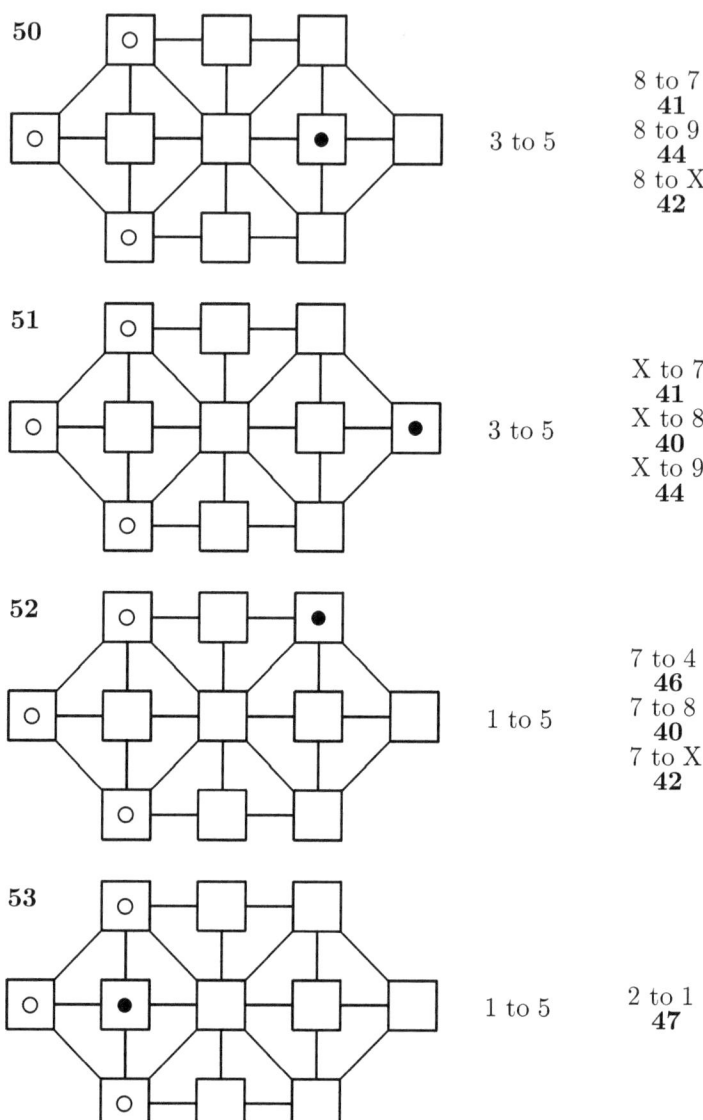

Figure 8.

MIX
Papier aus verantwortungsvollen Quellen
Paper from responsible sources
FSC® C105338

If you have any concerns about our products,
you can contact us on
ProductSafety@springernature.com

In case Publisher is established outside the EU,
the EU authorized representative is:
**Springer Nature Customer Service Center GmbH
Europaplatz 3, 69115 Heidelberg, Germany**

Printed by Libri Plureos GmbH
in Hamburg, Germany